Communications in Computer and Information Science 2500

Rationale

The CCIS series is devoted to the publication of proceedings of computer science conferences. Its aim is to efficiently disseminate original research results in informatics in printed and electronic form. While the focus is on publication of peer-reviewed full papers presenting mature work, inclusion of reviewed short papers reporting on work in progress is welcome, too. Besides globally relevant meetings with internationally representative program committees guaranteeing a strict peer-reviewing and paper selection process, conferences run by societies or of high regional or national relevance are also considered for publication.

Topics

The topical scope of CCIS spans the entire spectrum of informatics ranging from foundational topics in the theory of computing to information and communications science and technology and a broad variety of interdisciplinary application fields.

Information for Volume Editors and Authors

Publication in CCIS is free of charge. No royalties are paid, however, we offer registered conference participants temporary free access to the online version of the conference proceedings on SpringerLink (http://link.springer.com) by means of an http referrer from the conference website and/or a number of complimentary printed copies, as specified in the official acceptance email of the event.

CCIS proceedings can be published in time for distribution at conferences or as post-proceedings, and delivered in the form of printed books and/or electronically as USBs and/or e-content licenses for accessing proceedings at SpringerLink. Furthermore, CCIS proceedings are included in the CCIS electronic book series hosted in the SpringerLink digital library at http://link.springer.com/bookseries/7899. Conferences publishing in CCIS are allowed to use Online Conference Service (OCS) for managing the whole proceedings lifecycle (from submission and reviewing to preparing for publication) free of charge.

Publication process

The language of publication is exclusively English. Authors publishing in CCIS have to sign the Springer CCIS copyright transfer form, however, they are free to use their material published in CCIS for substantially changed, more elaborate subsequent publications elsewhere. For the preparation of the camera-ready papers/files, authors have to strictly adhere to the Springer CCIS Authors' Instructions and are strongly encouraged to use the CCIS LaTeX style files or templates.

Abstracting/Indexing

CCIS is abstracted/indexed in DBLP, Google Scholar, EI-Compendex, Mathematical Reviews, SCImago, Scopus. CCIS volumes are also submitted for the inclusion in ISI Proceedings.

How to start

To start the evaluation of your proposal for inclusion in the CCIS series, please send an e-mail to ccis@springer.com

Isabel Praça · Simona Bernardi ·
Pedro R.M. Inácio
Editors

Cybersecurity

9th European Interdisciplinary Cybersecurity Conference, EICC 2025
Rennes, France, June 18–19, 2025
Proceedings

Springer

Editors
Isabel Praça
Instituto Superior de Engenharia do Porto
Porto, Portugal

Simona Bernardi
Universidad de Zaragoza
Zaragoza, Spain

Pedro R.M. Inácio
Universidade da Beira Interior
Covilhã, Portugal

ISSN 1865-0929 ISSN 1865-0937 (electronic)
Communications in Computer and Information Science
ISBN 978-3-031-94854-1 ISBN 978-3-031-94855-8 (eBook)
https://doi.org/10.1007/978-3-031-94855-8

This Springer imprint is published by the registered company Springer Nature Switzerland AG
The registered company address is: Gewerbestrasse 11, 6330 Cham, Switzerland

Preface

The European Interdisciplinary Cybersecurity Conference (EICC) aims to establish a premium venue for the exchange of information and the sharing of research findings on cybersecurity and its many aspects between academics and practitioners in Europe. EICC encourages dialogue between computer scientists and researchers in all other fields related to cybersecurity, such as behavioral sciences, sociology, criminology, investigations, and law. Interdisciplinary contributions are especially welcome.

EICC 2025 was the 9th edition of the conference, which until 2019 was known as the Central European Cybersecurity Conference – CECC, being rebranded to underscore the interdisciplinarity of the conference and broaden its reach. This 9th edition of EICC was held at Inria, Rennes, France, on the 18th and 19th of June, 2025.

Three Special Session proposals were submitted to EICC 2025, and, after a thoughtful review process, all were accepted. The evaluation criteria for the selection of the special sessions included the relevance to the EICC, the timeliness and expected interest in the proposed topics, the ability of the organizers to lead a successful workshop, and their balance and synergy.

The EICC 2025 proceedings include the accepted papers from the main-track session and the following special sessions:

- Advances in Interdisciplinary Cybersecurity: Insights from Funded Research Projects – CyFRP;
- Complex Network Analysis for Cybersecurity – CNACYS;
- Medical Device Security and Privacy – MeDSec.

All sessions together received a total of 39 submissions from 23 different countries. Each session had an independent Program Committee, which was in charge of reviewing and selecting the papers submitted to the session. All the papers submitted to the main-track, CNACYS and MeDSec sessions underwent double-blind review, but two papers were discarded due to non-compliance with the double-blind policy. The papers submitted to the CyFRP Special Session underwent a single-blind review. Each paper received at least three reviews (139 reviews in total). Out of the 39 submissions, 23 papers were selected to be presented at EICC 2025 and included in these proceedings (21 full papers, 2 short papers), yielding an acceptance rate of 58.9%.

This edition of the conference featured a keynote by Miguel Pupo Correia, of the *Instituto Superior Técnico da Universidade de Lisboa*. He delivered a talk titled *Blockchain Interoperability: From Vulnerabilities to Attacks.*

Many people contributed to the success of the EICC 2025 program. We would like to express our gratitude to all those who supported this event: the special session organizers for their commitment, the authors who contributed to this volume, the reviewers for their help in the paper assessment, the keynote speaker and the EICC participants.

We would also like to thank all the members of the EICC Steering Committee, in particular to Simon Vrhovec, for his patience and continuous support in all the tasks, and to Organizing Chair, Jean-François Lalande.

Finally, many thanks are given to the Springer staff who provided professional support through all the phases that led to this volume.

June 2025

Isabel Praça
Simona Bernardi
Pedro R.M. Inácio

Organization

Conference Chair

Pedro R. M. Inácio	Universidade da Beira Interior, Portugal

Organizing Chair

Jean-François Lalande	CentraleSupélec/Inria, France

Program Committee Chairs

Simona Bernardi	University of Zaragoza, Spain
Isabel Praça	Polytechnic of Porto (ISEP/IPP), Portugal

Steering Committee

Igor Bernik	University of Maribor, Slovenia
Bela Genge	University of Medicine, Pharmacy, Sciences and Technology of Târgu Mureş, Romania
Jörg Keller	FernUniversität in Hagen, Germany
Blaž Markelj	University of Maribor, Slovenia
Wojciech Mazurczyk	Warsaw University of Technology, Poland
Aleksandra Mileva	Goce Delčev University of Štip, North Macedonia
Simon Vrhovec	University of Maribor, Slovenia

Program Committee

Habtamu Abie	Norwegian Computing Centre, Norway
Marios Anagnostopoulos	Democritus University of Thrace, Greece
Niyati Baliyan	National Institute of Technology Kurukshetra, India
João Paulo Barraca	University of Aveiro, Portugal
Giampaolo Bella	University of Catania, Italy
Luca Caviglione	CNR/IMATI, Italy

Richard Chbeir	University of Pau and the Pays de l'Adour, France
Michał Choraś	Bydgoszcz University of Science and Technology, Poland
Kovila Coopamootoo	King's College London, UK
Tiago Cruz	University of Coimbra, Portugal
Salvatore D'Antonio	University of Naples "Parthenope", Italy
Luca Davoli	University of Parma, Italy
Changyu Dong	Newcastle University, UK
Matthew Edwards	University of Bristol, UK
Pavlos Efraimidis	Democritus University of Thrace, Greece
Tobias Eggendorfer	TH Ingolstadt, Germany
Sergio Esposito	University of Catania, Italy
Virginia Franqueira	University of Kent, UK
Damjan Fujs	University of Ljubljana, Slovenia
Steven Furnell	University of Nottingham, UK
Dieter Gollmann	Hamburg University of Technology, Germany
Petra Grd	University of Zagreb, Croatia
Nils Gruschka	University of Oslo, Norway
Mordechai Guri	Ben-Gurion University of the Negev, Israel
Piroska Haller	Petru Maior University, Romania
Marko Hölbl	University of Maribor, Slovenia
Martin Gilje Jaatun	University of Stavanger/SINTEF Digital, Norway
Georgios Karopoulos	European Commission/Joint Research Centre, Italy
Sokratis Katsikas	Norwegian University of Science and Technology, Norway
Stefan Katzenbeisser	University of Passau, Germany
Shujun Li	University of Kent, UK
Wenjuan Li	Education University of Hong Kong, China
Shigang Liu	CSIRO's Data61, Australia
Brad Malin	Vanderbilt University, USA
Lorenzo Mannocci	University of Pisa, Italy
Stefano Marrone	University of Campania "Luigi Vanvitelli", Italy
Ibéria Medeiros	University of Lisbon, Portugal
David Megías	Universitat Oberta de Catalunya, Spain
Weizhi Meng	Lancaster University, UK
Rodrigo Miani	Federal University of Uberlândia, Brazil
Pal-Stefan Murvay	Politehnica University of Timisoara, Romania
Sebastian Pape	Social Engineering Academy GmbH, Germany
Sergio Pastrana	Universidad Carlos III de Madrid, Spain
Pedro Pinto	Instituto Politécnico de Viana do Castelo, Portugal
Kaja Prislan Mihelič	University of Maribor, Slovenia

Cigdem Sengul	Brunel University London, UK
Bernardo Sequeiros	University of Beira Interior, Portugal
Gerardo Simari	Universidad Nacional del Sur/CONICET, Argentina
Kai Simon	Fraunhofer-Gesellschaft, Germany
Michael Sirivianos	Cyprus University of Technology, Cyprus
Florian Skopik	Austrian Institute of Technology, Austria
Hung-Min Sun	National Tsing Hua University, Taiwan
Yuqiong Sun	Facebook, USA
Igor Tomičić	University of Zagreb, Croatia
Fernando Tricas	University of Zaragoza, Spain
Nicolas Tsapatsoulis	Cyprus University of Technology, Cyprus
Abhishek Vaish	Indian Institute of Information Technology Allahabad, India
Steffen Wendzel	University of Ulm, Germany
Christos Xenakis	University of Piraeus, Greece
Nicola Zannone	Eindhoven University of Technology, Netherlands

Special Sessions

Advances in Interdisciplinary Cybersecurity: Insights from Funded Research Projects – CyFRP 2025 Special Session

Chairs

Jean-François Lalande	CentraleSupélec/Inria, France
Simon Vrhovec	University of Maribor, Slovenia

Program Committee

Luca Caviglione	CNR/IMATI, Italy
Michał Choraś	Bydgoszcz University of Science and Technology, Poland
Kovila Coopamootoo	King's College London, UK
Tiago Cruz	University of Coimbra, Portugal
Pavlos Efraimidis	Democritus University of Thrace, Greece
George Drosatos	Athena Research Center, Greece
Damjan Fujs	University of Ljubljana, Slovenia
Petra Grd	University of Zagreb, Croatia

Piroska Haller	Petru Maior University, Romania
Marko Hölbl	University of Maribor, Slovenia
Shujun Li	University of Kent, UK
Ioannis Mavridis	University of Macedonia, Greece
David Megías	Universitat Oberta de Catalunya, Spain
Pal-Stefan Murvay	Politehnica University of Timisoara, Romania
Niels Nijdam	University of Geneva, Switzerland
Konstantinos Rantos	Democritus University of Thrace, Greece
Michael Sirivianos	Cyprus University of Technology, Cyprus
Aimilia Tasidou	Institut Polytechnique de Paris, France
Igor Tomičić	University of Zagreb, Croatia
Steffen Wendzel	University of Ulm, Germany
Christos Xenakis	University of Piraeus, Greece

Complex Network Analysis for Cybersecurity – CNACYS 2025 Special Session

Chairs

Martin Atzmueller	Osnabrück University/DFKI, Germany
Rushed Kanawati	Sorbonne Paris Nord University/CNRS, France

Program Committee

Martin Husák	Masaryk University, Czech Republic
Pierre Parrend	EPITA Strasbourg, France
Hamida Seba	Claude Bernard University Lyon 1, France

Medical Device Security and Privacy – MeDSec 2025 Special Session

Chairs

Habtamu Abie	Norwegian Computing Center, Norway
Martin Gilje Jaatun	SINTEF Digital, Norway
Dusko Milojevic	KU Leuven, Belgium
Nikos Papadakis	SPACE HELLAS S.A., Greece
Sandeep Pirbhulal	Norwegian Computing Center, Norway

Program Committee

Shaukat Ali	Simula Research Laboratory/Oslo Metropolitan University, Norway
Marios Anagnostopoulos	Democritus University of Thrace, Greece
Manos Athanatos	Foundation for Research and Technology Hellas/Technical University of Crete, Greece
Sabarathinam Chockalingam	Institute for Energy Technology, Norway
Sabine Delaitre	Bosonit group, Spain
Gencer Erdogan	SINTEF, Norway
Vasileios Gkioulos	Norwegian University of Science and Technology, Norway
Ilias Gkotsis	Inlecom Innovation, Greece
Dieter Gollmann	Hamburg University of Technology, Germany
Geir-Erlend Johansen	Norwegian Directorate of Health, Norway
Nesrine Kaaniche	Télécom SudParis, France
Sokratis Katsikas	Norwegian University of Science and Technology, Norway
Petros Malitas	PKNM Solutions, Switzerland
Carsten Maple	University of Warwick, UK
Luis Perez Pau	Futurs Health, Spain
Ankur Shukla	Institute for Energy Technology, Norway
Ali Hassan Sodhro	Kristianstad University, Sweden
Steve Taylor	University of Southampton, UK
Mohsen Toorani	University of South-Eastern Norway, Norway
Christos Xenakis	University of Piraeus, Greece
Shouhuai Xu	University of Colorado Colorado Springs, USA

Additional Reviewers

Vaios Bolgouras
Aristeidis Farao
Georgios Paparis
Nikolaos Pavlidis

Sponsors

Keynote

Blockchain Interoperability: From Vulnerabilities to Attacks

Miguel Pupo Correia

Instituto Superior Técnico, Universidade de Lisboa, Portugal

Blockchain —also referred to as Web3 or Distributed Ledger Technology— is a rapidly evolving field with many applications. Two aspects make it particularly relevant for our discussion. First, cybersecurity is foundational in blockchain, enabling the development of high-value distributed applications. A prime example is the emergence of cryptocurrencies and Decentralized Finance (DeFi), which together represent a market exceeding 2 trillion euros. Second, blockchain is inherently interdisciplinary. It sits at the intersection of computer science, economics, law, sociology, and more, creating a rich space for cross-domain innovation and challenges. While blockchain platforms such as Ethereum, BNB Chain, and Polygon serve as infrastructures that support a wide range of decentralized applications, there is increasing interest in enabling interoperability between these systems. However, this trend has introduced a new and significant attack surface —one that has already resulted in hundreds of millions of euros in losses. In this talk, I will explore the concept of blockchain interoperability, focusing on the specific vulnerabilities introduced by Bridges and Bridge Aggregators. We will examine how these components work, why they are targeted, and what can be done to secure them.

Contents

Main Track

An Automated Home Secretary: Evaluating the Ability of Large Language Models to Understand Extremist Conversations ... 3
James Stevenson, Luke Gassmann, and Matthew Edwards

PARAG: Proactive Answering Framework Integrating LLMs with Retrieval-Augmented Generation ... 20
Bikash Saha, Nanda Rani, Joheen Chakraborty, Divyanshu Singh, Soumyo V. Chakraborty, and Sandeep Kumar Shukla

Evaluating LLaMA 3.2 for Software Vulnerability Detection ... 38
José Gonçalves, Miguel Silva, Bernardo Cabral, Tiago Dias, Eva Maia, Isabel Praça, Ricardo Severino, and Luís Lino Ferreira

Power Consumption Analysis for Reverse Engineering Digital Modulation: A Novel Approach to Physical Layer Attacks on Communication Systems ... 52
Pablo Pareja, Yehya Nasser, Ahmed Masmoudi, and Sebastien Houcke

Analysis of Permissions and Trackers in Android Apps Targeting Children Under Thirteen ... 63
Jamie Fitzpatrick and Virginia N. L. Franqueira

Understanding Corporate Ransomware – A Case Study in the Nordics ... 82
Emil Larsson and Meiko Jensen

Exploring the Cybercrime Potential of LLMs: A Focus on Phishing and Malware Generation ... 98
Orçun Çetin, Baturay Birinci, Çağlar Uysal, and Budi Arief

The Fragmented Research Space of Cybersecurity: A Map of the Territory ... 116
Michael Lang

Hiding Information in Short Network Packets ... 133
Jörg Keller and Sebastian Litzinger

Examining the Influence of Cultural Diversity in Cybersecurity Culture ... 151
Mohammed Aljohani, Steven Furnell, Xavier Carpent, and Nicholas Gervassis

A Framework to Enable Cheating Detection in Multiplayer Online Games 166
Salman Shaikh, Hugo Bertin, Marc Dacier, and Yérom-David Bromberg

Framework and Roadmap for Secure Design and Development of Applications in the Cloud and Mobile Ecosystem 182
Francisco T. Chimuco, João B. F. Sequeiros, Mário M. Freire, and Pedro R. M. Inácio

Enhancing JavaScript Malware Detection Through Weighted Behavioral DFAs 201
Pedro Pereira, José Gonçalves, João Vitorino, Eva Maia, and Isabel Praça

Advances in Interdisciplinary Cybersecurity: Insights from Funded Research Projects - CyFRP 2025 Special Session

Maritime Cyber Resilience: Bridging Cybersecurity and Regulatory Frameworks 217
Sandra Lielbārde, Agnė Brilingaitė, Linas Bukauskas, Evita Roponena, Elizabete Citskovska, and Rūta Pirta

Evaluating KASLR Break on RISC-V Using Gem5: Microarchitectural Side-Channel Analysis of Page-Table Walks 229
Mahreen Khan, Maria Mushtaq, Renaud Pacalet, and Ludovic Apvrille

Complex Network Analysis for Cybersecurity -CNACYS 2025 Special Session

Spectral Graph Analysis of Bipartite Graphs for Advanced Attack Detection ... 239
Majed Jaber, Nicolas Boutry, and Pierre Parrend

Hypergraph-Based Model for Adversary Threat Intelligence Analysis 255
Juan Manuel Matalobos, Regino Criado, and Santiago Moral

Medical Device Security and Privacy - MeDSec 2025 Special Session

IoT-Driven Security Solutions for Resource-Constrained Wearable Medical Devices 275
Emmanouil Bakiris and George Oikonomakis

Cyber-Risk Indicators for Connected Medical Devices 288
Simeon Tverdal, Gencer Erdogan, Andrea Neverdal Skytterholm, Samuel M. Senior, Steve Taylor, and Laura Carmichael

A Secure Data Transmission Protocol for Cardiac Implants to Mitigate Intrusive Cyber Attacks 307
Anisha Mitra and Dipanwita Roy Chowdhury

Strengthening the MDCG Guidance on Cybersecurity for Medical Devices: A Legal Approach 324
Oguzhan Yesiltuna, Ruoxin Su, Ashwinee Kumar, and Paul Quinn

MDCG 2019-16 Guidelines: Case Study-Based Assessment and Path Forward 338
Christos Androutsos, Steve Taylor, Karin Bernsmed, Andrea Neverdal Skytterholm, Gregory Epiphaniou, Nabil Moukafih, Theodoros N. Arvanitis, Sotiris Messinis, Nikos Papadakis, Marco Fruscione, Andrés Castillo, Dusko Milojevic, Dimitrios S. Karas, Nikolaos Fotos, Max Ostermann, Oscar Freyer, Stephen Gilbert, Vasilis Pezoulas, Lambros Athanasiou, George Gkois, and Dimitrios I. Fotiadis

Threshold Multiparty Private Set Intersection via Federated Graph Neural Networks 356
Sotirios Messinis, Nicholas E. Protonotarios, Dimitris Kalogeras, and Nikolaos Doulamis

Correction to: Analysis of Permissions and Trackers in Android Apps Targeting Children Under Thirteen C1
Jamie Fitzpatrick and Virginia N. L. Franqueira

Author Index 373

Main Track

An Automated Home Secretary: Evaluating the Ability of Large Language Models to Understand Extremist Conversations

James Stevenson(✉), Luke Gassmann, and Matthew Edwards

Bristol Cyber Security Group, School of Computer Science, University of Bristol, Bristol, UK
{my19303,dn21518,matthew.john.edwards}@bristol.ac.uk

Abstract. Large Language Models (LLMs) offer promising opportunities for automating online extremism detection, yet challenges remain in capturing nuanced rhetoric and complex social dynamics. In this study, we evaluate state-of-the-art LLMs (GPT-4o, LLaMA, Gemini, and Mixtral) using the updated UK government definition of extremism, analysing 500 user examples. GPT-4o notably achieved an accuracy of 80% in extremism detection with a Macro-F1 score of 71%. Fine-tuned approaches similarly produced 82% accuracy and 82% Macro-F1. However, performance dropped significantly when distinguishing nuanced social roles, particularly within contexts characterised by complex emotional tones or negative sentiment. These findings highlight both the strengths and current limitations of LLMs in automated content moderation, pointing to critical areas for further improvement in the analysis of extremist discourse.

Keywords: Extremism detection · Large Language Models · Content moderation · Social roles · Far-right networks · Cyber Security

1 Introduction

The rise of online extremism presents urgent challenges for societies seeking to curtail the influence of extremist rhetoric and behaviour, which can fuel radicalisation, incite violence, and destabilise social cohesion. Given the increasing ubiquity of social media and encrypted platforms such as Telegram, extremist groups now have unprecedented avenues to propagate their ideologies, recruit followers, and organise actions. Research within social and computer sciences has highlighted the need for scalable, automated tools that can detect and monitor such online activity with the precision needed to capture subtle contextual cues and complex social dynamics [14].

In March 2024, the UK government updated its definition of extremism to address the evolving tactics of radical groups and the social harm they can cause.

I. Praça et al. (Eds.): EICC 2025, CCIS 2500, pp. 3–19, 2025.
https://doi.org/10.1007/978-3-031-94855-8_1

The new definition clearly defines the scope of extremism to the "promotion or advancement of an ideology based on violence, hatred, or intolerance," specifically targeting actions that negate fundamental rights, undermine democratic institutions, or foster environments permissive of extremist outcomes [9]. This study seeks to evaluate whether LLMs, with their capacity for natural language understanding and contextual analysis, can effectively apply this new three-bullet-point definition to identify extremist content in diverse online discussions. By examining if LLMs can detect the targeted behaviours—such as promoting intolerance or undermining democratic norms—outlined in the updated definition, we aim to assess their potential for real-world extremism detection.

This study builds on existing approaches by evaluating whether advanced LLMs can match human-level accuracy in the detection of extremism under the UK's new guidelines. Additionally, we investigate the ability of LLMs to classify user roles-such as influencers, supporters, and bystanders-in online extremist communities, aiming to create a nuanced understanding of how these roles contribute to online extremism dynamics. Previous studies have demonstrated that specific language markers, like sentiment and motivational appeals, are associated with mobilisation and group influence in online extremist circles [17]. However, the extent to which LLMs can parse these markers in real-world datasets remains under-explored.

Our primary research questions (RQs) are as follows:

- **RQ1:** Can large language models accurately detect extremism according to new government definitions at a human-equivalent level?
- **RQ2:** Are large language models effective at identifying user roles within online extremist discourse with comparable accuracy to human labelling?
- **RQ3:** How does the granularity of label categories impact model accuracy in community role detection?
- **RQ4:** How do emotional and contextual content features influence model accuracy in assessing social dynamics?

To address these questions, we employed the Pushshift Telegram Dataset [3], which provides an extensive corpus of Telegram messages from public channels, enabling detailed analysis of extremist discourse across real-world conversations.

2 Related Work

2.1 Detecting Extremism

From a social science perspective, efforts to detect extremism have focused on understanding the sociopolitical dynamics that lead to radicalisation. Social scientists have long emphasised the importance of identifying radical ideologies and the contexts in which they flourish. For example, Bailard et al. [2] examined the Proud Boys' online discourse on platforms like Telegram, highlighting how online rhetoric often precedes offline violent actions. Their study illustrates the importance of detecting not only extremist language but also motivational appeals that serve to mobilise individuals into action. Similarly, Smith et al. [17] analysed the digital traces of far-right groups, showing how socially validated online interactions can translate into offline mobilisation. These works demonstrate that merely analysing language as text content is insufficient and more complex approaches are required to understand and classify radical language.

In contrast, data science approaches have focused on automating the detection of extremist content, often using supervised learning and natural language processing (NLP) techniques. For instance, Alatawi et al. [1] applied domain-specific word embeddings and deep learning models, including BERT, to detect white supremacist hate speech. Their study demonstrated that advanced language models could detect nuanced and coded extremist language, which is often difficult to identify manually, with high F1 scores. This showcases the potential of machine learning models in scaling the detection of extremist content across vast datasets. Similarly, Stevenson et al. [18] analysed far-right extremist activities on Parler, applied classical binary classification random forest classifiers to detecting extremism on the network, and scored mid to high accuracy scores with this approach.

The detection of extremist recruitment activities has also been a key focus in computational extremism detection research. Scanlon and Gerber [15] explored using supervised learning techniques, such as support vector machines (SVMs), to detect recruitment efforts on jihadist forums. Their research demonstrated that machine learning models could classify recruitment-related content with high accuracy, highlighting the potential of automated systems to perform real-time monitoring of online extremist activities and prevent the spread of radical ideologies.

This research seeks to contribute to this literature by assessing how off-the-shelf and fine-tuned large language models handle the extremism classification task, and if they can handle nuances in extremist rhetoric and speech identified.

2.2 Detecting Social Roles

Online social networking platforms provide users with abundant interaction opportunities. With its continuing development, social media has extended its purpose from simply providing conversation, to online users serendipitously finding business opportunities, topics, news and products [13,19]. This abundance of information and objectives has facilitated role-related research across social network analysis and natural language processing.

Transformer architecture has accelerated the pace of leadership and influence role identification using online content [5]. In an attempt to identify leadership roles, for example, Oueslati et al. [11] used conversational feature extraction techniques to develop a new Dynamic Detection of Opinion Leaders framework which built upon established work [10]. The framework itself comprised two variables spanning both domains of natural language processing and network topology. To assess the recipients' attitudes to the content, the first variable was retrieved using sentiment feature extraction via term frequency. Terms were based upon three dictionaries indicating positive, negative and neutral. The second variable used platform attributes (such as likes, follows and reposts) to assess the after-effects of posted content. Both these variables were weighted before being passed to a BERT transformer which was then trained for classifying content for opinion leader status. By introducing content-based variables such as sentiment, Oueslati et al. [11] was able to increase framework precision from 80.13% to 94.67%, further indicating the benefit of conversational features on leadership role detection.

Alongside detecting traits of leadership, Micallef et al. [8] researched the impact of online roles and their behaviours during periods of misinformation in the COVID-19 pandemic. The researchers first identified misinformation roles by assessing social media posts with a fine-tuned BERT model across three labels: misinformation spreader, mitigator, and bystander. By identifying user roles across the network, and using feature extraction, the researchers assessed the types of language being used between groups. The results showed that aggressive and negative emotional traits were more common amongst opinion posts which did not source information. They also found that 96% of negators originated from non-official fact-checking accounts and were instead concerned citizens. Reflecting on these studies and the surrounding literature, the motivation to identify leadership and online roles is well rooted in social network analysis. The introduction of content-based toolsets such as transformers has accelerated the application of content. Because of this, we aim to explore the capabilities of this resource linking both extremism and online roles.

3 Methodology

3.1 Data Selection

Telegram is a particularly effective platform for extremism research due to its combination of encrypted communication and large open channels that allow extremist groups to disseminate propaganda, recruit members, and organise activities with minimal oversight, making it an ideal environment for studying both covert and public extremist behaviour [2].

For this research, we utilised the existing "Radical Agenda" channel from the Pushshift Telegram Dataset [3], which contained 127,404 posts from 630 unique users, spanning from 24 March 2018 to 9 September 2019. This channel was selected by first aggregating keywords from previous research by Stevenson et al. [18] and extracting channels with a high occurrence of these keywords in their 'about' sections. These keywords, such as 'genocidal', 'civil war', and 'armed', are strongly associated with violent extremism and were key in identifying relevant channels. After this keyword filter, the list of channels was then further filtered to channels flagged for violent content (i.e., channels violating Apple's App Store guidelines). Following this, the labelling process confirmed that the selected channel contained extreme content.

To facilitate analysis, the dataset was chunked into segments of 10 posts, with a sliding window of 3 posts to maintain context across overlapping conversations.

This approach ensured that multiple messages from the same users, as well as interactions between different users, could be captured within each chunk, providing a richer context for analysis.

3.2 Data Labelling and Extremism Definitions

As well as measuring a model's ability to predict extremism and social roles, instruction comprehension was an important trait we wanted to compare against human labellers. To do this, we created a set of rules for the human and model labellers to follow before any labelling took place [20]. By providing the same prompt rules to the humans and model labellers, we provide a fairer environment to evaluate model capabilities as outlined in past related research [6]. This rule set highlighted the definitions for each of the seven social roles as well as the binary extremist definition. Three definitions were used for extremism. These definitions were the full (Sects. 2, 3, and 4) 2024 UK Government definition of extremism [9], the short (Sect. 2) 2024 UK Government definition of extremism, and the old 2011 UK Government definition of extremism [16]. The majority of this research focuses on the short 2024 definition, however, to identify the most effective definition, an initial exploration of the long and old definitions were evaluated.

After labelling with the short definition, extremism labels had the following distribution: Extremist: 66 users, Non-Extremist: 434 users.

Role definitions were based on roles observed in social media based on the categories *leader, engager and bystander*. These categories were then subdivided into the following set of roles and definitions. The role **People Leader** directs, recruits and mobilises conversation members either virtually or in the real world. The **Influencer Leader** directs the conversation as a knowledge source and/or gatekeeper, with other members reflecting their attitudes. The **Engager Negator** exhibits negative or berating interactions in an attempt to reduce discussion or offer a counterargument against a fundamental principle of the group. The **Engager Supporter** positively interacts with the topic, encouraging or promoting future discussion and ideological success. The **Engager Neutral** demonstrates neutral topic interaction to learn or socially interact with members. A **Bystander** does not engage with the main discussion but remains within the discussion chunk. The role **NATTC** (Not Applicable to the Conversation) is used when a user is at the beginning or end of a chunk discussing a topic that the content is not engaged with.

For both extremism and role classification, two researchers independently labelled 50 random sets of user conversations, reviewing all posts from each user in a chunk and assigning two labels: one for whether the post was extremist or not, and another for the user's role within the conversation. After labelling and mapping (i.e. changing resolution of the data), the Cohen's Kappa scores were calculated to measure inter-rater reliability:

- Cohen's Kappa Score for Extremist/Not Extremist: 0.85
- Cohen's Kappa Score for Role Labels (No Mapping): 0.67

– Cohen's Kappa Score for Role Labels (With Mapping): 0.71

Previous research into extremism classification techniques has seen Cohen's Kappa scores vary between 0.43 and 0.76 [4]. It is hypothesised that this score varies drastically due to the complex variety of extremism definitions and the thin line between hate-speech and extremism. With the definitions of extremism used in this research a comparatively high Cohen's Kappa score has been achieved, indicating strong agreement.

3.3 Prompting Large Language Models

For the analysis, the IBM WatsonX API [7] was utilised to send the prepared dataset to a set of pre-trained language models for evaluation. This facilitated the communication between our data pipeline and the following models: LLaMA, Mixtral, GPT-3.5, GPT-4o, Gemini, and LLaMA 3.1. The models were provided with each 10-post chunk and asked to label each user in that chunk with an extremism and role label. Standard prompt templates were used, with minor modifications for extremism classification depending on the definition under test [20].

Table 1. Role Label distribution and combination of labels used in label resolution sets. *E* stands for Engager. *L* stands for Leader.

	LLM Role Classification Labels						
Categories	NATTC	Bystander	E Supporter	E Neutral	E Negator	People L	Influencer L
Distribution	33	69	186	39	33	22	118
Detailed	NATTC	Bystander	E Supporter	E Neutral	E Negator	People L	Influencer L
Moderate	Passive		E Supporter	E Neutral	E Negator	Leading	
Simple	Passive		Engaging			Leading	

Given the complexity of the role labels, we expected that there would be some difficulty for models to differentiate between them. To reduce this issue, we created three resolution categories. As shown in Table 1, these are mapped into the following categories: Simple (containing three labels - Bystander, Engager and Leader) and Detailed (containing all original seven labels). We also include a Moderate resolution category where only the Engager categories are not grouped.

3.4 Fine-Tuning of Large Language Models for Extremism Classification

Previous research has demonstrated that LLM fine-tuning can lead to performance improvements and provide a means to achieve these improvements while reducing large training dataset requirements [12]. We explored if fine-tuning a base LLM would provide a demonstrable performance increase for this task. For this, the Unsloth Meta LLaMA 3.1-8B model was used. To address the lack of pre-labelled data a teacher-student approach was adopted. ChatGPT-4o was used to label the posts of users between 2018-03-24 to 2018-04-23 (31 days) - totally 47 users and 5,878 posts. With the short extremism definition, ChatGPT labelled these as 5,493 False and 385 True - these labels became the fine-tuning dataset. Then, a Low-Rank Adaptation (LoRA) fine-tuning approach with 60 steps, a 2e-4 learning rate, and a 0.01 weight decay was employed with these labelled data points to fine-tune the LLaMA 3.1 base model. The effectiveness of this model was evaluated against the same human-labelled dataset as the other base models - this was done with a temperature of 0.1.

4 Results

4.1 Assessing Extremism Predictions (RQ1)

In this section, we evaluate the capability of each model to accurately classify extremist content. As shown in Table 2, the results demonstrate that GPT-4o significantly outperforms the other base models, achieving a Macro-F1 score of 0.71 and an accuracy of 80.8%. The fine-tuned LLaMA 3.1 model scores above this with an 82% accuracy. These results are particularly notable when compared to the classical random forest models used in previous research by Stevenson et al. [18], which showed lower performance in identifying extremist content (with an accuracy of 78.9%). Equally, in regards to fine-tuning, we can see a marked increase in the accuracy of the fine-tuned LLaMA 3.1 8b model compared to the base LLaMA 3.1 8b model of approximately 24%, in turn surpassing the accuracy and F1 of the GPT-4o model – despite the fact that this fine-tuned model was trained using output from the GPT-4o model. This suggests that fine-tuning can produce better task-specific results than even a generally more powerful model.

In comparison, other models such as LLaMA 3.0 and Mixtral showed lower performance, with Macro-F1 scores of 0.657 and 0.573, respectively. Although these models provided some accurate classifications, their rationales were less consistent, often missing key context or failing to follow the prompt rules, a problem that led to hallucinations in several cases. Despite these shortcomings, the LLaMA 3.0 70B model still achieved a respectable accuracy of 73.8%, while Mixtral followed closely with 62%.

These results highlight that large language models (LLMs) such as GPT-4o can be highly effective in real-world applications for identifying and classifying extremist content. These results can be seen in Table 2 and Cohen Kappa scores can be seen in Table 5. Equally, it can be seen that the fine-tuning of a base LLM can drastically improve a model's accuracy and F1 in extremism classification.

In addition to comparing overall model performance, it is important to assess how the models fared under different definitions of extremism, particularly the long [9] and old UK definitions [16]. In this research we compared three permutations of extremism definitions:

Table 2. Macro-F1 and accuracy for each LLM predicting extremist' labels using the short UK definition.

Model	Macro-F1	Accuracy
LLaMA 3.0 70B	0.657	0.738
Mixtral 8 × 7B	0.573	0.62
GPT 3.5	0.573	0.552
GPT-4o	**0.71**	**0.808**
Gemini	0.675	0.778
LLaMA 3.1 8B	0.6426	0.5760
Fine-Tuned LLaMA 3.1 8B	**0.8211**	**0.82**

- **2024 UK Definition Short** The main definition used for this research. This definition uses section '2. The definition' of the 2024 UK definition of extremism. It is a short definition featuring approximately 100 words. We refer to this definition as the 'short' definition. This definition's results are covered in Table 2.
- **2024 UK Definition Long** A comparison definition taken from the 2024 UK Definition, covering sections '2. The definition', '3. Behaviour that could constitute extremism', and '4. Further context' - totalling upward of 1103 words. We refer to this definition as the 'long' definition. This definition's results are covered in Table 3.
- **2011 UK Definition** A high-level, older definition, totalling approximately 50 words. We refer to this definition as the 'old' definition. This definition's results are covered in Table 3.

To first verify the most optimal extremism definition, a comparison between the 'short', 'old', and 'long' definitions was performed using LLaMA 3.0 70B as a base model. The results demonstrated a trend where LLaMA performed best when using the short version of the new definition, and it's performance declined when using either the long or old versions. Due to this, for this research, the 'short' definition was used.

The short definition provided models like LLaMA with clearer criteria for extremism classification, contributing to its superior performance (as shown in Table 2). In contrast, when applying the long or old definitions, we see a drop in performance. For example, LLaMA 3.0's agreement score dropped from 0.75 with the short definition to 0.592 with the long definition. Due to the rationales provided by the models, we hypothesise that this drop was due to hallucinations and misclassification caused by the broader and more complex nature of the long definition, leading to incorrect labels being created, such as 'unknown'. Similarly, when using the old definition, LLaMA 3.0's agreement score fell from 0.75 (short) to 0.623 (old), as shown in Table 3 - this is hypothesised to being down to the shorter and vaguer 2011 definition.

These findings underscore the importance of definition clarity in extremism detection tasks. We hypothesise that the new short definition presented a sweet spot of detail and clarity that allowed the models to perform optimally by reducing ambiguity and focusing on specific, actionable elements of extremist behaviour. Conversely, it could be seen that both the long and old definitions potentially introduce complexities with the definition, which may have lead to the reduction in the overall accuracy and consistency of the models' predictions.

Table 3. Performance Comparison Between Short, Long, and Old Extremism Definitions

Model Accuracy	Short Definition	Old Definition	Long Definition
LLaMA 3.0 70 B	0.738	0.623	0.592

4.2 Assessing Community Role Predictions (RQ2, RQ3)

Alongside extremism, each model was tasked to identify each user's role within the conversation chunk. This prediction would be based on their output and the reaction from their community. As shown in Table 1, the models' labels were divided into three resolution categories: *Simple, Moderate* and *Detailed.*

As mentioned before, we measured model performance across different label resolutions, with *Detailed* containing the most role types (7 categories) and *Simple* containing the fewest (3 categories). As expected, in most cases, resolution is directly related to the accuracy of the models. This can be seen by a decline in both accuracy and Macro-F1 scores across each model, with the best-performing model achieving a Macro-F1 score of 0.604 in the simplest resolution but a Macro-F1 of 0.302 at a detailed resolution (see Table 4 and Cohen Kappa scores can be seen in Table 5).

This being the case, all models underperformed to our expectations, showing clear signs of weakness in social reasoning and theory of mind skillsets. When evaluating model accuracy and Macro-F1 results we can see consistent outliers depending on the metric we use to assess the results which may be due to the real world imbalances of role occurrence. Reflecting on Table 4, we can assess each role category with equal weight (regardless of appearance) by evaluating models based on their Macro-F1 scores. In this case, GPT-4o and Mixtral are the best-performing models in simple and moderate resolutions. Extending this to the detailed resolution, Mixtral rapidly declines in its position amongst the other models, whilst GPT-4o comparatively improves and is capable of outperforming the GPT-3.5 moderate label. Whilst this could be expected given the generational jump between GPT-3.5 and GPT-4o, the difficulty level for a detailed resolution model to outperform a moderate resolution model is notable. Whilst degrading in position amongst the other models as resolution increases, LLaMA 3.1 performs considerably better than initially expected. Given its parameter size being considerably smaller than many of the other models it was expected that scaling laws could have a considerable effect on its Macro-F1 score. However, LLaMA 3.1 (8b parameters) manages to outperform LLaMA 3.0 (70B parameters), GPT-3.5 (175b parameters) and Mixtral (8×7b parameters). This shows the generational jump for LLaMA 3.1.

By assessing the accuracy we can see a different picture developing across the label resolutions, providing insight into model detection capabilities with roles weighted by their occurrence in the dataset. For example, GPT-4o performed well when we assess Macro-F1 scores, however, with accuracy its performance degrades considerably faster than other models moving from 0.604 (simple resolution) to 0.278 in the detailed resolution. This surprising drop highlights the rarity of some labels in the dataset. Because of this, we surmise that GPT-4o would have performed considerably better if the dataset had not reflected a real-world occurrence of roles and that future work should assess GPT-4o based on the social task and the relevance of its label distribution. This effect in degrading performance can also be noted for LLaMA 3.1.

Table 4. Macro-f1 and accuracy for each LLM predicting the roles present in online conversations, at different label resolutions. Bold represents the highest-performing model in each row.

Model	LLaMA 3.0	Mixtral	GPT 3.5	GPT-4o	Gemini	LLaMA 3.1
	Simple					
Macro-F1	0.438	0.463	0.371	**0.501**	0.363	0.453
Accuracy	0.588	0.578	0.54	**0.604**	0.574	0.568
	Moderate					
Macro-F1	**0.319**	0.237	0.306	0.268	0.293	0.236
Accuracy	**0.348**	0.256	0.31	0.322	0.304	0.298
	Detailed					
Macro-F1	0.236	0.205	0.185	**0.266**	0.207	0.213
Accuracy	0.298	0.284	0.238	0.278	**0.302**	0.256

Table 5. Cohen Kappa of model classifications compared against human classification.

Classification Category	LLaMA	Mixtral	GPT-3.5	GPT-4	Gemini	LLaMA 3.1	Fine-Tuned LLaMA
Extremist	0.3507	0.1650	0.1373	0.4418	0.3785	0.1246	0.4102
Role Simple	0.2120	0.2201	0.1304	0.2794	0.1584	0.2202	N/A
Role Moderate	0.1736	0.1843	0.1161	0.1745	0.1803	0.1512	N/A
Role Detailed	0.1520	0.1322	0.1084	0.1552	0.1634	0.1137	N/A
Average	0.2221	0.1754	0.1230	0.2627	0.2202	0.1524	N/A

4.3 Content Features and LLM Challenges (RQ4)

Assessing Extremism Prediction Reasoning. In this section, we analyse how different LLMs - GPT-4o, GPT-3.5-turbo, Meta LLaMA-3 (70B), and Mixtral (8×7B) classify and reason about posts labelled as either *extremist* or *non-extremist*.

As part of this analysis, we can review the responses from these LLMs and review the reasoning they generated as part of their classification. While this reasoning is not representative of the actual LLM decision-making, it provides insights into its understanding of the prompt at hand.

In one example, the LLMs were provided with a post speaking about knife violence, hate speech, and dehumanising language - which the human researchers labelled as extremist, stating that it fell into the third category of intentionally creating a permissive environment for others to achieve extremist results. The below exhibits the LLM response on classification and reasoning:

- **GPT-4o**: (Label) Extremist
  (Reasoning) This post promotes violence (*use a knife when I need to*) and dehumanising language (*pedos, hebes*). It creates a permissive environment for violence and intolerance, fitting the updated definition of extremism by encouraging hatred and indirectly justifying violent actions.
- **GPT-3.5-turbo**: (Label) Non-extremist
  (Reasoning) While the user mentions violent language, there is no clear intent to promote actual violence or incitement of hatred against a specific group.
- **LLaMA-3 (70B)**: (Label) Extremist
  (Reasoning) The user's post promotes violence (*use a knife*) and hateful language (*hebes*), which aligns with extremist behaviour by fostering a culture of violence and dehumanisation.
- **Mixtral** (8×7B): (Label) Non-extremist
  (Reasoning) Although the post contains aggressive language, it lacks a direct call to action or explicit incitement to violence.
- **Fine-Tuned LLaMA-3.1 (8B)**: (Label) Non-extremist
  (Reasoning) The user's post contains derogatory language and stereotypes, but it does not promote or advance an ideology rooted in violence, hatred, or intolerance. It does not aim to negate or destroy fundamental rights and freedoms, undermine or replace the UK's democratic system, or encourage others to create conditions that lead to these outcomes. The emotional intensity is low, and there is no call to action. The post does not evoke anger, hatred, or fear, and it does not have a significant potential for impact.

Of the base models, GPT-4o performs the strongest evaluation by accurately identifying extremist content, highlighting the post's violent language and dehumanising terms, which contribute to an environment of extremism. While disagreeing, the fine-tuned LLaMA model also provides a strong case for why the post does not fall into the extremist definition. In contrast to GPT-4o, GPT-3.5-turbo, on the other hand, misses the implicit promotion of violence by focusing too much on the lack of explicit intent. LLaMA-3 (70B) also correctly identifies the post as extremist but provides a less detailed explanation than GPT-4o. Meanwhile, Mixtral (8×7B) focuses only on the absence of a direct call to action. One reason some LLMs may struggle with the nuances of applying the UK definition could be that, while they understand the attitudes being expressed correctly, their training data or system prompts may lead them to more heavily favour other definitions of extremism, and as such they are applying built-in topic knowledge and ignoring the direction given about labelling extremism using the UK definition.

Next, in this example the author highlights the benefits of violent 'revolutionary tactics' and how democracy cannot be relied on to 'save wester civilization' - this in turn, serves as a clear example of a user directly advocating political violence.

- **GPT-4o**: (Label) Extremist
  (Reasoning) User Cs posts advocate for revolutionary tactics and dismiss reliance on democracy, indicating an intent to undermine and overturn democracy. Additionally, User Cs use of racial slurs suggests an ideology based on hatred and intolerance.
- **GPT-3.5-turbo**: (Label) Extremist
  (Reasoning) User C's posts include language promoting violence and hatred towards certain racial groups, advocating for the overthrow of democracy, and suggesting the use of revolutionary tactics, which align with the definition of Extremist provided.
- **LLaMA-3 (70B)**: (Label) Extremist
  (Reasoning) The user promotes the idea of overthrowing democracy and uses derogatory language towards certain groups, which aligns with the definition of Extremist.
- **Mixtral** (8×7B): (Label) Extremist
  (Reasoning) User C is promoting an ideology based on hatred and intolerance. This is an attempt to negate or destroy the fundamental rights and freedoms of others, specifically those who are not white or do not share the same interests as the user. User C is also attempting to create a permissive environment for others to achieve the same results by using derogatory language and encouraging intolerance.
- **Fine-Tuned LLaMA-3.1 (8B)**: (Label) Extremist
  (Reasoning) The user promotes an ideology rooted in violence and intolerance, aiming to undermine the UK's democratic system and rights. It encourages individuals to abandon democratic processes and instead adopt revolutionary tactics, which can be interpreted as promoting actions that negate or destroy fundamental rights and freedoms. The use of derogatory language ('cuckolds') further indicates a high level of hostility and intolerance. The call to action, 'Say it with me', suggests a level of social validation and potential for amplification, which could increase the post's impact.

Here all models correctly identify the post as extremist and highlight not only the inclusion of racial slurs but also the intent to overthrow democracy with 'revolutionary tactics.' All models agree that this matches the provided definition of extremism.

Overall while models succeed at detecting explicit extremist content, they frequently fail to detect subtle forms of extremism, such as implied violence or content that creates a permissive environment for extremism. Both GPT-3.5-turbo and Mixtral struggle in this area, often missing the broader context or underlying intentions behind the content, which are key to identifying extremism under the 2024 guidelines. This focus on explicit calls to violence limits their accuracy in cases where extremism is more covert or implicit.

Equally, with the fine-tuned model, we also see a more verbose alignment with the 2024 UK definition of extremism, linking its decision to *"an ideology rooted in violence and intolerance, aiming to undermine the UK's democratic system and rights"*.

Among the evaluated models, GPT-4o performs the best of the base models, with an accuracy of 80.8% and a Macro-F1 score of 71%. It consistently provides detailed and nuanced explanations that align well with the UK's extremism definition. Its reasoning mirrors human judgement more closely, particularly when identifying less obvious forms of extremist rhetoric, such as when posts intentionally create a permissive environment for others to achieve the results, which contributes to its higher accuracy. Equally, of all of the models, the fine-tuned LLaMA model performs the best with an accuracy of 82% and an F1 of 82%. This model also provides strong justifications and reasoning for posts while highlighting the difference between proposed violence and extremism intent.

Feature Comparisons. A benefit of analysing content for extremism is that we can review the association between conversational features and model accuracy. To do this, we retrieve 33 features from each person's contribution to the conversation using a set of BERT models fine-tuned for toxicity, sentiment and emotion feature extraction. Then, based on their correlations with model accuracy, we reduced these to the most significant relationships. We also condensed role resolutions by averaging the results.

Reflecting on the features present in Fig. 1, there is a notable pattern separating positive and negative correlations. This separation can be best seen by toxicity features, where traits relating to toxicity have a high correlation with a model's accuracy for predicting extremism (in particular with LLaMA 3, GPT-4o and Gemini). This is not unexpected given the clear relationship between extremism and toxic behaviour, however, these same traits show the same consistent correlation when the model is tasked to predict social roles. Since social roles have little thematic association with either positive or negative traits, the content is likely having an effect on the model's reasoning skills rather than these being traits associated with the roles. More broadly, positive traits improve model accuracy for both extremist and role average predictions. The most notable characteristics include Positive Sentiment, Neutral Sentiment, Admiration, Amusement and Approval. Meanwhile, traits such as Anger, Annoy-

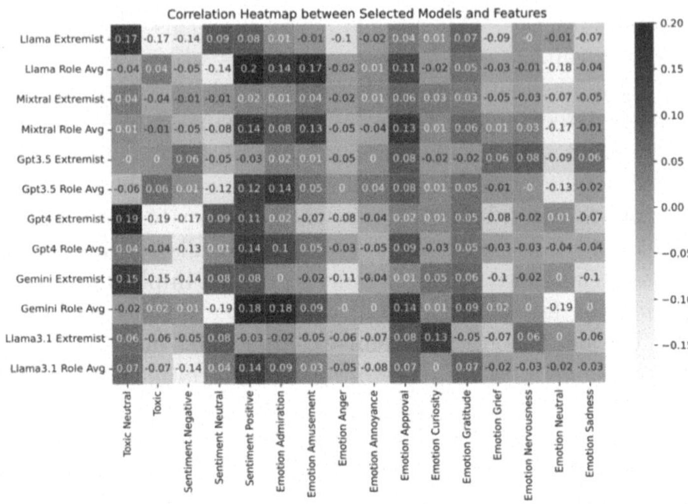

Fig. 1. A heatmap showing the correlation between models and tasks and the accuracy of predictions. Role resolutions are averaged into one cell.

ance and Grief contribute negatively to model prediction, surprisingly also when models predict extremism labels.

We argue that this is an extension of the Sally-Anne test noted earlier, since we similarly find that models have greater difficulty predicting the presence of negative traits in content. This could be directly related to the problem that models have a bias towards Engager Negator. This is reinforced by the higher levels of accuracy when positive traits are present. It would be more unusual to require agent and environment separation if the two users show positive affirming behaviour about a topic in which they disagree strongly (unlike the reverse), especially in an extremist setting. This would suggest that in the presence of positive emotions, the models can show less ability in social theory of mind whilst still being correct.

5 Conclusion

In this study, we have explored the capabilities of large language models (LLMs) in detecting extremism and identifying community roles within online extremist conversations, using new government definitions. Our findings indicate that while advanced models like GPT-4o demonstrate a high level of accuracy in extremism detection they exhibit notable weaknesses in accurately identifying user roles, particularly as the complexity of role definitions increase and in certain emotional environments (such as when displaying negative emotions). We have also seen the ability for models to greatly improve their accuracy in extremism classification tasks after fine-tuning, and for a greater level of reasoning after this fine-tuning

- which suggests further ability to build models in this space that can capture this nuance.

We addressed our first research question (RQ1) by evaluating whether LLMs can accurately detect extremism at a human level. The results showed that GPT-4o outperformed other models, achieving a Macro-F1 score of 0.71 and an accuracy of 80.8%, indicating that LLMs can effectively identify extremist content when provided with clear and concise definitions. For RQ2 and RQ3, concerning the detection of users' community roles and the impact of label resolution on model accuracy, we found that all models struggled with social reasoning tasks. The accuracy of role prediction deteriorated as the resolution of label categories increased, highlighting a limitation in the models' ability to navigate nuanced social dynamics. Furthermore, we found that across both extremism and role categorisation, models found emotional environments a factor for prediction accuracy. For instance, content displaying positive emotions (toxic neutral, sentiment positive, admiration) correlated to notably better model performance across the library of LLMs. Negative emotions, however, were consistently more difficult for the model to distinguish categories perhaps due to our niche topic-specific definitions. It is also hypothesised that experimenting with hyperparameters such as temperature will unlock higher and more stable classification results.

Overall our results show that despite some models being unsuitable for extremism classification, state-of-the-art LLMs have made some improvements to meet human-level categorisation, most notably seen between GPT generations and in fine-tuning approaches. Furthermore, we show that the definition of extremism announced by the UK government in 2024 is easier for LLMs to apply than its predecessor. Despite these benefits, there are still issues that language models will need to overcome before they reach reliable human standards for reasoning about complex extremist environments, but these could be realistically addressed in future work.

References

1. Alatawi, H.S., Alhothali, A.M., Moria, K.M.: Detecting white supremacist hate speech using domain specific word embedding with deep learning and BERT. IEEE Access **9**, 106363–106375 (2021)
2. Bailard, C.S., Tromble, R., Zhong, W., Bianchi, F., Hosseini, P., Broniatowski, D.: "Keep your heads held high boys!": examining the relationship between the proud boys' online discourse and offline activities. Am. Polit. Sci. Rev., 1–18 (2024)
3. Baumgartner, J., Zannettou, S., Squire, M., Blackburn, J.: The Pushshift Telegram dataset. In: Proceedings of the International AAAI Conference on Web and Social Media, vol. 14, pp. 840–849 (2020)
4. Gaikwad, M., Ahirrao, S., Phansalkar, S., Kotecha, K.: Online extremism detection: a systematic literature review with emphasis on datasets, classification techniques, validation methods, and tools. IEEE Access **9**, 48364–48387 (2021). https://doi.org/10.1109/ACCESS.2021.3068313
5. Gassmann, L., McConville, R., Edwards, M.: Predicting interpersonal influence from conversational features. In: 2023 10th International Conference on Behavioural and Social Computing (BESC) (2023)

6. Gassmann, L., Campbell, J., Edwards, M.: Influence reasoning capabilities of large language models in social environments. In: Proceedings of the AAAI Symposium on AI Trustworthiness and Risk Assessment for Challenging Contexts (ATRACC). AAAI Press, United States (2024). https://sites.google.com/view/aaai-atracc. aI Trustworthiness and Risk Assessment for Challenged Contexts : AAAI 2024 Fall Symposium, ATRACC ; Conference date: 07-11-2024 Through 09-11-2024
7. IBM Corporation: IBM WatsonX API (2024). https://www.ibm.com/docs/en/watsonx/saas?topic=started-watsonx-apis. Accessed 23 Nov 2024
8. Micallef, N., He, B., Kumar, S., Ahamad, M., Memon, N.: The role of the crowd in countering misinformation: a case study of the COVID-19 infodemic. In: 2020 IEEE International Conference on Big Data (Big Data), pp. 748–757 (2020). https://doi.org/10.1109/BigData50022.2020.9377956
9. Ministry of Housing, Communities and Local Government: New definition of extremism (2024). https://www.gov.uk/government/publications/new-definition-of-extremism-2024/new-definition-of-extremism-2024. Accessed 28 Oct 2024
10. Oueslati, W., Arrami, S., Dhouioui, Z., Massaabi, M.: Opinion leaders' detection in dynamic social networks. Concurr. Comput.: Pract. Experience **33**(1), e5692 (2021)
11. Oueslati, W., Mejri, S., Al-Otaibi, S., Ayouni, S.: Recognition of opinion leaders in social networks using text posts' trajectory scoring and users' comments sentiment analysis. IEEE Access **11**, 123589–123609 (2023). https://doi.org/10.1109/ACCESS.2023.3329049
12. Parthasarathy, V.B., Zafar, A., Khan, A., Shahid, A.: The ultimate guide to fine-tuning LLMS from basics to breakthroughs: an exhaustive review of technologies, research, best practices, applied research challenges and opportunities. Technical report, CeADAR: Ireland's Centre for AI, University College Dublin, Belfield, Dublin, Ireland (2024)
13. Piao, S., Whittle, J.: A feasibility study on extracting Twitter users' interests using NLP tools for serendipitous connections. In: 2011 IEEE Third International Conference on Privacy, Security, Risk and Trust and 2011 IEEE Third International Conference on Social Computing, pp. 910–915 (2011). https://doi.org/10.1109/PASSAT/SocialCom.2011.164
14. Prentice, S., Rayson, P., Taylor, P.J.: The language of Islamic extremism: towards an automated identification of beliefs, motivations and justifications. Int. J. Corpus Linguist. **17**(2), 183–204 (2012)
15. Scanlon, J.R., Gerber, M.S.: Automatic detection of cyber-recruitment by violent extremists. Secur. Inf. **3**(1), 1–10 (2014). https://doi.org/10.1186/s13388-014-0005-5
16. Secretary of State for the Home Department: Prevent Strategy. The Stationery Office, London (2011). https://www.official-documents.gov.uk. presented to Parliament by the Secretary of State for the Home Department by Command of Her Majesty
17. Smith, L.G., Piwek, L., Hinds, J., Brown, O., Chen, C., Joinson, A.: Digital traces of offline mobilization: a study of far-right extremist groups. J. Pers. Soc. Psychol. **125**(3), 496–518 (2023)
18. Stevenson, J., Edwards, M., Rashid, A.: Analysing the activities of far-right extremists on the Parler social network. In: 2023 IEEE/ACM International Conference on Advances in Social Networks Analysis and Mining (ASONAM), pp. 392–399. IEEE/ACM (2023)

19. Teutle, A.R.M.: Twitter: network properties analysis. In: 2010 20th International Conference on Electronics Communications and Computers (CONIELECOMP), pp. 180–186 (2010). https://doi.org/10.1109/CONIELECOMP.2010.5440773
20. user1342: Extremism classification LLM prompt (2025). https://gist.github.com/user1342/71c51bc82fcb5be0013b77a1e3eb89e6. Accessed 22 Mar 2025

PARAGAG: Proactive Answering Framework Integrating LLMs with Retrieval-Augmented Generation

Bikash Saha[1(✉)], Nanda Rani[1], Joheen Chakraborty[2], Divyanshu Singh[3], Soumyo V. Chakraborty[3], and Sandeep Kumar Shukla[1]

[1] Indian Institute of Technology Kanpur, Kanpur, India
{bikash,nandarani,sandeep}@cse.iitk.ac.in
[2] Massachusetts Institute of Technology, Cambridge, USA
joheen@mit.edu
[3] C3i Hub, IIT Kanpur, Kanpur, India

Abstract. Navigating organizational security policies is challenging due to their expansive and continuously evolving nature. LLMs can help streamline this process by understanding and interpreting complex policy language. This paper benchmarks this possibility and develops PARAG (Proactive Answering framework using Retrieval-Augmented Generation). It enhances policy comprehension by generating contextually accurate responses using a RAG-based approach. By structuring policy documents into semantically meaningful representations, the framework enables efficient indexing and retrieval of relevant information. This structured retrieval process allows the system to extract the most pertinent policy details, ensuring precise, policy-grounded answers without requiring extensive fine-tuning. To assess its effectiveness, we introduce an organization-agnostic dataset comprising 100 security policy-related question-answer pairs and evaluate the framework using policies from two real-world organizations. A comparative analysis of nine LLMs, including both black-box and open-source models, across 11 performance metrics demonstrates that GPT-4o-mini outperforms other black-box models, while Mistral-7B leads among open-source alternatives. These findings highlight the potential of RAG-based systems in transforming policy accessibility, enabling organizations to bridge the gap between policy documentation and practical compliance. Both the dataset and the framework code are hosted on the PARAG GitHub repository.

Keywords: Retrieval-Augmented Generation (RAG) · Large Language Models (LLMs) · Security Policy Automation · Policy Compliance · Question Answering · Cyber Policy · Policy Management · Security Management · LLM for Cybersecurity · Cybersecurity

1 Introduction

Organizational security policies are the bedrock of enterprise information and system protection. These documents typically detail access controls, data protec-

I. Praça et al. (Eds.): EICC 2025, CCIS 2500, pp. 20–37, 2025.
https://doi.org/10.1007/978-3-031-94855-8_2

tion standards, incident response procedures, and compliance requirements [4]. They serve as vital references for security teams, compliance officers, and employees seeking clarity on protocols. Yet, their extensive, intricate, and frequently updated nature can make specific information difficult to find [5,12]. Recent advancements in large language models (LLMs) present new opportunities for intelligent policy querying and automated question answering [36,39]. LLMs can help simplify this process by understanding complex policy language and providing quick, context-driven responses.

However, standard LLM-based approaches struggle with organization-specific policy queries due to a lack of training on proprietary documents, making them unaware of unique policy structures and compliance requirements [9]. Additionally, security policies are dynamic, often updated in response to regulatory changes and organizational needs. To maintain relevance, traditional LLM-based question-answering systems rely on frequent fine-tuning, which is both computationally expensive and resource-intensive [15,25,26,31]. Traditional LLM-based answering systems introduce significant maintenance overhead and limit real-time adaptability, making them impractical for enterprises requiring up-to-date security compliance information.

This paper explores the feasibility of using LLMs for security policy question-answering through a structured benchmarking approach. To enable this, we develop PARAG (Proactive Answering framework using Retrieval-Augmented Generation), an LLM-powered RAG framework designed to handle queries related to organization-specific security policies. PARAG offers a scalable and adaptable solution that seamlessly updates as policy documents evolve, ensuring continuous relevance and accuracy.

Our framework builds a vector database containing organization-specific security policies embedded in a vector format. At query time, the system retrieves relevant policy segments, appends them to the user's query, and creates a context-aware prompt for the LLM. This ensures that the LLM provides responses aligned with organization-specific policies and security guidelines. Maintaining a structured vector database facilitates efficient updates whenever policies are modified, eliminating the need for costly fine-tuning while preserving the framework's core structure and scalability.

To evaluate the feasibility of using LLMs for this task, we curate a benchmark dataset of 100 diverse security policy-related questions, each paired with human-written answers to establish a ground truth for comparison. We systematically test the framework using both black-box and open-source LLMs, assessing their effectiveness in retrieving and answering policy-based queries. To ensure a comprehensive analysis, we evaluate the system using 11 diverse performance metrics, offering a holistic comparison of each model's capability in generating accurate, reliable, and policy-grounded responses. The contributions of this paper are summarized as follows:

1. We benchmark the use of LLMs for organization-specific security policy question-answering by developing PARAG (Proactive Answering framework

using Retrieval-Augmented Generation), an efficient and scalable RAG-based LLM framework.

2. To support benchmarking, we present a benchmark organization-agnostic security policy question-answer pair dataset. This dataset comprises 100 diverse questions spanning 10 distinct categories, enabling a standardized assessment of organization-specific policy question-answering systems.[1]
3. We comprehensively evaluate the framework using nine black-box and open-source LLMs, applying it to security policies from two distinct industry sectors, and measuring performance with 11 diverse evaluation metrics for a thorough comparative analysis.
4. We also develop a prototype chatbot system based on our framework to support real-world adoption. The system allows organizations to seamlessly integrate their security policies and select their preferred LLM, enabling easy deployment without extensive customization.

The rest of the paper is structured as follows: Sect. 2 discusses existing retrieval-augmented approaches in cybersecurity. Section 3 details the proposed framework in detail. Section 4 describes the experimental setup and evaluation metrics. Section 5 presents the results, key insights, and analysis of real-world security policy applications, along with identified limitations and future directions. Section 6 discusses ethical considerations and responsible AI use. Finally, Sect. 7 summarizes the findings and concludes the study.

2 Related Work

Several research studies have explored the role of Retrieval-Augmented Generation (RAG) in cybersecurity. By incorporating external, task-specific knowledge retrieval, RAG enhances the generative capabilities of LLMs. In this section, we discuss key related work that underpins this approach.

CADDIE [18] is a RAG-based application for governance and compliance, integrating policy analysis and real-time regulatory updates. It leverages a dense vector index of regulatory documents to provide compliance officers with factually grounded responses, significantly reducing manual effort in governance, risk, and compliance (GRC) frameworks. Similarly, ProveRAG [7] uses an automated RAG-based LLM pipeline to analyze security vulnerabilities and retrieve mitigation strategies from structured sources like the National Vulnerability Database (NVD) and Common Weakness Enumeration (CWE) repositories.

Another RAG-powered question-answering model, CyberRAG [38], is designed for cybersecurity education. It integrates an ontology-aware RAG system that enhances factual accuracy by validating LLM-generated responses against an ontology-based knowledge graph. Similarly, research on hybrid RAG architectures [22] demonstrates that combining multiple retrieval techniques, such as Sentence-Window and Parent-Child methodologies, significantly

[1] The complete set of questions along with framework code is provided at: https://github.com/bikasaha/PARAG-Proactive-Answering-Framework-Integrating-LLMs-with-Retrieval-Augmented-Generation.

enhances response fidelity and contextuality. These approaches leverage re-ranking mechanisms to optimize retrieval, ensuring that policy-related queries receive the most relevant and fact-grounded responses.

CyRAG and GraphCyRAG [27], designed for cybersecurity threat intelligence, integrate RAG with Neo4j-based knowledge graphs and vector databases to provide policy analysts with contextualized threat insights. These systems improve security posture analysis by retrieving interconnected attack patterns, vulnerabilities, and compliance policies. Additionally, research on self-reflective RAG architectures, such as CRAG (Corrective Retrieval-Augmented Generation) [35], demonstrates improvements in policy compliance models by iteratively verifying retrieved responses against an evidence-based retrieval evaluation framework, reducing errors and enhancing accuracy.

Unlike previous works that focus on generalized policy retrieval or cybersecurity threat intelligence, this work benchmarks the effectiveness of LLMs in answering security policy queries through PARAG. It builds on traditional RAG-based approaches by incorporating structured document chunking, which ensures that the most relevant policy segments are retrieved and processed to generate accurate responses. By systematically evaluating diverse LLMs on real-world policy datasets, this work provides a comprehensive understanding of how LLMs can be leveraged for policy comprehension and compliance automation, offering insights into their strengths and limitations in handling complex security policies.

3 PARAG

3.1 System Design

PARAG consists of three integrated system modules designed to facilitate the RAG-based answer generation process. These modules are: (i) Database Module - stores organization-specific security policy documents, (ii) Query Processing Module - processes user queries and retrieves relevant contextual documents from the database, and (iii) Answer Generation Module - generates responses based on the retrieved contextual documents and the user's query. The details of each module are explained in this section, and an overview of PARAG's design is shown in Fig. 1.

Database Module. This module serves as the central repository for storing organization-specific security policy documents. These documents are preprocessed and structured to enable efficient retrieval, ensuring that relevant policy information can be accessed with minimal latency. This module contains three different sub-modules discussed below:

1. **Pre-processing:** The input to database module consists of one or more PDF files, each containing organizational security policies that often vary in structure-ranging from paragraphs and bullet points to tables and headers [3]. Given this complexity, preprocessing becomes a critical step. We extract textual content using a robust PDF parsing library. Throughout the extraction

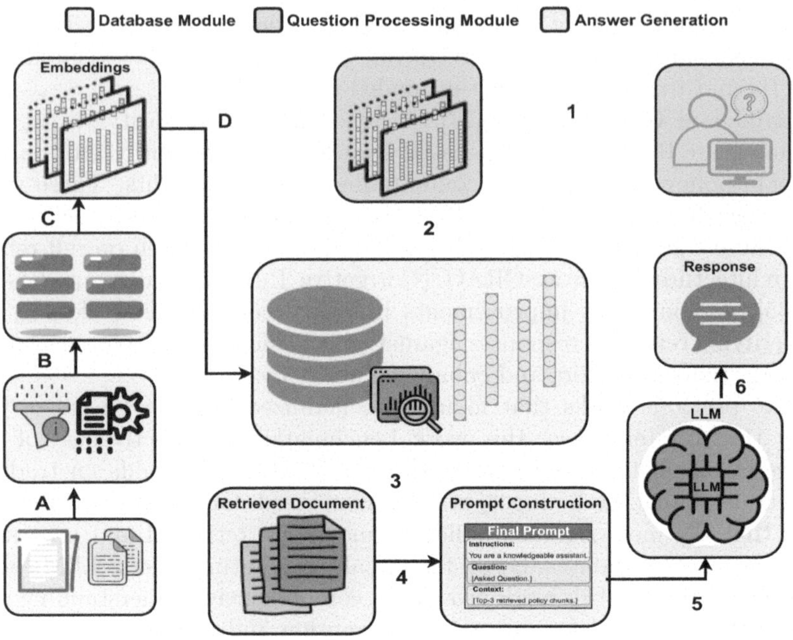

Fig. 1. PARAG Workflow Overview

process, we preserve structural markers such as headings and subheadings, ensuring that the logical flow of the policy remains intact.

2. **Embeddings:** Given the varying lengths of policy documents, ranging from short to extensive, it is required to divide them into smaller, manageable chunks for efficient processing. Due to the dense and technical nature of these documents, each PDF document is segmented into chunks of up to 512 tokens. This chunk size is carefully chosen to maintain a balance between preserving contextual coherence and ensuring computational efficiency.
   To prevent the loss of crucial information at chunk boundaries and keep chunks connected, we employ an overlapping mechanism, wherein the last 50 tokens of a chunk are repeated at the beginning of the next chunk within the same document. This ensures smooth context transitions and prevents abrupt breaks in information flow. This overlap mechanism is strictly applied within individual documents to preserve their internal narrative structure while optimizing retrieval accuracy.
   Further, each chunk is converted into a fixed-size vector representation using a Sentence Transformer [28], enabling efficient semantic retrieval. This transformation ensures that even contextually complex queries can be accurately matched to the most relevant policy segments.
3. **Store:** To enable fast and efficient retrieval, the generated chunk embeddings, along with their associated metadata, are stored in a vector database. Each entry in the database comprises three key components: the embedding, a d-dimensional vector representing the semantic encoding of the chunk; the

text data, which includes the original chunk text or a reference to its location within the policy document; and metadata, which provides additional contextual details such as document ID, chunk index, section headers, and other relevant identifiers. This structured storage approach ensures that policy segments can be retrieved with high accuracy while preserving the integrity and traceability of the original documents.

Query Processing Module. Once the vector database is populated with all policy chunks, the system can respond to user queries using a RAG approach [15]. This module consists of two sub-module discussed below:

1. **Question Embedding:** When a user poses a query (e.g., "What are the guidelines for password construction?"), the system encodes the question text into a query embedding using the *same* sentence transformer as it embeds in database module. This consistency ensures that the user query and the policy chunks lie in the same semantic embedding space.
2. **Similarity-based Retrieval:** To retrieve the most relevant policy segments, the query embedding undergoes a *similarity search* within the vector database. We employ similarity matching to measure the closeness between the query and stored chunk embeddings, ensuring precise matching. PARAG retrieve the *top-k most similar chunks* to output them as relevant contextual documents, prioritizing those with the highest similarity scores. This approach enhances the quality of the final language model prompt by filtering out irrelevant content and ensuring that only the most contextually relevant information contributes to the generated response.

Answer Generation. Once the question-relevant contextual documents are retrieved from the database, it gets passed to answer generation module which contains LLM module to construct the prompt and generate the answer. This module includes two sub-module discussed below:

1. **Prompt Construction:** The retrieved contextual document act as supporting document, such as hint for LLM model to provide response specific to

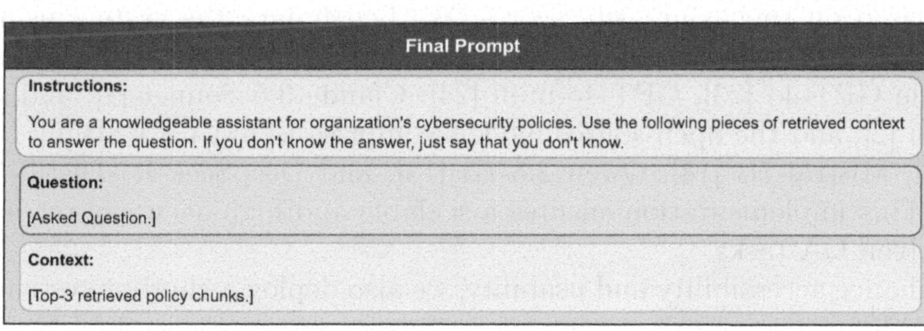

Fig. 2. Final Prompt Template After Retrieval.

organization policies. The retrieved context gets appended to prompt template of instruction and user prompt to integrate organization specific relevant knowledge in the prompt. We refer the updated prompt as enhanced prompt. An example is shown in Fig. 2.

2. **Response Generation:** The prepared prompt ensures that the model synthesizes an answer that is directly aligned with the provided policy context, minimizing the risk of hallucinations or inaccuracies. By restricting the model's knowledge to the retrieved chunks, the generated response remains faithful to the original policy text. The final answer is then formatted and presented to the user.

Our methodology seamlessly integrates dense retrieval with generative question answering (QA), specifically designed to handle the complexities of security policy documents. By employing overlapping chunks, we preserve crucial boundary information and ensure that important context is not lost during retrieval. Utilizing the sentence-transformer allows the system to accurately capture nuanced terminology and regulatory language, improving the relevance of retrieved information. By focusing the LLM's context on the most relevant retrieved chunks, we enhance factual accuracy and alignment with the source text. This scalable pipeline helps organizations maintain up-to-date policy comprehension, supporting compliance and decision-making as security protocols evolve.

3.2 Implementation

The PARAG system is designed with a robust pipeline that integrates efficient document processing, vector storage, and retrieval mechanisms. The preprocessing phase utilizes PyPDF2[2] to extract text from security policy documents, ensuring structured and accurate content extraction. To generate dense vector representations of both policy document chunks and user queries, we employ Sentence Transformer, which encodes textual data into meaningful embeddings. The embeddings are stored and managed using Milvus Lite[3], a high-performance vector database optimized for Approximate Nearest Neighbor (ANN) search.

For similarity matching, we use Cosine Similarity to retrieve the most relevant policy segments, ensuring contextually accurate responses. The retrieval process is fine-tuned using a top-3 search mechanism, selecting the three most relevant chunks based on their similarity scores. We benchmark the system using nine diverse LLMs, including black-box and open-source models. The black-box models include GPT-4o [23], GPT-4o-mini [24], Claude 3.5 Sonnet [1], and Claude 3.5 Haiku [2], and the open-source models comprise LLaMA 3.1-8B [6], LLaMA 2-7B [32], Mistral-7B [14], Qwen 2.5-7B [13], and DeepSeek-R1-Distill-Llama-8B [11]. This implementation ensures a scalable and high-accuracy retrieval for policy-driven QA tasks.

To enhance accessibility and usability, we also deploy a chatbot system based on the PARAG framework, maintaining the same retrieval and LLM integra-

[2] https://pypi.org/project/PyPDF2/.

[3] https://milvus.io/docs/v2.2.x/milvus_lite.md.

tion settings. This chatbot enables a portable security policy-related question-answering system, allowing relevant users to efficiently query organizational policies in real-time. The deployment enables seamless policy queries across organizations, enhancing accessibility and compliance. A glimpse of the deployed system can be seen in Fig. 3.

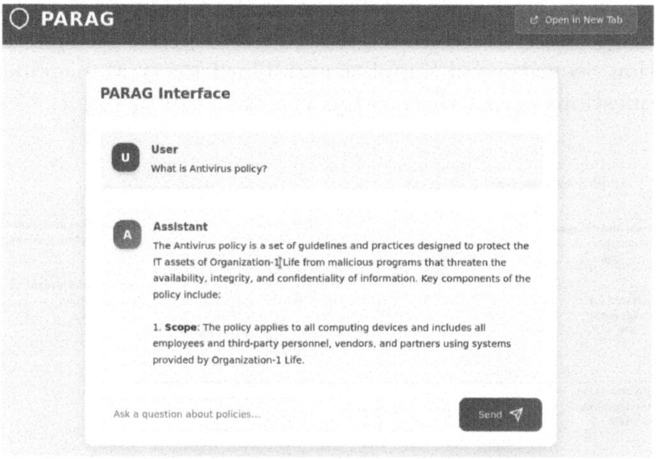

Fig. 3. Deployed system based on PARAG framework.

4 Experimental Setup

To assess the efficiency of PARAG, we adopt a comprehensive evaluation approach, testing it on real-world organizational policies as discussed in Sect. 3.2. This section discusses all essential components, including datasets, evaluation metrics, and system details.

4.1 Dataset

Policy Documents: To assess the effectiveness of PARAG, we conduct experiments using security policy documents from two real-world organizations, referred to as Organization-1 and Organization-2. Organization-1's dataset comprises 24 security policies, while Organization-2's dataset includes 31 security policies, enabling a diverse evaluation across different policy structures. We preprocess each organization's security policy documents as discussed in Sect. 3.1. Since our evaluation includes closed-source LLMs, we prioritized privacy by masking and anonymizing all organization-specific identifiers within the documents before chunking. This ensures that no proprietary or sensitive information is exposed during processing while still preserving the integrity and structure of the policy content. Each dataset is independently stored and processed, ensuring that retrieval and question answering were performed within the context of a single organization's policies at a time.

Question-Answer Pairs: To systematically evaluate our proposed framework, we developed a comprehensive set of 100 benchmark questions distributed across 10 distinct categories. These questions are organization-agnostic and serve as a standardized evaluation set, enabling the assessment of security policies for any organization. The categories include Access and Usage Policies, Information and Cybersecurity Policies, Incident and Crisis Management, Corrective Actions and Documentation, Data and Asset Management, Antivirus and Patch Management, Change and Release Management, Privacy and Compliance, Software and Application Security, and Supplier and Third-Party Management. The distribution of questions across these categories is shown in Fig. 4.

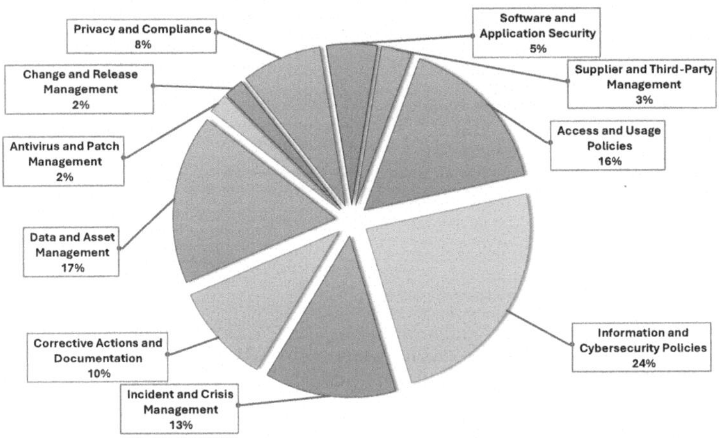

Fig. 4. Categorization of question types.

Each question set includes a combination of real-world security inquiries and synthetic questions designed to explore edge cases or less frequently referenced policy aspects, such as encryption key rotation frequency. To ensure a rigorous evaluation, we meticulously reviewed all 100 questions per organization, thoroughly examined the corresponding policy documents, and crafted detailed human-written answers aligned with each organization's security policies. The prepared responses serve as a benchmark, enabling a direct comparison between model-generated and human-authored answers. This comparison allows for a deeper analysis of accuracy, completeness, and factual consistency. By incorporating domain expertise in answer preparation, we establish a robust benchmark for evaluating the retrieval and answer-generation capabilities of our framework across diverse security domains.

4.2 Evaluation Metrics

We employ multi-dimensional evaluation approach [30] to compare the generated responses with ground truth to ensure PARAG's effectiveness. The evaluation framework includes *lexical metrics* (ROUGE [17]), *semantic similarity metrics* (BERTScore [37], Semantic Similarity), *structural and readability metrics* (Flesch-Kincaid [8], Distinct-N [16]), and *content focus metrics* (Keyphrase Matching [29]).

Lexical metrics assess word and sequence overlap between generated and reference responses, with ROUGE measuring n-gram overlap and emphasizing recall to capture key information while maintaining alignment. BERTScore computes contextual similarity using pre-trained models like BERT by evaluating the cosine similarity of token embeddings, ensuring robustness to paraphrasing. Structural and readability metrics evaluate coherence, fluency, and uniqueness, with Distinct-N measuring n-gram diversity to ensure responses remain engaging and meaningful. Content focus metrics verify the model's ability to capture and retain critical information relevant to security policies.

For a comprehensive evaluation, we employ 11 performance metrics, which are explained as explained in Table 1.

Table 1. Performance Metrics Details

Metrics	Key Detail
ROUGE-1 (R-1)	Measures unigram (single-word) overlap to assess basic word-level content relevance
ROUGE-2 (R-2)	Measures bigram (two consecutive words) overlap, capturing contextual and sequential alignment
ROUGE-L (R-L)	Valuates the longest common subsequence (LCS) to assess structural similarity and text fluency
Semantic Similarity (SS)	Computes cosine similarity between sentence embeddings to assess the semantic closeness of responses
BERTScore-Precision (BS-P)	Measures how much of the generated content aligns with the reference answer
BERTScore-Recall (BS-R)	Assesses how well the generated response captures all key points from the reference
BERTScore-F1-Score (BS-F1)	Provides a balanced view of BERTScore precision and recall
Flesch-Kincaid Readability (FKR)	Valuates readability based on sentence length and syllable count, ensuring generated responses are accessible and suitable for the target audience
Distinct-1 (D-1)	Measures the proportion of unique unigrams (words) in the response to assess lexical diversity
Distinct-2 (D-2)	Measures the proportion of unique bigrams (two-word sequences) to quantify repetition and variation
Keyphrase Matching (KM)	Evaluates keyphrase overlap using KeyBERT [10] to assess core idea alignment.

The scores for all the mentioned performance metrics range from 0 to 1, except for the Flesch-Kincaid Readability Score, which falls within a range of 0 to 100. The combination of these metrics provides a holistic evaluation of response quality, covering *relevance, coherence, fluency, diversity, and readability.* This comprehensive assessment allows for a nuanced understanding of PARAG's performance, facilitating targeted improvements to enhance policy-driven question-answering capabilities.

5 Results and Discussion

We evaluate PARAG on datasets from both organizations using a series of LLMs, as discussed in Sects. 3.2 and 4. This section presents an analysis of PARAG's performance, compares the employed LLMs across selected evaluation metrics, and highlights key observations. Additionally, we discuss the system's limitations and outline potential directions for future improvements.

5.1 Discussion

We evaluate PARAG on policy documents from both organizations using the prepared question-answer pairs. For each organization's document set, we generate responses using nine different LLMs, treating the respective policy documents as the retrieval database. Each question in the dataset is processed through PARAG, and responses are generated across all LLMs. Responses from all models are compared with human-written ground truth to compute performance metrics from Sect. 4.2.

Black-Box LLM Performance Analysis. The evaluation results for black-box LLMs across both organizations, as shown in Table 2, indicate that GPT-4o-mini delivers the best overall performance for Organization-1, leading in 9 out of 11 metrics. In contrast, GPT-4o emerges as the top-performing model for Organization-2, outperforming others in 5 out of 11 metrics. GPT-4o ranks second-best in Organization-1 with the highest score in 2 metrics, while GPT-4o-mini and Claude 3.5 Haiku share the second-best position for Organization-2, each excelling in 3 metrics.

Table 2. Performance Comparison of Black-box Models

Met-rics	GPT-4o-mini		GPT-4o		Claude 3.5 Sonnet		Claude 3.5 Haiku	
	Org1	Org2	Org1	Org2	Org1	Org2	Org1	Org2
R-1	**0.442**	0.298	0.441	0.348	0.396	0.349	0.396	**0.373**
R-2	**0.227**	0.134	0.208	**0.142**	0.170	0.125	0.171	0.133
R-L	**0.315**	0.199	0.294	**0.218**	0.248	0.207	0.244	0.210
BS-P	**0.709**	**0.679**	0.680	0.676	0.626	0.644	0.620	0.647
BS-R	0.650	0.547	**0.674**	0.599	0.651	0.594	0.663	**0.615**
BS-F1	**0.676**	0.602	0.674	**0.633**	0.635	0.617	0.638	0.629
SS	0.792	0.714	**0.802**	0.779	0.782	0.776	0.792	**0.795**
FKR	**27.00**	**31.98**	22.90	26.09	15.93	19.23	09.84	13.326
D-1	**0.790**	**0.820**	0.732	0.780	0.708	0.728	0.679	0.708
D-2	**0.941**	0.927	0.940	**0.946**	0.938	0.932	0.931	0.938
KM	**0.296**	0.23	0.272	**0.252**	0.234	0.228	0.242	0.250

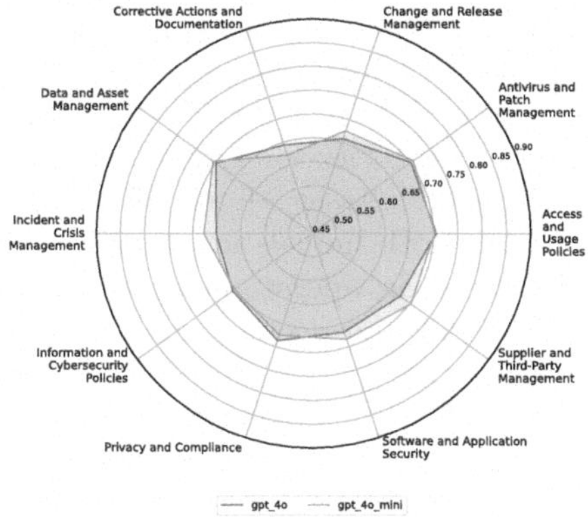

(a) Comparison of F1-scores on Organization 1 documents

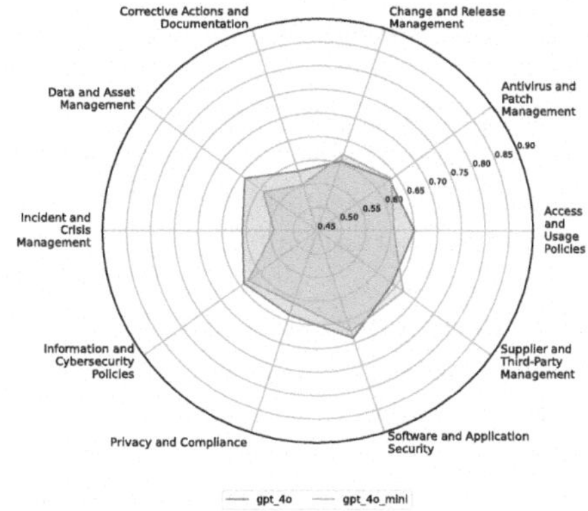

(b) Comparison of F1-scores on Organization 2 documents

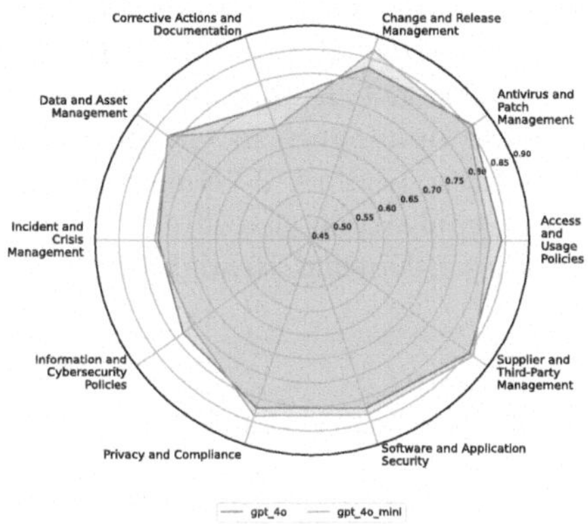

(c) Comparison of semantic similarity on Organization 1 documents

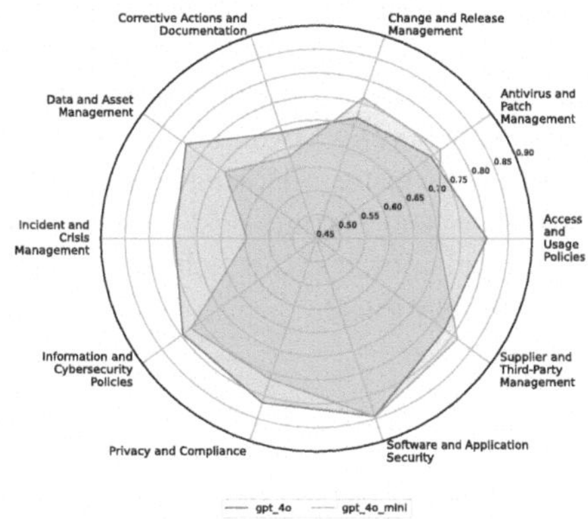

(d) Comparison of semantic similarity on Organization 2 documents

Fig. 5. Question category-wise comparison of black-box models.

Analyzing specific aspects, GPT-4o-mini excels in retrieval precision and lexical diversity for Organization-1, with superior ROUGE-1, ROUGE-2, and Distinct-N scores. It also achieves better readability (FKR) across both organizations, suggesting clearer and more digestible responses. However, for Organization-2, GPT-4o surpasses others in semantic similarity and BERTScore Recall, highlighting its strength in capturing policy-specific context and nuances.

Claude 3.5 models perform reasonably well in terms of semantic preservation, but their lower precision and lexical diversity scores suggest more generalized or less informative responses compared to the GPT-4o series. Based on these observations, the analysis was later narrowed to focus specifically on GPT-4o and GPT-4o-mini, as they consistently ranked among the top two performers across both organizations and offered a meaningful trade-off between performance and efficiency.

Supporting these findings, the radar plots in Fig. 5 show that GPT-4o consistently outperforms GPT-4o-mini in most policy categories for Organization-2, particularly in semantically rich areas like Data and Asset Management and Corrective Actions and Documentation. Although both models are comparable in Organization-1-with GPT-4o-mini slightly ahead in operational categories-GPT-4o demonstrates stronger contextual alignment and deeper semantic representation, especially as captured by cosine similarity.

In conclusion, while GPT-4o-mini is highly efficient and performs best in precision-driven metrics for simpler contexts, GPT-4o stands out as the most reliable and context-aware model overall, making it better suited for handling complex, policy-intensive tasks in real-world cybersecurity governance.

Open-Source LLM Performance Analysis. The evaluation results for open-source LLMs across both organizations are summarized in Table 3. Mistral-7B emerged as the top-performing model for Organization-1, leading in 7 out of 11 metrics, whereas Qwen2.5-7B demonstrated clear dominance in Organization-2, outperforming all others in 10 out of 11 metrics. LLaMA 2-7B consistently ranked as the second-best model across both organizations, excelling in a few individual metrics and achieving the highest Flesch-Kincaid Readability Score in Organization-2.

Table 3. Performance Comparison of Open-Source Models

Met-rics	LLama3.1		LLama2		Mistral		Qwen2.5		DeepSeek-LL	
	Org1	Org2	Org1	Org2	Org1	Org2	Org1	Org2	Org1	Org2
R-1	0.259	0.271	0.255	0.194	**0.325**	0.270	0.287	**0.386**	0.287	0.326
R-2	0.094	0.080	0.085	0.045	**0.127**	0.077	0.112	**0.122**	0.099	0.097
R-L	0.158	0.156	0.154	0.121	**0.193**	0.158	0.158	**0.181**	0.162	0.164
BS-P	0.546	0.579	0.567	0.563	**0.573**	0.579	0.551	**0.613**	0.570	0.600
BS-R	0.619	0.591	0.546	0.494	0.643	0.585	**0.670**	**0.668**	0.632	0.618
BS-F1	0.576	0.583	0.547	0.520	**0.602**	0.578	0.601	**0.638**	0.594	0.605
SS	0.724	0.733	0.638	0.570	**0.749**	0.710	0.748	**0.770**	0.737	0.743
FKR	31.88	35.50	**34.71**	**40.36**	25.72	39.45	22.41	22.57	28.84	31.67
D-1	0.366	0.306	**0.617**	0.574	0.578	0.336	0.561	**0.584**	0.562	0.520
D-2	0.570	0.469	0.757	0.660	0.849	0.483	**0.881**	**0.892**	0.837	0.795
KM	0.186	0.186	0.176	0.112	**0.228**	0.204	0.227	**0.206**	0.220	0.188

For Organization-1, Mistral-7B achieved the highest scores in ROUGE-1, ROUGE-2, and ROUGE-L, indicating strong fluency and contextual relevance in its responses. Qwen2.5-7B, though slightly behind in some surface-level precision metrics, showed competitive performance in semantic similarity and lexical diversity. In Organization-2, Qwen2.5-7B significantly outperformed other models in Distinct-2 and Keyphrase Matching, reflecting its ability to generate more informative and policy-aligned responses. It also led in BERTScore Recall and semantic similarity, highlighting its superior capability to retain contextual depth.

Although LLaMA 2-7B produced highly readable outputs, its lower semantic alignment suggests that readability alone may not ensure policy-specific relevance. In contrast, Qwen2.5-7B and Mistral-7B demonstrated a stronger balance between accuracy, diversity, and contextual understanding. Based on this comprehensive performance across multiple dimensions, our analysis was later

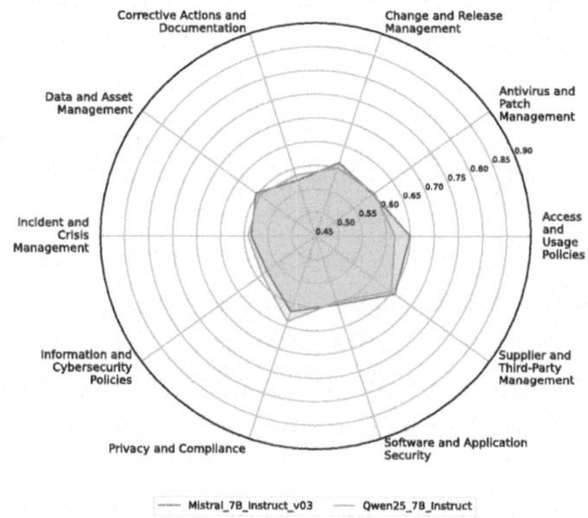

(a) Comparison of F1-score on Organization 1 documents

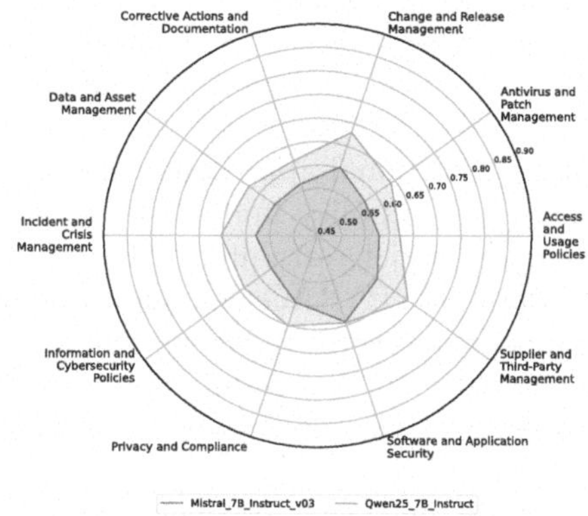

(b) Comparison of F1-score on Organization 2 documents

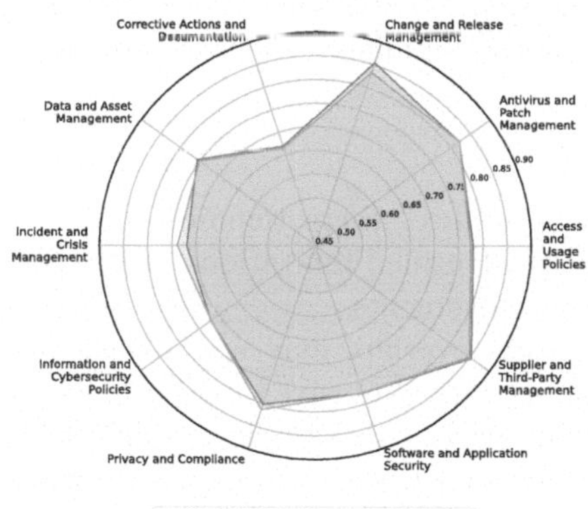

(c) Comparison of semantic similarity on Organization 1 documents

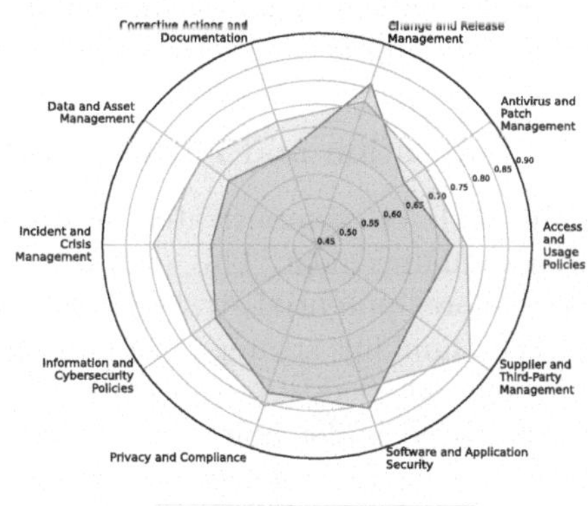

(d) Comparison of semantic similarity on Organization 2 documents

Fig. 6. Question category-wise comparison of open-source models.

narrowed to focus on Qwen2.5-7B and Mistral-7B, as they consistently emerged as the top-performing open-source models across both organizations.

The radar plots shown in Fig. 6 further support these findings. In Organization-1, Mistral-7B slightly outperformed Qwen in BERTScore F1 in categories such as Access and Usage Policies and Change and Release Management. However, Qwen exhibited stronger semantic similarity in areas like Information and Cybersecurity Policies and Incident and Crisis Management. In Organization-2, Qwen2.5-7B clearly outshone Mistral across most policy categories, particularly in complex domains such as Supplier and Third-Party Management, Privacy and Compliance, and Incident Response. These results affirm Qwen's robust generalization and deeper semantic retention.

In conclusion, while Mistral-7B remains highly effective for scenarios requiring concise and relevant answers, Qwen2.5-7B stands out as the overall best-performing open-source model, excelling in semantic depth, policy alignment, and response diversity. This makes it especially suitable for complex cybersecurity policy analysis across varying organizational contexts.

Unlike traditional approaches that require frequent fine-tuning, PARAG adapts seamlessly to policy updates by simply refreshing the database, eliminating the need for model retraining. Through extensive evaluation across multiple policy datasets, PARAG demonstrates its ability to generate precise and policy-grounded answers, reducing reliance on direct LLM inference and mitigating hallucinations. The results demonstrate that integrating a structured retrieval process with advanced language models significantly improves response relevance and coherence.

By systematically benchmarking multiple black-box and open-source LLMs, our findings highlight key trends in retrieval effectiveness, response diversity, and readability across different policy structures. The study demonstrates how retrieval-augmented approaches make organizational policies more accessible and actionable, improving real-world decision-making alignment.

5.2 Limitation and Future Work

While PARAG demonstrates promising performance, there remains scope for further enhancement. This section outlines the identified limitations of PARAG and explores potential strategies for future improvements.

The chunking strategy, though incorporating overlap, may not always preserve context effectively, especially in documents with extensive tables, diagrams, or structured bullet points, potentially causing fragmented retrieval. The interpretability of LLM-generated answers remains another challenge, as reformulated or abstracted responses may lack direct traceability to original policy texts, necessitating stricter extractive summarization for high-stakes decisions.

Future enhancements include implementing adaptive chunking and hierarchical retrieval, enabling the system to dynamically adjust chunk sizes based on semantic boundaries for more precise information retrieval. Incorporating explainability features, such as confidence scores and policy phrase highlighting, can further improve compliance traceability and user trust. Additionally, integrating vision-language models would allow to process multi-modal policy documents, including security diagrams, flowcharts, and structured tables, expanding its applicability beyond textual policies. These advancements would make PARAG a more robust, transparent, and adaptable solution for managing evolving security policies.

6 Ethics and Responsible Use

Deploying PARAG for security policy retrieval requires careful attention to confidentiality, privacy, accuracy, and fairness. Given that security policies may contain sensitive information, ensuring secure storage, controlled access, and regulatory compliance (e.g., GDPR, HIPAA) is essential [19,33]. While RAG improves factual accuracy, LLMs can still generate misleading responses, necessitating human oversight, audit trails, and disclaimers for accountability [34].

To address bias and fairness, continuous output monitoring and policy reviews help minimize unintended biases from both policy data and external LLM training sources [20]. User privacy is safeguarded through anonymization, secure logging, and encrypted query handling to prevent data exposure. Additionally, AI-driven policy retrieval may influence compliance workflows and decision-making, requiring employee training and adaptive security protocols [21]. Balancing automation and human oversight ensures reliability, while ongoing auditing and refinement mitigates misinformation, bias, and AI overreliance.

7 Conclusion

This study benchmarks the effectiveness of LLMs in handling security policy queries through PARAG, a retrieval-augmented generation (RAG) framework designed to enhance organizational security policy comprehension. By integrating an efficient retrieval mechanism with advanced LLMs, PARAG ensures that responses remain factually grounded, policy-aligned, and semantically relevant without requiring frequent model fine-tuning. Through extensive evaluation on security policies from two real-world organizations, we benchmarked nine LLMs across 11 performance metrics, demonstrating that retrieval-augmented approaches significantly enhance policy accessibility and response reliability. GPT-4o-mini and GPT-4o emerged as the best-performing black-box models, while Mistral-7B and Qwen2.5-7B led among open-source alternatives, highlighting the impact of policy structure on retrieval effectiveness. The results confirm that PARAG's structured retrieval process improves policy comprehension, reduces hallucinations, and provides a scalable, adaptable, and security-aware question-answering system for enterprise use. This work establishes a structured benchmarking framework for evaluating LLMs in policy-based question-answering, offering insights into the strengths and limitations of retrieval-augmented approaches in managing complex security policies.

References

1. Anthropic: Introducing claude 3.5 sonnet (2024). https://www.anthropic.com/news/claude-3-5-sonnet
2. Anthropic: Introducing the next generation of claude (2024). https://www.anthropic.com/news/claude-3-family
3. Beltagy, I., Peters, M.E., Cohan, A.: LongFormer: the long-document transformer. arXiv preprint: arXiv:2004.05150 (2020)
4. Beuchelt, G.: Information technology security management. In: Computer and Information Security Handbook, pp. 475–508. Elsevier (2025)
5. De Bruijn, H., Janssen, M.: Building cybersecurity awareness: the need for evidence-based framing strategies. Gov. Inf. Q. **34**(1), 1–7 (2017)
6. Dubey, A., et al.: The llama 3 herd of models. arXiv preprint: arXiv:2407.21783 (2024)
7. Fayyazi, R., Trueba, S.H., Zuzak, M., Yang, S.J.: ProveRAG: provenance-driven vulnerability analysis with automated retrieval-augmented LLMS. arXiv preprint: arXiv:2410.17406 (2024)
8. Flesch, R.: A new readability yardstick (1948). https://doi.org/10.1037/h0057532
9. Gao, Y., et al.: Retrieval-augmented generation for large language models: a survey. arXiv preprint: arXiv:2312.10997 (2023)
10. Grootendorst, M.: KeyBERT: minimal keyword extraction with BERT. (2020).https://doi.org/10.5281/zenodo.4461265
11. Guo, D., et al.: DeepSeek-R1: Incentivizing reasoning capability in LLMS via reinforcement learning. arXiv preprint: arXiv:2501.12948 (2025)
12. Harknett, R.J., Stever, J.A.: The new policy world of cybersecurity. Public Adm. Rev. **71**(3), 455–460 (2011)
13. Hui, B., et al.: Qwen2.5-coder technical report. arXiv preprint: arXiv:2409.12186 (2024)
14. Jiang, A.Q., et al.: Mistral 7B. arXiv preprint: arXiv:2310.06825 (2023)
15. Lewis, P., et al.: Retrieval-augmented generation for knowledge-intensive NLP tasks. In: Advances in Neural Information Processing Systems, vol. 33, pp. 9459–9474 (2020)
16. Li, J., Galley, M., Brockett, C., Gao, J., Dolan, B.: A diversity-promoting objective function for neural conversation models. In: Proceedings of the 2016 Conference of the North American Chapter of the Association for Computational Linguistics: Human Language Technologies (NAACL-HLT), pp. 110–119. Association for Computational Linguistics, San Diego (2016). https://aclanthology.org/N16-1014. Accessed 12 Feb 2024
17. Lin, C.Y.: Rouge: A package for automatic evaluation of summaries. In: Text Summarization Branches Out, pp. 74–81. Association for Computational Linguistics, Barcelona, Spain (2004). https://aclanthology.org/W04-1013. Accessed 02 Dec 2024
18. Lodge, B.: Rage against the machine with BERT for proactive cybersecurity posture. In: 2024 IEEE International Conference on Big Data (BigData), pp. 3579–3588. IEEE (2024)
19. McCallister, E.: Guide to Protecting the Confidentiality of Personally Identifiable Information. Diane Publishing (2010)
20. Mitchell, S., Potash, E., Barocas, S., D'Amour, A., Lum, K.: Algorithmic fairness: choices, assumptions, and definitions. Ann. Rev. Statist. Appl. **8**(1), 141–163 (2021)

21. Novelli, C., Taddeo, M., Floridi, L.: Accountability in artificial intelligence: what it is and how it works. AI Soc. **39**(4), 1871–1882 (2024)
22. Omrani, P., Hosseini, A., Hooshanfar, K., Ebrahimian, Z., Toosi, R., Akhaee, M.A.: Hybrid retrieval-augmented generation approach for LLMS query response enhancement. In: 2024 10th International Conference on Web Research (ICWR), pp. 22–26. IEEE (2024)
23. OpenAI: Gpt-4o (2024). https://platform.openai.com/docs/models#gpt-4o
24. OpenAI: Gpt-4o mini (2024). https://platform.openai.com/docs/models#gpt-4o-mini
25. Ovadia, O., Brief, M., Mishaeli, M., Elisha, O.: Fine-tuning or retrieval? Comparing knowledge injection in LLMS. arXiv preprint: arXiv:2312.05934 (2023)
26. Prabhune, S., Berndt, D.J.: Deploying large language models with retrieval augmented generation. arXiv preprint: arXiv:2411.11895 (2024)
27. Rahman, M., et al.: Retrieval augmented generation for robust cyber defense. Technical report, Pacific Northwest National Laboratory (PNNL), Richland, WA (United States) (2024)
28. Reimers, N.: Sentence-BERT: sentence embeddings using Siamese BERT-networks. arXiv preprint: arXiv:1908.10084 (2019)
29. Rose, S., Engel, D., Cramer, N., Cowley, W.: Automatic keyword extraction from individual documents. Text Min.: Appl. Theory., 1–20 (2010).https://doi.org/10.1002/9780470689646.ch1
30. Saha, B., Rani, N., Shukla, S.K.: MaLAware: automating the comprehension of malicious software behaviours using large language models (LLMS) (2025). https://arxiv.org/abs/2504.01145
31. Soudani, H., Kanoulas, E., Hasibi, F.: Fine tuning vs. retrieval augmented generation for less popular knowledge. In: Proceedings of the 2024 Annual International ACM SIGIR Conference on Research and Development in Information Retrieval in the Asia Pacific Region, pp. 12–22 (2024)
32. Touvron, H., et al.: Llama 2: open foundation and fine-tuned chat models. arXiv preprint: arXiv:2307.09288 (2023)
33. Voigt, P., Von dem Bussche, A.: The EU General Data Protection Regulation (GDPR). A Practical Guide, 1st ed., vol. 10, no. 3152676, pp. 10–5555. Springer International Publishing, Cham (2017)
34. Weidinger, L., et al.: Ethical and social risks of harm from language models. arXiv preprint: arXiv:2112.04359 (2021)
35. Yan, S.Q., Gu, J.C., Zhu, Y., Ling, Z.H.: Corrective retrieval augmented generation. arXiv preprint: arXiv:2401.15884 (2024)
36. Yang, X., Wang, Z., Wang, Q., Wei, K., Zhang, K., Shi, J.: Large language models for automated q&a involving legal documents: a survey on algorithms, frameworks and applications. Int. J. Web Inf. Syst. **20**(4), 413–435 (2024)
37. Zhang, T., Kishore, V., Wu, F., Weinberger, K.Q., Artzi, Y.: BERTScore: evaluating text generation with BERT. In: Proceedings of the International Conference on Learning Representations (ICLR) (2020). https://arxiv.org/abs/1904.09675. Accessed 02 Dec 2024
38. Zhao, C., et al.: Ontology-aware rag for improved question-answering in cybersecurity education. arXiv preprint: arXiv:2412.14191 (2024)
39. Zhu, Y., et al.: Large language models for information retrieval: a survey. arXiv preprint: arXiv:2308.07107 (2023)

Evaluating LLaMA 3.2 for Software Vulnerability Detection

José Gonçalves[1], Miguel Silva[1(✉)], Bernardo Cabral[2], Tiago Dias[1], Eva Maia[1], Isabel Praça[1], Ricardo Severino[2], and Luís Lino Ferreira[2]

[1] GECAD, ISEP, Polytechnic of Porto, rua Dr. António Bernardino de Almeida, 4249-015 Porto, Portugal
{jpsgs,mdgsa,tiada,egm,icp}@isep.ipp.pt

[2] ISEP/INESC TEC, Polytechnic of Porto, rua Dr. António Bernardino de Almeida, 4249-015 Porto, Portugal
{bemac,sev,llf}@isep.ipp.pt

Abstract. Deep Learning (DL) has emerged as a powerful tool for vulnerability detection, often outperforming traditional solutions. However, developing effective DL models requires large amounts of real-world data, which can be difficult to obtain in sufficient quantities. To address this challenge, DiverseVul dataset has been curated as one of the largest datasets of vulnerable and non-vulnerable C/C++ functions extracted exclusively from real-world projects. Its goal is to provide high-quality, large-scale samples for training DL models. Nevertheless, during our study several inconsistencies were identified in the raw dataset while applying pre-processing techniques, highlighting the need for a refined version. In this work, we present a refined version of DiverseVul dataset, which is used to fine-tune a large language model, LLaMA 3.2, for vulnerability detection. Experimental results show that the use of pre-processing techniques led to an improvement in performance, with the model achieving an F1-Score of 66%, a competitive result when compared to our baseline, which achieved a 47% F1-Score in software vulnerability detection.

Keywords: Artificial intelligence · Cybersecurity · Large language models · Software testing · Vulnerability detection

1 Introduction

As software adoption grows, so do concerns about its security. In modern software development, these risks are worsened by the use of AI-powered code assistants that can generate insecure code that can be deployed in production environments [21,31]. Thus, it is of paramount importance to ensure the quality of software, largely due to the fact that it serves as the backbone of many critical operations [6,19,26].

One of the key approaches to maintain software quality is software testing, which provides insight into the development process to assess whether the software is well developed [18]. This field encompasses a wide range of methods,

I. Praça et al. (Eds.): EICC 2025, CCIS 2500, pp. 38–51, 2025.
https://doi.org/10.1007/978-3-031-94855-8_3

each designed to evaluate different quality attributes [5]. Among these methods, Software Vulnerability Detection (SVD) plays a critical role in identifying security vulnerabilities. Static analysis, a common technique for SVD, examines source code for vulnerabilities without executing it, relying on methods such as rule/template-based analysis, static symbolic execution, and code similarity detection [6,19,25,26,37]. In order to reduce the time and cost associated with manual vulnerability analysis, extensive research has been conducted on automating SVD using Deep Learning (DL). Studies have explored different models and feature representations using structural, syntactic, and semantic information extracted from source code [32,37]. Despite achieving high performance, DL-based SVD still faces several challenges, mainly due to the complexity of source code and the scarcity of high-quality labeled datasets [1,26].

To address the complexity of source code, several studies have explored multiple feature representations at different levels of granularity. Most modern approaches reduce data complexity by pre-processing and extracting implicit structural, syntactic and semantic features from the source code [1]. Successful methods remove non-essential elements such as comments and variable names [38], while normalising code components that contribute to vulnerabilities, thus simplifying the samples. The presence of programmer-defined elements can also affect model performance. Experimental results [33] indicate that Artificial Intelligence (AI) models often rely on identifier names rather than the underlying code semantics associated with vulnerabilities. This reliance introduces a security risk, as malicious actors can create adversarial samples to manipulate the model's output by changing identifier names while preserving the actual vulnerability. Graph representations have also been explored, specifically Abstract Syntax Trees (ASTs), Control Flow Graphs, and Code Property Graphs, to effectively represent the semantic and syntactic information hidden in source code [29,32,37].

In recent years, many research works have attempted to fine-tune Large Language Models (LLMs) on source code for SVD, as will be described in Sect. 2. However, the most promising results are achieved using synthetic datasets, which do not capture the full complexity of real-world source code [9]. As a result, these models often underperform when applied to real-world source code. Driven by the lack of valuable real-world data, a dataset named DiverseVul [4], containing vulnerable and non-vulnerable C/C++ source code collected from several real-world vulnerable projects, was proposed. The authors of the dataset analyzed it in combination with other well-known ones, and even though they present discouraging results, they mention that the combination of the datasets could have turned some data points miscellaneous, which would affect performance [4].

This work aims to study the DiverseVul dataset in the context of LLMs for SVD, while also exploring the influence of pre-processing techniques. LLaMA 3.2, a recent model released by Meta, was fine-tuned for SVD using the DiverseVul dataset. This text-only lightweight model allows for the development of solutions that can be executed in lower end devices, while achieving good results. The performance of this model for SVD was analyzed both before and after the application of source code pre-processing techniques, to assess if the model is influenced by non-relevant code elements. This pre-processing was conducted using the Source Code Processing Engine (SCoPE) [7], a C/C++ data processing framework that operates at the function level. During the data processing phase, several erroneous data points in DiverseVul were identified and removed, resulting in the creation of a refined version of the original dataset.

In summary, the main contributions of this work include:

- a refined version of the DiverseVul dataset processed by the SCoPE framework without the erroneous entries. The refined version of the dataset is publicly available on Zenodo [8].
- an assessment of the influence of pre-processing on vulnerability detection with LLMs.
- a performance assessment of newer and lighter models, such as LLaMA 3.2, for SVD.

This work is organized as follows. Section 2 provides a literature review of existing LLMs applications for SVD. Section 3 presents the experimental setup, describing the dataset and its flaws, the data pre-processing and the LLaMA 3.2 model. Section 4 presents and discusses the results achieved before and after the data processing phase. Section 5 addresses the potential threats to validity. Finally, Sect. 6 summarises the work with the main conclusions drawn from it and describes new research lines to be explored in future work.

2 State of the Art

SVD is the process of identifying weaknesses or flaws within software systems. It can be performed via two different analysis methods: (i) static and (ii) dynamic [37]. Static analysis consists of analyzing the source code to detect vulnerabilities. This method has high code coverage but is usually more prone to false positives [37]. Dynamic analysis is performed by running a set of test cases in the target system and observing its behavior in real-time to ensure compliance with specified requirements [37]. This method has higher miss rate and achieves less coverage, but it is generally more accurate and raises fewer false alarms [37]. In comparison to one another, dynamic analysis requires a substantial amount of computational power to generate and run multiple test cases in the System Under Test (SUT), whereas static analysis does not. However, static analysis requires a great amount of knowledge and attention, which can be very laborious. To circumvent this process and decrease the amount of expertise necessary to perform this static vulnerability detection, several research works have attempted to leverage DL to automate the vulnerability detection process [14,30].

The literature shows that DL has been widely adopted to perform static analysis of source code, experimenting several feature representations for ingestion and algorithms, ranging from the most simple Neural Networks (NN) to more complex model architectures, such as Transformers in the case of LLMs. For instance, Russell et al. [24] used multiple types of NN to detect software vulnerabilities from C/C++ source code extracted from Debian, Github, and SATE IV Juliet Test Suite repositories represented by token embeddings. The Convolution Neural Network (CNN) had the best performance on the SATE IV dataset, and the CNN combined with the random forest showed the most promising results on the Debian and Github datasets. Their work proved that machine learning is more effective than the use of static analyzers, which are typically rule-based. Similarly, Bilgin et al. [3] decided to experiment with the CNN, but instead of using the token representation, the authors chose to leverage ASTs to represent the source code. The AST ensured a structured format suitable for the model, with the source code transformed into a binary tree of uniform depth for consistent function size. The binary tree was then converted into a depth-ordered token vector and further transformed into a three-dimensional numeric vector using token embeddings. The study, conducted on a dataset with predetermined vulnerabilities, yielded promising results, noting variations in the difficulty of detecting different vulnerability types.

Other works continued to explore more data representation of source code that would contain more information than the one provided by ASTs. Li et al. [15] explored the impact of incorporating semantic code information in detecting vulnerabilities associated with specific library/API calls in source code. They compared the effectiveness of classification using only data dependency features with the simultaneous use of control dependency and data dependency features. The Joern tool was employed to analyze programs, extract features, identify library/API function calls, and generate code gadgets from program slices. These gadgets were vectorized and tested with different NNs. The study found that combining control and data dependency features is more effective than relying solely on data dependency. The best performance was achieved by a Bidirectional Long Short-Term Memory (BLSTM) NN, with an F1-Score of 92.4% on a dataset of 68353 code gadgets, including 13686 labelled as vulnerable. The authors tested various methods to handle imbalanced data and found that not using any sampling produced the best results.

Zou et al. [38] proposed μVulDeePecker, a deep learning-based system for multi-classification of software vulnerabilities. Inspired by VulDeePecker [16], the first binary vulnerability detection system based on deep learning, μVulDeePecker refines the code gadget concept. Unlike VulDeePecker, which proved ineffective for multiclass vulnerability detection, μVulDeePecker introduces two variants. The first, VulDeePecker+, modifies the NN architecture directly. The second trains separate VulDeePecker classifiers for each vulnerability type. However, the latter variant faces scalability issues and is less effective with small sample sizes. μVulDeePecker comprises six modules, including data pre-processing for cleaning and normalizing input data, classification of collected data into vulnerability classes, and an automatic program based on lexical analysis to analyze tokens. Vectors representing normalized code gadgets are generated using a word-to-vector model. The BLSTM NN architecture, built using Keras, is trained separately for global and local feature learning to prevent interference between networks. Using the same dataset as Li et al. [16], μVulDeePecker achieved a maximum F1-Score of 96.28%, demonstrating its effectiveness and improvement over its predecessor.

Zhou et al. [36] introduced Devign, a model utilizing graph structures to learn vulnerability patterns from source code functions through Graph Neural Networks (GNNs). To mitigate dataset inaccuracies resulting from static tool classifications, they produced their own dataset containing labelled C functions from open-source projects like Linux Kernel, QEMU, Wireshark, and FFmpeg. The Devign architecture includes a graph layer representing the semantics of composite code, a Gated Graph Recurrent layer learning node characteristics from neighboring nodes, and a Conv module extracting significant node representations for graph-level prediction. The results show that Devign, with comprehensive graph-encoded semantics, outperforms most modern vulnerability identification methods. It achieved a maximum F1-Score of 84.97% on Linux function data following best practices. For the FFmpeg dataset, Devign obtained a lower maximum F1-Score of 34.92%, while on the Qemu dataset, it achieved a higher F1-Score of 41.12%.

With the advances in AI, LLMs have also gained some popularity in the vulnerability detection field. For instance, Khare et al. [11] evaluate the performance of LLMs from the GPT and LLaMA families on 5 different datasets spanning two programming languages and including code samples from synthetic and real-world projects. The authors obtained great results on the synthetic datasets, with smaller fine-tuned LLMs performing better than larger models and lower results on the real-world projects datasets which include CVEFixes Java and C/C++ subsets. Nevertheless, their results show that LLMs can outperform existing static analysis and deep learning vulnerability detection tools by identifying the vulnerable data flows in the code. Chen et al. [4], responsible for the DiverseVul dataset, also combined multiple datasets including their own, and fine-tuned LLMs of four different families to assess their quality. The results achieved by the authors show that NatGen, CodeGPT, CodeBERT, and ReVeal were the best-performing algorithms of each family, with NatGen achieving the highest F1-Score of 47.2%. However, their approach is prone to errors since the combination of datasets tailored differently might result in miscellaneous data. Additionally, the authors do not seem to perform any data balancing, which could explain the high accuracy and low F1-Score. Mamede, Pinconschi, and Abreu [20] fine-tuned JavaBERT from the BERT model architecture family to perform multi-label classification. The authors use the Juliet Test Suite for Java, composed of synthetic data covering 21 different vulnerability types, to fine-tune the model and its evaluation presents an accuracy of 98.9%.

Though the results seem promising, multiple works have shown that synthetic datasets in the field do not represent the code implemented in real projects [2,10,17,28]. Driven by the lack of valuable data, Kuang et al. [12] mutated user-defined identifiers in the source code while maintaining its syntactic and semantic information to generate counterfactual training data. They implement their data processing method on VulDeePecker, Draper, and CodeXGLUE datasets and evaluate them on CodeBERT. Their results show that their method performed better on Draper and VulDeePecker datasets, achieving 94% and 85% F1-Score, respectively. However, these are both semi-synthetic datasets.

Alternatively, some works have also leveraged prompt engineering techniques to perform vulnerability detection. For instance, Steenhoek et al. [28] evaluated the performance of 11 well-known LLMs and compared their performance on vulnerability detection on a dataset curated from real-world data collected from Common Vulnerabilities and Exposures (CVE) records. The results show that the LLMs achieved 50% to 63% F1-Score. Purba et al. [22] decided to use CVEFixes for fine-tuning LLMs for vulnerability detection on two types of attacks, SQL injection and buffer overflow. However, their results using CodeGen, DaVinci and GPT-4 and GPT3.5 show low performance, similar to other literature works. Zhang et al. [34] planned to explore further prompt engineering for SVD using specifically ChatGPT and proceeded to complement previous works in the field by improving the design of the prompts utilized. They experimented on two datasets containing Java and C/C++ source code. In regards to vulnerability detection in the C/C++ code, their chain-of-thought prompts improved performance to 74% F1-Score.

3 Experimental Setup

In this paper, we explore the DiverseVul dataset in the field of SVD by fine-tuning a pre-trainned model for vulnerability detection, while also analysing the impact of the data pre-processing. The experiments were conducted on a laptop running Windows 11 OS, equipped with a 12th Gen I7-12650HX 2.1 GHz CPU, 64GB of RAM, and an RTX A4500 GPU. The following subsections describe the most relevant aspects of the experiment's setup.

3.1 Dataset and Data Pre-processing

DiverseVul is a comprehensive vulnerability dataset that includes 18945 vulnerable C/C++ across 150 Common Weakness Enumerations (CWEs). Complementing these are 330492 non-vulnerable functions sourced from 7514 vulnerability-fixing commits. The dataset's curation involved the crawling of two security issue websites and the extraction of data from the vulnerability fix commits of 797 vulnerable projects. Commits implying vulnerability introduction were excluded using regex and manual verification for commits with more than 10 functions altered. Projects related to these commits were cloned, and their vulnerable functions were extracted and labelled as vulnerable or non-vulnerable before and after the commits, respectively. Duplicate entries were removed via MD5 hash comparison. Functions lacking specific CWE labels were cross-referenced with CVE numbers from the titles, aligning them with respective CWEs in the National Vulnerability Database (NVD). The final dataset is composed of 8 features that are described in Table 1.

Table 1. DiverseVul Dataset Features.

Feature	Description
func	Contains the C/C++ function code
target	Is the function vulnerable or not
cwe	List of respective CWEs present in the function
project	Project from where the function was extracted
commit_id	Identifier of the commit
size	Size of the function
message	Commit message

AI models may rely on programmer-controlled elements to predict whether a function is vulnerable, potentially overlooking the actual vulnerability. To address this, the SCoPE framework was used to process all source code from DiverseVul dataset. SCoPE facilitates code standardization by leveraging regular expressions. It efficiently removes redundant white spaces and comments while substituting specific strings with generic tokens. It also uses token substitution to streamline both function and variable declarations, significantly reducing code size. This ensures a consistent representation of the data, free from superfluous elements that could impact the model's performance. To apply the transformations, the code is first converted into a representative tree, which is used to detect variables and function names. Then, token replacement and code normalization are performed using regular expressions, resulting in a normalized representation of the source code.

Following the application of SCoPE to DiverseVul, 268253 entries were successfully processed without problems, while 62239 entries were processed with errors due to the use of macros in the code. Subsequently, an additional step was taken to identify functions sharing identical declarations. This precautionary measure aimed to validate the integrity of the data curation process. As a result, 7901 duplicated entries were identified, and faulty data was detected. Examples of faulty data included entries containing only comments or entries that, despite having identical code, differed only in function names while having opposite labels. One of the faulty entries can be observed in Fig. 1, where two entries have the same code with different function names.

```
void schi_box_del(GF_Box *s)
{
        gf_free(s);
}
```

```
void jp2h_box_del(GF_Box *s)
{
        gf_free(s);
}
```

Fig. 1. Example of Duplicated Entries.

After identifying and removing erroneous entries, all entries containing only comments were eliminated. Duplicate entries were also identified and removed to ensure that there was only one unique example for each entry. After this pre-processing, a new refined version of the DiverseVul dataset, now free of such entries, was created and is publicly available on Zenodo for further use and exploration.

Given the significant class imbalance in the dataset, as shown in Fig. 2, a balanced subset of 10000 entries was selected for training and evaluation of our model in the SVD task, consisting of 5000 samples from each class. This selection aimed to optimize computational efficiency while maintaining the representativeness of the dataset. To evaluate model performance, the holdout method was used to randomly partition the data into training, test and validation sets. The training set contained 70% of the samples, while the remaining 30% were split between the validation and test sets.

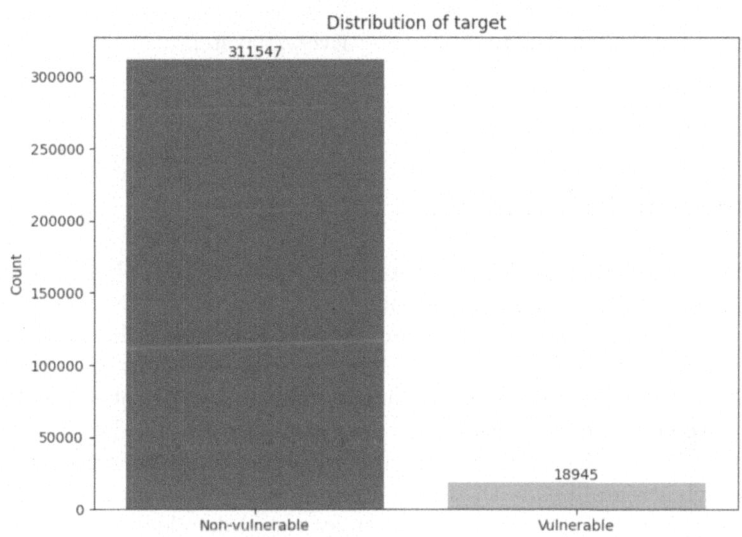

Fig. 2. DiverseVul Target Distribution.

3.2 Model and Fine-Tuning

This study aims to assess if LLaMA 3.2 with 1.23 billion parameters is effective at SVD, both before and after optimizing the input data. Despite the original purpose of the LLaMA 3.2 model was sequence-to-sequence generation, originally pre-trained on a large multilingual dataset for general language tasks, in this study, the model was fine-tuned to perform binary classification on code samples, categorizing them as either vulnerable or not vulnerable. Due to computational constraints, low-rank adaptation (LoRA) was used in the fine-tuning process.

LoRA is a Parameter-Efficient Fine-Tuning (PEFT) technique that reduces the number of trainable parameters, significantly reducing training time while maintaining model performance [23].

To optimize the model for vulnerability detection, multiple fine-tuning iterations were performed to identify the best hyperparameter configuration. The final set of hyperparameters that achieved the best performance is detailed in Table 2.

Table 2. Model Hyperparameters.

Hyperparameter	Value
Epochs	4
Batch Size	32
Learning Rate	2×10^{-5}
Weight Decay	0.01
Evaluation Interval	Every 100 steps

4 Results and Discussion

One of the main challenges faced by some LLMs in SVD is their limited token capacity, which limits the amount of code they can process at a time [35]. When the length of the analyzed source code exceeds this limit, the model's ability to capture interrelationships between different code segments may be compromised [13]. To address this issue, reducing the number of tokens in the input not only mitigates this limitation, but also improves the model's performance, allowing it to more effectively detect vulnerabilities in larger segments of code. This approach streamlines both the training and inference processes [27].

In this context, since the dataset had an average token length of 374.13 tokens per function, SCoPE was used to reduce the token length, resulting in 300.38 tokens per function after processing. Although the model used in the experiments supports significantly larger context windows (up to 128,000 tokens), a limit of 1024 tokens was set in the tokenizer for the maximum size allowed per function. This limit was set to reduce the need for padding in smaller functions, optimizing the use of computational resources during training. Only 5.48% of the functions in the original dataset exceeded this limit, and this proportion was further reduced with SCoPE processing. Reducing the average length of the functions minimized the cases of truncation in the tokenization process.

The model was initially trained on just two epochs to quickly assess whether SCoPE would have a positive impact without the computational cost of full fine-tuning. Due to the long training time, full tuning of all setups was impractical, so only the best performing setup was fine-tuned to refine its performance.

Table 3. Obtained results for DiverseVul.

Processed	Accuracy	Precision	Recall	F1-Score
No	62%	63%	62%	62%
Yes	63%	63%	63%	63%

Table 3 summarizes the results, showing that pre-processing slightly improved the model's performance. This improvement can be attributed to the reduced average token length or to the fact that SCoPE removed all function and variable names. This forced the model to learn without relying on these potentially misleading identifiers, which are often defined by programmers and may not reliably reflect the intent of the underlying code. As a result, the model focused on understanding the root cause of vulnerabilities rather than relying on token names, resulting in more robust and generalized learning.

Building on this, LLaMA 3.2 was further fine-tuned on the processed dataset for four epochs, resulting in performance improvements. These improved results are summarized in Table 4. Although not originally designed for code-specific tasks, LLaMA 3.2 demonstrates competitive performance when fine-tuned for vulnerability detection.

Table 4. Performance of NatGen and LLaMA 3.2.

Model	Accuracy	Precision	Recall	F1-Score
NatGen [4]	92%	52%	43%	47%
LLaMA 3.2	66%	65%	67%	66%

We compared our results with the best AI model reported by the authors of the DiverseVul dataset, who used NatGen for the same purpose as our model, SVD. While the differences in models, datasets, and pre-processing prevent a direct comparison, this provides a baseline for performance. LLaMA achieves lower accuracy, however, it outperforms NatGen in terms of precision, recall, and F1-Score. Since accuracy is heavily influenced by the number of classes tested and the imbalance of the original dataset, as shown in Fig. 2, it cannot be used as a reliable metric without proper balancing. The authors of the DiverseVul dataset noted that errors may have occurred during the merging of multiple datasets, which might have introduced label noise. In our study, we used SCoPE to remove erroneous data points from the DiverseVul dataset, which may explain the improved model performance.

5 Threats to Validity

The threats to the results' validity can essentially be found in the data utilized to train the models. Since the process of identifying erroneous entries is automated, some unidentified faulty data points may still exist in the dataset. Additionally, this study did not consider if the selected LLM was pre-trained on data contained in the dataset utilized for fine-tuning. If this is the case, it could hurt performance.

6 Conclusions and Future Work

This study explores the usage of DiverseVul in the context of LLMs for SVD. The dataset was curated using the SCoPE framework, which played a critical role in identifying and eliminating erroneous data points. Those points were removed, and a refined version of the dataset was built and will be made publicly available. To the best of the authors' knowledge, the erroneous data identified in DiverseVul has not been reported in previous research, making this study a valuable contribution to the field.

While evaluating the performance of LLaMA 3.2 on SVD, the impact of various pre-processing techniques was explored. Fine-tuning the model on the processed dataset demonstrated an improvement in detection performance. This pre-processing step allowed the model to focus more effectively on the semantic structure of the code, improving its ability to identify vulnerabilities and minimizing distractions caused by irrelevant elements. This success motivated further experimentation, which included increasing the number of training epochs, as the initial training approach did not fully stabilize the model's loss. This adjustment led to a 3% improvement, leading to a 66% of F1-Score, demonstrating that LLaMA 3.2 can achieve competitive performance in SVD tasks.

Several promising directions for future work can be explored. One direction is to expand the scope of the study by using a larger portion of the DiverseVul dataset, as only a subset was used in this initial research. Extending the training time and increasing the number of epochs could further optimize the model's performance, as resource constraints limited the potential for refinement in this study. Another direction could be to explore the robustness of the model, as the results suggest that programmer-controlled elements may influence its predictions.

Given recent advances in large language models, it would also be valuable to compare the performance of other state-of-the-art models, such as DeepSeek-r1, on SVD task. Such comparisons could provide deeper insights into their relative strengths and weaknesses. In addition, advanced techniques such as prompt engineering and retrieval-augmented generation could be explored to further improve both the performance and explainability of the model when applied to SVD tasks.

Acknowledgments. This work was done and funded in the scope of the BEHAVIOR project (NORTE2030-FEDER-00576300 no. 14391). This work was also supported by UIDB/00760/2020.

Disclosure of Interests. The authors have no competing interests to declare that are relevant to the content of this article.

References

1. Akimova, E.N., et al.: A survey on software defect prediction using deep learning. Mathematics **9**(11) (2021). https://doi.org/10.3390/math9111180
2. Alexopoulos, N., Brack, M., Wagner, J.P., Grube, T., Mühlhäuser, M.: How long do vulnerabilities live in the code? A large-scale empirical measurement study on FOSS vulnerability lifetimes. In: 31st USENIX Security Symposium (USENIX Security 22), pp. 359–376. USENIX Association, Boston (2022). https://www.usenix.org/conference/usenixsecurity22/presentation/alexopoulos
3. Bilgin, Z., Ersoy, M.A., Soykan, E.U., Tomur, E., Çomak, P., Karaçay, L.: Vulnerability prediction from source code using machine learning. IEEE Access **8**, 150672–150684 (2020). https://doi.org/10.1109/ACCESS.2020.3016774
4. Chen, Y., Ding, Z., Alowain, L., Chen, X., Wagner, D.: DiverseVul: a new vulnerable source code dataset for deep learning based vulnerability detection. In: Proceedings of the 26th International Symposium on Research in Attacks, Intrusions and Defenses. RAID 2023. ACM (2023).https://doi.org/10.1145/3607199.3607242
5. Ganney, P.S., Pisharody, S., Claridge, E.: Chapter 9 - software engineering. In: Taktak, A., Ganney, P.S., Long, D., Axell, R.G. (eds.) Clinical Engineering (Second Edition), pp. 131–168. Academic Press, second edition edn. (2020). https://doi.org/10.1016/B978-0-08-102694-6.00009-7
6. Ghaffarian, S.M., Shahriari, H.R.: Software vulnerability analysis and discovery using machine-learning and data-mining techniques: a survey. ACM Comput. Surv. **50**(4), 1–36 (2017). https://doi.org/10.1145/3092566
7. Gonçalves, J., Dias, T., Maia, E., Praça, I.: Scope: evaluating LLMS for software vulnerability detection. In: Mehmood, R., et al. (eds.) Distributed Computing and Artificial Intelligence, Special Sessions I, 21st International Conference, pp. 34–43. Springer Nature Switzerland, Cham (2025)
8. Gonçalves, J.P., et al.: RDiverseVul: Refined diversevul (2025). https://doi.org/10.5281/zenodo.15051277
9. Jain, R., Gervasoni, N., Ndhlovu, M., Rawat, S.: A code centric evaluation of C/C++ vulnerability datasets for deep learning based vulnerability detection techniques. In: Proceedings of the 16th Innovations in Software Engineering Conference. ISEC '23, Association for Computing Machinery, New York (2023). https://doi.org/10.1145/3578527.3578530
10. Katsadouros, E., Patrikakis, C.: A survey on vulnerability prediction using GNNs. In: Proceedings of the 26th Pan-Hellenic Conference on Informatics. PCI 2022, ACM. https://doi.org/10.1145/3575879.3575964
11. Khare, A., Dutta, S., Li, Z., Solko-Breslin, A., Alur, R., Naik, M.: Understanding the effectiveness of large language models in detecting security vulnerabilities (2023). https://doi.org/10.48550/ARXIV.2311.16169
12. Kuang, H., Yang, F., Zhang, L., Tang, G., Yang, L.: Leveraging user-defined identifiers for counterfactual data generation in source code vulnerability detection. In: 2023 IEEE 23rd International Working Conference on Source Code Analysis and Manipulation (SCAM), pp. 143–150 (2023). https://doi.org/10.1109/SCAM59687.2023.00024
13. Li, T., Zhang, G., Do, Q.D., Yue, X., Chen, W.: Long-context LLMs struggle with long in-context learning (2024). https://arxiv.org/abs/2404.02060
14. Li, Z., Zou, D., Tang, J., Zhang, Z., Sun, M., Jin, H.: A comparative study of deep learning-based vulnerability detection system. IEEE Access **7**, 103184–103197 (2019). https://doi.org/10.1109/access.2019.2930578

15. Li, Z., Zou, D., Xu, S., Jin, H., Zhu, Y., Chen, Z.: SySeVR: a framework for using deep learning to detect software vulnerabilities (2018). https://doi.org/10.48550/ARXIV.1807.06756
16. Li, Z., et al.: VulDeePecker: a deep learning-based system for vulnerability detection (2018). https://doi.org/10.48550/ARXIV.1801.01681
17. Lipp, S., Banescu, S., Pretschner, A.: An empirical study on the effectiveness of static C code analyzers for vulnerability detection. In: Proceedings of the 31st ACM SIGSOFT International Symposium on Software Testing and Analysis. ISSTA '22, ACM. https://doi.org/10.1145/3533767.3534380
18. M. Altaie, A., Gh. Alsarraj, R., H. Al-Bayati, A.: Verification and validation of a software: a review of the literature. Iraqi J. Comput. Inf. **46**(1), 40-47 (2020). https://doi.org/10.25195/ijci.v46i1.249
19. Malhotra, R., Gupta, S., Singh, T.: A systematic review on application of deep learning techniques for software quality predictive modeling. In: 2020 International Conference on Computational Performance Evaluation (ComPE). IEEE. https://doi.org/10.1109/compe49325.2020.9200103
20. Mamede, C., Pinconschi, E., Abreu, R.: A transformer-based ide plugin for vulnerability detection. In: Proceedings of the 37th IEEE/ACM International Conference on Automated Software Engineering. ASE '22, Association for Computing Machinery, New York (2023). https://doi.org/10.1145/3551349.3559534
21. Perry, N., Srivastava, M., Kumar, D., Boneh, D.: Do users write more insecure code with AI assistants? In: Proceedings of the 2023 ACM SIGSAC Conference on Computer and Communications Security. CCS '23, ACM. https://doi.org/10.1145/3576915.3623157
22. Purba, M., Ghosh, A., Radford, B.J., Chu, B.: Software vulnerability detection using large language models. In: 2023 IEEE 34th International Symposium on Software Reliability Engineering Workshops (ISSREW), pp. 112–119. IEEE Computer Society, Los Alamitos (2023). https://doi.org/10.1109/ISSREW60843.2023.00058
23. Razuvayevskaya, O., et al.: Comparison between parameter-efficient techniques and full fine-tuning: a case study on multilingual news article classification. PLOS ONE **19** (2023). https://api.semanticscholar.org/CorpusID:260886825
24. Russell, R.L., et al.: Automated vulnerability detection in source code using deep representation learning (2018). https://doi.org/10.48550/ARXIV.1807.04320
25. Shafiq, S., Mashkoor, A., Mayr-Dorn, C., Egyed, A.: A literature review of using machine learning in software development life cycle stages. IEEE Access **9**, 140896-140920 (2021). https://doi.org/10.1109/access.2021.3119746
26. Shen, Z., Chen, S.: A survey of automatic software vulnerability detection, program repair, and defect prediction techniques. Secur. Commun. Netw. **2020**, 1–16 (2020). https://doi.org/10.1155/2020/8858010
27. Song, W., et al.: Hierarchical context merging: better long context understanding for pre-trained LLMs (2024). https://arxiv.org/abs/2404.10308
28. Steenhoek, B., Rahman, M.M., Roy, M.K., Alam, M.S., Barr, E.T., Le, W.: A comprehensive study of the capabilities of large language models for vulnerability detection (2024). https://doi.org/10.48550/ARXIV.2403.17218
29. Sun, H., Cui, L., Li, L., Ding, Z., Li, S., Hao, Z., Zhu, H.: VDTriplet: vulnerability detection with graph semantics using triplet model. Comput. Secur. **139**, 103732 (2024). https://doi.org/10.1016/j.cose.2024.103732, https://www.sciencedirect.com/science/article/pii/S0167404824000336

30. Tang, G., Yang, L., Zhang, L., Kuang, H., Wang, H.: MRC-VulLoc: software source code vulnerability localization based on multi-choice reading comprehension. Comput. Secur. **141**, 103816 (2024). https://doi.org/10.1016/j.cose.2024.103816, https://www.sciencedirect.com/science/article/pii/S0167404824001172
31. Tóth, R., Bisztray, T., Erdodi, L.: LLMs in web-development: evaluating LLM-generated PHP code unveiling vulnerabilities and limitations (2024). https://doi.org/10.48550/ARXIV.2404.14459
32. Wang, J., Huang, M., Nie, Y., Li, J.: Static analysis of source code vulnerability using machine learning techniques: a survey. In: 2021 4th International Conference on Artificial Intelligence and Big Data (ICAIBD), pp. 76–86 (2021). https://doi.org/10.1109/ICAIBD51990.2021.9459075
33. Yang, Z., Shi, J., He, J., Lo, D.: Natural attack for pre-trained models of code. In: Proceedings of the 44th International Conference on Software Engineering, pp. 1482–1493. ICSE '22. Association for Computing Machinery, New York (2022). https://doi.org/10.1145/3510003.3510146
34. Zhang, C., Liu, H., Zeng, J., Yang, K., Li, Y., Li, H.: Prompt-enhanced software vulnerability detection using ChatGPT (2024)
35. Zhang, Y., Li, J., Liu, P.: Extending LLMs' context window with 100 samples (2024). https://arxiv.org/abs/2401.07004
36. Zhou, Y., Liu, S., Siow, J., Du, X., Liu, Y.: Devign: effective vulnerability identification by learning comprehensive program semantics via graph neural networks (2019). https://doi.org/10.48550/ARXIV.1909.03496
37. Zhu, Y., Lin, G., Song, L., Zhang, J.: The application of neural network for software vulnerability detection: a review. Neural Comput. Appl. **35**(2), 1279–1301 (2022). https://doi.org/10.1007/s00521-022-08046-y
38. Zou, D., Wang, S., Xu, S., Li, Z., Jin, H.: μvuldeepecker: a deep learning-based system for multiclass vulnerability detection. IEEE Trans. Dependable Secure Comput. **18**(5), 2224–2236 (2021). https://doi.org/10.1109/TDSC.2019.2942930

Power Consumption Analysis for Reverse Engineering Digital Modulation: A Novel Approach to Physical Layer Attacks on Communication Systems

Pablo Pareja(✉), Yehya Nasser, Ahmed Masmoudi, and Sebastien Houcke

IMT Atlantique, Lab-STICC, UMR CNRS 6285, Brest, France
{pablo.pareja,yehya.nasser,ahmed.masmoudi, sebastien.houcke}@imt-atlantique.fr
https://labsticc.fr/en

Abstract. Physical layer security is a critical aspect of embedded wireless communication devices, particularly in military, civil, intelligence, and security applications. Ensuring secure communication at this layer is essential to prevent unauthorized access and exploitation of vulnerabilities. While many existing techniques focus on the transmitted signal, they often overlook potential side-channel vulnerabilities that arise when an attacker gains access to the device itself. In this work, we propose a novel approach to enhancing physical layer security by demonstrating the feasibility of differentiating modulation schemes through power consumption analysis (PCA). Using a Field Programmable Gate Array (FPGA), we implemented various modulation schemes and measured the associated power consumption traces. Based on these measurements, we developed a classification algorithm using the gradient-boosted decision tree method. Our approach achieved a classification accuracy of approximately 99%, highlighting the potential of power analysis as both a tool for identifying vulnerabilities and strengthening security in embedded wireless communication devices.

Keywords: Side-channel attack · Power Consumption Analysis · Digital Modulations · Automatic Modulation Classification · Machine Learning · Boosted Trees · Field-Programmable Gate Array · HDL · Physical Layer Security

1 Introduction

Signal identification from unknown or partially known sources has always played an important role in security and military applications, where analysis can reveal crucial information about signal parameters such as carrier frequency, bandwidth, modulation type, coding scheme, frequency spreading method, or the number of carriers. This task is vital for signal interception, radio surveillance, interference identification and mitigation, and electronic warfare. A key aspect

I. Praça et al. (Eds.): EICC 2025, CCIS 2500, pp. 52–62, 2025.
https://doi.org/10.1007/978-3-031-94855-8_4

of this process is modulation classification [8], where the modulation scheme used is unknown to a potential attacker. However, attackers can exploit various modulation classification methods to deduce the modulation scheme and other parameters. These methods typically focus on analyzing the intercepted radio-frequency signal, in cases where an attacker gains physical access to the device, side-channel information may be leaked, enabling an attacker to extract critical information. Moreover, in scenarios where only a transceiver is available and the signal quality is very low, a method that does not rely on the signal-to-noise ratio (SNR) would be highly advantageous. In military applications, the inability to identify the modulation scheme can prevent receivers from recovering the message [21]; without this knowledge, demodulating the signal and accessing the transmitted information is impossible. Additionally, knowing the modulation scheme could enable the generation of a jamming signal using a similar modulation method.

Previous research on analytical power models has shown that dynamic power consumption is dependent on factors such as application, technology, operating frequency, and data that are being processed [13]. Furthermore, as indicated in [12], different modulation schemes exhibit different power consumption patterns. With this insight, this study aims to prove, for the first time, that it is possible to differentiate between various digital modulation schemes based on their power consumption profiles. We simulate on an FPGA various modulations and subsequently measure the power consumption of each modulation method. We then statistically analyze the results to observe differences among the power traces obtained. Finally, we will implement a boosted tree classification model to classify each modulation method by extracting representative features from the power traces and evaluating the classification results.

This work first examines the state of the art in modulation classification methods and PCA. Next, we present the methodology, detailing the measurement setup, the proposed model, and the data review process. Finally, we analyze the results and discuss the corresponding conclusions.

2 State of the Art

2.1 Modulation Classification

Automatic Modulation Classification (AMC) is a critical component of digital communication systems, with applications spanning military, intelligence, and commercial sectors [11]. Existing AMC algorithms can be broadly categorized into two main approaches: likelihood-based methods and feature-based methods [6].

While recent advancements have improved AMC performance, classification accuracy significantly degrades at low SNRs [10,20]. Specifically, most models achieve less than 60% accuracy when the SNR is less or equal than 0 dB, although some exceed 90% accuracy at SNR values above 0 dB. These performance variations depend heavily on the classification methods employed and the diversity of modulation types considered.

Moreover, the complexity of AMC increases with higher modulation orders due to the intricacy of the associated constellation diagrams, which makes accurate classification more challenging. A further limitation is the vulnerability of existing AMC techniques to advanced physical layer security methods, such as directional modulation (DM), spatial modulation (SM), covert communications, and intelligent reflecting surface (IRS)-aided communication [17]. These techniques effectively obfuscate signals, complicating efforts by eavesdroppers to intercept and identify the modulation schemes.

2.2 Power Side-Channel Attack

Given that AMC requires high SNR, exploring alternative approaches that are less dependent on signal quality is crucial for developing new security mechanisms, particularly in areas with potential unexplored vulnerabilities. Side-Channel Attack (SCA) can pose a significant threat if the physical implementation of an algorithm unintentionally exposes exploitable information [18]. Examples of such leaks include power consumption, electromagnetic radiation, timing analysis, and acoustic emissions.

SCAs that exploit these vulnerabilities are well documented in the cryptography field, particularly in the field of power analysis attack techniques, such as simple power analysis (SPA), differential power analysis (DPA), template attacks, and correlation power analysis (CPA), among others, have been extensively studied [16]. Since many SCA challenges can be framed as classification problems, machine learning models offer a powerful tool for SCA [7,9]. By extracting relevant features, these models can be trained to analyse power consumption data, thereby revealing critical information about the underlying processes.

3 Methodology

3.1 Measurement Setup

We employed an FPGA kit from ChipWhisperer (CW) [14] to capture power consumption traces without the need for an oscilloscope. The kit provides various features that facilitate SCA, with one of the most significant being its ability to perform synchronous sampling. The target board used in this study is the CW305, while the measurement board is the CW-Lite. The specifications for both boards are available on their respective websites [1,2].

The FPGA on the CW305 board is an Artix-7, which supports USB interfaces running on an ARM microcontroller. This setup allows for FPGA configuration, VCC-INT adjustment, PLL configuration, and writing to the data bus. Python is the primary language for interacting with this platform, and we utilized Jupyter notebooks to manage configurations and run experiments. A dynamic configuration script was developed to facilitate running experiments with varying FPGA configuration settings, which will be detailed further in the text.

The CW-Lite measurement board integrates hardware for power analysis, programming, glitching, and serial communications. It features a 10-bit Analog-to-digital converter (ADC) with a sampling rate of up to 105 MS/s and a flexible clocking mechanism that enables synchronous power capture. The importance of synchronous sampling has been demonstrated in prior research [15], where it consistently outperformed asynchronous sampling during CPA attacks. However, the CW-Lite has limitations: synchronous sampling is restricted to frequencies that are either one or four times the FPGA clock frequency, which means the sampling rate is inherently tied to the FPGA clock and cannot be freely adjusted.

For power SCA, direct access to the device's power lines eliminates noise interference commonly found in traditional AMC methods. The most challenging scenario arises when clock synchronization is not achievable, necessitating asynchronous sampling. To account for both scenarios, we conducted experiments using both asynchronous and synchronous sampling.

Another parameter we modified during the experiments was the FPGA clock frequency, which served as a key variable. Notably, the FPGA clock frequency, like other FPGA configurations, must be set prior to algorithm execution and cannot be dynamically adjusted during runtime. While methods to protect against SCPA attacks exist [19], their application outside the cryptographic domain remains largely unexplored. Our study aims to first establish the presence of information leakage to justify the adoption of such countermeasures in other domains.

3.2 Proposed Model

To measure the power consumption of different modulation schemes while leveraging the CW development environment, we selected a Vivado project from the official CW repository [3] that implements a basic 1-round/cycle AES core as the starting point for our project. This choice allowed us to utilize the existing infrastructure for capturing power traces while making the necessary modifications for our specific scenario. Since the communication logic with the Python API operates independently of the algorithm running on the FPGA, we replaced the AES algorithm block with a digital modulation block.

We developed two additional blocks required for data processing. First, a **Serial Transmitter** was designed to receive a specified number of samples. Once all data were received, the transmitter sent one input at a time to the modulator. The modulated data were subsequently stored in a **RAM block**, enabling all data processing to occur at FPGA speed without being constrained by the Python API's communication speed. After the modulation process was completed, we retrieved both the modulated data and the corresponding power traces.

The following modulation schemes were implemented in this study:

- **Frequency-Shift Keying (FSK):** 4-FSK, 16-FSK.
- **Amplitude-Shift Keying (ASK):** OOK, 4-ASK, 16-ASK.

- **Quadrature Amplitude Modulation (QAM):** 4-QAM, 16-QAM, 64-QAM, 256-QAM.

For QAM, we implemented two distinct methods for calculating the I and Q symbols:

1. Using a **Look-Up Table (LUT)**, referred to as QAM.
2. Using **arithmetic operations**, referred to as QAMv2.

This dual implementation was designed to evaluate whether different methods for calculating the same constellation could affect the algorithm's classification accuracy. We verified that our AMC approach is effective for both implementations. In this work, we focus on analyzing the power consumption from the transmitter side to identify the modulation. The data acquisition model is depicted in Fig. 1, where the trigger signal initiates the measurement process as soon as the serial transmitter starts transmitting data to the modulator. We implemented only the transmitter side of the communication system because of constraints of time and the laboratory setup available, it was then decided to only implement this system. The data acquisition model is described in Fig. 1. Here, the trigger signal is the signal for starting the measurement and it starts when the serial transmitter starts transmitting the data to the modulator.

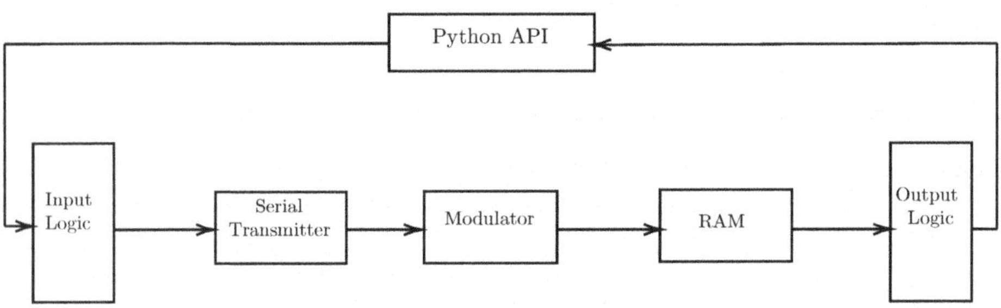

Fig. 1. Model for data acquisition.

3.3 Power Characterization

The developed serial transmitter accepts chunks of 128 bits, with a total of 3072 samples of 8 bits, we send from the python API 16 symbols at the time. Once all samples are received, the transmitter begins transmitting the data to the modulator and subsequently to the RAM block. For this purpose, we generated random inputs of 3072 symbols to be sent to the FPGA. For the measurement, we chose 12 FPGA clock frequencies: 4 MHz, 6 MHz, 8 MHz, 10 MHz, 12 MHz, 14 MHz, 16 MHz, 18 MHz, 20 MHz, 22 MHz, 24 MHz and 26 MHz. We perform 10 runs of data collection for each FPGA clock frequency and modulation type. The idea behind using different clock frequencies is to build a model that is resilient to devices that could change their clock frequency, this happen very frequently in IoT devices with dynamic clock adjustments.

For each run of measurements, we process 3072 symbols, taking 3072 samples of power traces in the case that we sample at one sample per clock cycle, considering that the processing of each input takes one cycle. We made sure of this point by doing validation of each modulation model and recollecting the outputs of each symbol processed. The amount of samples increases proportionally to the sampling speed; this means that if we sample at twice the speed of the target, we take 6144 samples, with the purpose of measuring the same working number of cycles. For synchronous sampling, we create two datasets, one with one sample per clock cycle and the other with four, because they are the only two cases that the capture board allows us to do synchronously. For asynchronous sampling, we can freely choose the sampling rate, as long as we stay inside the limits of the hardware. We then take two and four samples per clock cycle and we take two samples instead of one because we try to compensate for the inaccuracy in asynchronous sampling.

3.4 Data Review

From the measured data, we compute the first four moments of the power trace as input to the AMC algorithm: mean, variance, skewness, and kurtosis, using Scipy. Moments are a useful way to study a distribution; they are well explained in [4]. For the case of the variance, we instead use the standard deviation to have the same scale as the original data. We show that each modulation has distinctive central moments, which justifies the use of these values as features. Although there are alternative feature extraction methods, this approach suffices for an initial analysis.

Following feature extraction using moments, we develop a boosted tree model using the XGBoost algorithm [5], which is widely used by data scientists to achieve state-of-the-art results. We create different models using the four datasets previously explained, and then compare the accuracy, recall, precision and F1 score to evaluate the performance of each model. Accuracy measures the proportion of correct predictions out of all predictions made, recall (or sensitivity) assesses the model's ability to correctly identify positive instances, precision measures the ratio of correctly predicted positives observations to the total predicted positives, and the F1 score provides a balance between precision and recall by calculating their harmonic mean. The ultimate goal is to determine whether these models can accurately classify all the different modulations based solely on power consumption, thereby allowing us to validate or refute our initial hypothesis.

4 Results

4.1 Statistical Analysis

The primary objective of this study is to determine whether there are notable differences in power consumption among the different modulation schemes and

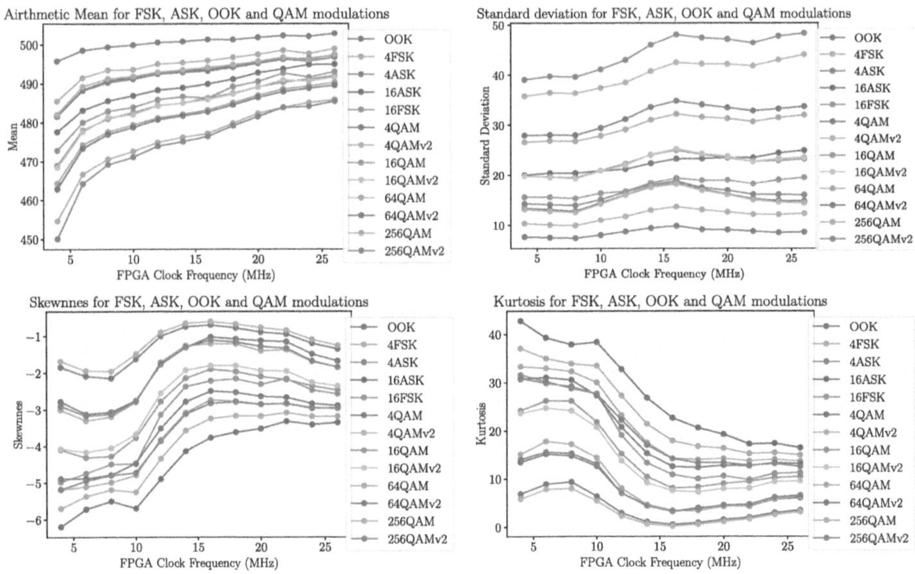

Fig. 2. Mean, standard deviation, skewness and kurtosis for different modulation methods with synchronous sampling for one sample per clock cycle.

between the asynchronous and synchronous datasets. To address this, we analyzed the standard deviation of the power traces for various modulation schemes, as shown in Figs. 3 and 2.

In both datasets, the standard deviation reveals distinct patterns for different modulation schemes. However, in the asynchronous sampling dataset, despite having twice the number of samples, the traces exhibit significantly higher variability, resulting in overlaps between some modulation schemes. This issue is notably mitigated in the synchronous sampling dataset, though some overlap persists for modulation schemes of similar order.

These observations suggest that while the standard deviation provides valuable information, it is insufficient on its own to fully classify all modulation schemes. Moreover, as shown in Fig. 2, additional statistical features such as the mean, skewness, and kurtosis demonstrate distinct and consistent behaviors across modulation schemes and clock frequencies. This consistency is particularly important for analyzing devices with variable clock speeds, a scenario that this experiment seeks to replicate.

4.2 Classification Model

Experimentally, the extracted features-mean, standard deviation, skewness, and kurtosis-demonstrated consistent differences depending on the modulation scheme and clock frequency. Based on this observation, we trained classification models using these features and evaluated their performance. Table 1 summarizes the accuracy, recall, and F1-score for each model.

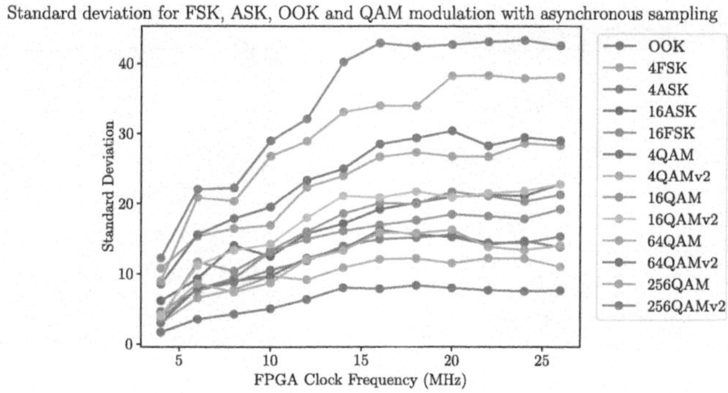

Fig. 3. Standard Deviation for different modulation methods with asynchronous sampling with two samples per clock cycle.

The worst-performing model corresponds to asynchronous sampling with two samples per clock cycle. Figure 4a illustrates the confusion matrix, which shows the probabilities of classifying each modulation scheme given the actual modulation. The matrix reveals that certain modulations, such as 16-ASK and 256-QAM, are distinguishable, whereas errors arise in distinguishing 16-FSK from 16-QAM and 4-ASK from 4-QAM or 4-FSK. These errors can be attributed to the similarity in the power consumption patterns of these modulation schemes, as well as the increased variability observed in asynchronous sampling traces.

When the asynchronous sampling model is adjusted to take four samples per clock cycle, as shown in Fig. 4b, the overall classification accuracy improves significantly.

In contrast, the best-performing models, as shown by the confusion matrices in Figs. 4c and 4d, exhibit excellent classification performance. These models demonstrate that in synchronous sampling, the number of samples taken is less critical than in asynchronous sampling. Additionally, these models can differentiate between modulations of the same order with very high accuracy, underscoring the significant differences in power consumption patterns among modulation schemes.

These results validate our initial hypothesis: modulation schemes can indeed be differentiated solely based on power consumption. By extracting features and training models on datasets acquired under varying sampling conditions, we achieved results that strongly support the feasibility of this approach. Furthermore, the results highlight the superior performance of synchronous sampling and its critical importance in SCA.

When compared to traditional AMC methods, the results of our approach align with the performance of algorithms operating at high SNR. However, since the traditional AMC methods are used in entirely different contexts, a direct and definitive comparison is not feasible.

Notably, even in the worst-case scenario of asynchronous sampling, the achieved accuracy remains considerably high and improves further when more

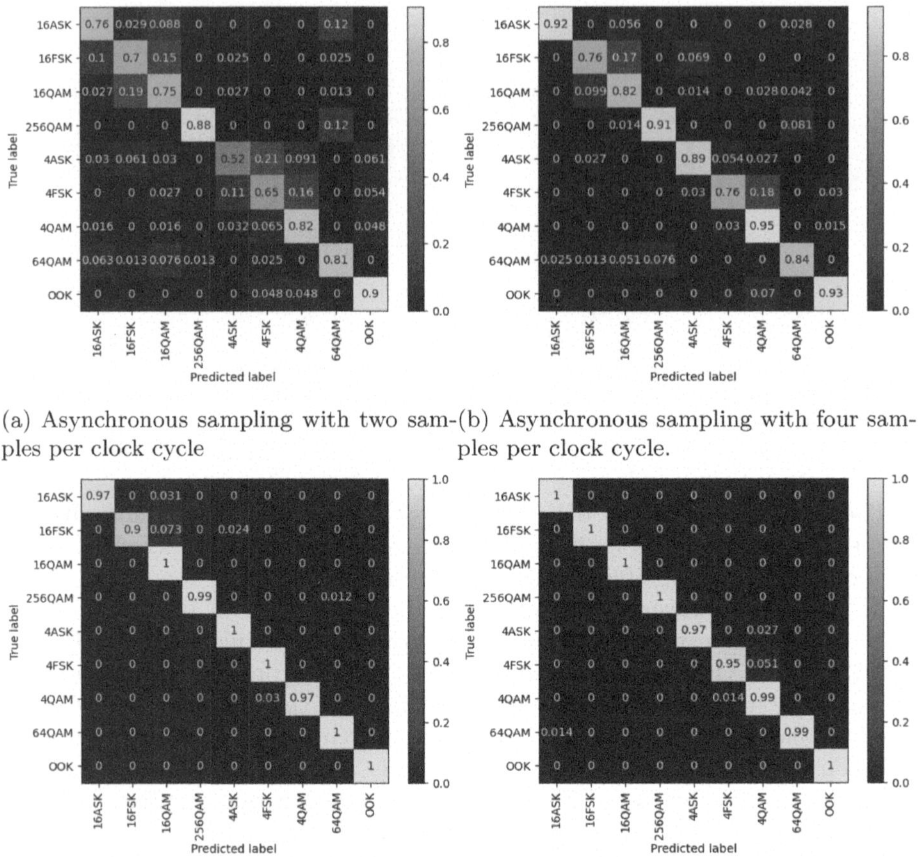

(a) Asynchronous sampling with two samples per clock cycle

(b) Asynchronous sampling with four samples per clock cycle.

(c) Synchronous sampling with one sample per clock cycle.

(d) Synchronous sampling with four samples per clock cycle.

Fig. 4. Confusion matrices for modulation classification using a XGBoost classifier for four different testing scenarios.

Table 1. Classification model performances

Dataset	Accuracy	Recall	Precision	F1 Score
Asynchronous sampling 2x	0.77	0.77	0.78	0.751
Asynchronous sampling 4x	0.87	0.87	0.87	0.865
Synchronous sampling 1x	0.98,	0.98	0.98	0.982
Synchronous sampling 4x	0.99,	0.99	0.99	0.989

samples are taken. This results emphasizes the broader objective of this research: exploring alternative methods for deducing key elements of wireless digital communication by leveraging potential information leakage through material characteristics, such as power consumption.

5 Conclusion

This work demonstrates the feasibility of revealing information about digital communication transceivers through power analysis methods. We achieved a classification accuracy of approximately 99% across a wide range of modulation schemes, highlighting the potential of power consumption as a distinguishing feature for reverse engineering the digital modulation.

Considering the advantages and limitations of this approach, we conclude that alternative methods of modulation classification, such as the one proposed in this study, can be particularly beneficial in specific scenarios. The same logic used here for modulation classification could potentially be extended to demodulation tasks. Additionally, in scenarios involving reverse engineering of wireless devices, this analysis could uncover not only the modulation scheme but also other critical parameters of the device.

Future research should focus on investigating more complex systems to validate and extend these findings. Moreover, combining side-channel information, such as power consumption, with transmitted radio frequency signals could offer deeper insights into wireless communication systems. Such approaches may also facilitate more effective attacks on transceivers, expanding the scope of side-channel analysis in digital communication security and hardware vulnerability studies. Finally, we propose that this analysis could be extended to other, less invasive SCA techniques, such as electromagnetic side-channel analysis. Combining such techniques with established signal processing methods may further aid in reverse engineering wireless communication devices.

References

1. CW1173 ChipWhisperer-Lite - NewAE Hardware Product Documentation. https://rtfm.newae.com/Capture/ChipWhisperer-Lite/
2. CW305 Artix FPGA Target - NewAE Hardware Product Documentation. https://rtfm.newae.com/Targets/CW305ArtixFPGA/
3. Chipwhisperer (2024). https://github.com/newaetech/chipwhisperer
4. Blitzstein, J.K., Hwang, J.: Introduction to Probability, Second Edition. CRC Press, Boca Raton (February 2019)
5. Chen, T., Guestrin, C.: XGBoost: A Scalable Tree Boosting System (2016). https://doi.org/10.48550/ARXIV.1603.02754
6. Ghasemzadeh, P., Banerjee, S., Hempel, M., Sharif, H.: Accuracy analysis of feature-based automatic modulation classification with blind modulation detection. In: 2019 International Conference on Computing, Networking and Communications (ICNC), pp. 1000–1004 (2019). https://doi.org/10.1109/ICCNC.2019.8685638

7. Hettwer, B., Gehrer, S., Güneysu, T.: Applications of machine learning techniques in side-channel attacks: a survey. J. Cryptogr. Eng. **10**(2), 135–162 (2019). https://doi.org/10.1007/s13389-019-00212-8
8. Hoque, N., Rahbari, H.: Circumventing the defense against modulation classification attacks. In: Proceedings of the 16th ACM Conference on Security and Privacy in Wireless and Mobile Networks, pp. 377–382. WiSec '23, Association for Computing Machinery, New York, NY, USA (2023).https://doi.org/10.1145/3558482.3590197
9. Hospodar, G., Gierlichs, B., De Mulder, E., Verbauwhede, I., Vandewalle, J.: Machine learning in side-channel analysis: a first study. J. Cryptogr. Eng. **1**(4), 293–302 (2011). https://doi.org/10.1007/s13389-011-0023-x
10. Huang, S., Lin, C., Zhou, K., Yao, Y., Lu, H., Zhu, F.: Identifying physical-layer attacks for IoT security: an automatic modulation classification approach using multi-module fusion neural network. Phys. Commun. **43**, 101180 (2020). https://doi.org/10.1016/j.phycom.2020.101180
11. Huynh-The, T., et al.: Automatic modulation classification: a deep architecture survey. IEEE Access **9**, 142950–142971 (2021). https://doi.org/10.1109/ACCESS.2021.3120419
12. Lorandel, J., Prévotet, J.C., Hélard, M.: Fast power and performance evaluation of fpga-based wireless communication systems. IEEE Access **4**, 2005–2018 (2016). https://doi.org/10.1109/ACCESS.2016.2559781
13. Nasser, Y., et al.: Neupow: a cad methodology for high-level power estimation based on machine learning. ACM Trans. Des. Autom. Electron. Syst. **25**(5) (2020). https://doi.org/10.1145/3388141
14. O'Flynn, C., Chen, Z.D.: Chipwhisperer: an open-source platform for hardware embedded security research. In: Prouff, E. (ed.) Constructive Side-Channel Analysis and Secure Design, pp. 243–260. Springer International Publishing, Cham (2014)
15. O'Flynn, C., Chen, Z.: Synchronous sampling and clock recovery of internal oscillators for side channel analysis and fault injection. J. Cryptogr. Eng. **5**(1), 53–69 (2014). https://doi.org/10.1007/s13389-014-0087-5
16. Randolph, M., Diehl, W.: Power side-channel attack analysis: a review of 20 years of study for the layman. Cryptography **4**(2) (2020).https://doi.org/10.3390/cryptography4020015, https://www.mdpi.com/2410-387X/4/2/15
17. Shi, W., et al.: Physical layer security techniques for future wireless networks (2021). https://arxiv.org/abs/2112.14469
18. Spreitzer, R., Moonsamy, V., Korak, T., Mangard, S.: Systematic classification of side-channel attacks: a case study for mobile devices. IEEE Commun. Surv. Tutor. **20**(1), 465–488 (2018). https://doi.org/10.1109/COMST.2017.2779824
19. Sundaresan, V., Rammohan, S., Vemuri, R.: Defense against side-channel power analysis attacks on microelectronic systems. In: 2008 IEEE National Aerospace and Electronics Conference, pp. 144–150 (2008). https://doi.org/10.1109/NAECON.2008.4806536
20. Yalcinkaya, B., Coruk, R.B., Kara, A., Tora, H.: Hierarchical classification of analog and digital modulation schemes using higher-order statistics and support vector machines. Wirel. Pers. Commun. **136**(2), 827–847 (2024). https://doi.org/10.1007/s11277-024-11285-y
21. Zhu, Z., Nandi, A.K.: Automatic Modulation Classification: Principles, Algorithms, and Applications. John Wiley & Sons Inc., Hoboken (2015)

Analysis of Permissions and Trackers in Android Apps Targeting Children Under Thirteen

Jamie Fitzpatrick and Virginia N. L. Franqueira(✉)

Institute of Cyber Security for Society (iCSS), School of Computing, University of Kent, Canterbury CT2 7NP, UK
v.franqueira@kent.ac.uk

Abstract. As mobile phone and app use among children increases, ensuring their privacy and safety has become more crucial. This study examines 27 popular Android apps for children under 13, available in the UK via Google Play Store, by analysing the permissions they request and the trackers they share data with. The aim is to determine whether these practices are justified by the apps' functionalities, and comply with regulatory standards to adequately protect children. Using a traffic light system inspired by the ICO audits, each app is rated green, amber, or red taking into account dangerous permissions and runtime trackers in use. Only one app reached an overall green rating, while 10 apps achieved an overall red rating. The majority of tracker companies (20 out of 21) are based outside the UK or EU, including the US, Russia, China, Israel and India, and two trackers are linked to profiling and fingerprinting of users. Overall, findings suggest that apps often prioritise functionality and user experience over privacy and safety of children.

Keywords: Apps compliance · tracking behaviours · dangerous permissions · children privacy and safety · ICO Children's code · traffic light

1 Introduction

The Office of Communications (Ofcom) revealed that, as of 2024, 96% of children aged 3–17 went online and, by age 11, nine of ten children had their own mobile phones, with some children using digital devices as young as three [26]. The Coronavirus (COVID-19) pandemic further amplified children's use of technology, as represented by a longitudinal study which spanned from pre-pandemic to restrictions being lifted [16]. Therefore, this current prevalence of technology is urging a focus on protecting young people and ensuring that risks associated with its use do not overshadow the potential benefits.

There has been a considerable increase in the use of digital technologies to facilitate education, particularly during and following the pandemic [4]. Such

The original version of the chapter has been revised. The figure 2 is replaced with correct image. A correction to this chapter can be found at https://doi.org/10.1007/978-3-031-94855-8_24

I. Praça et al. (Eds.): EICC 2025, CCIS 2500, pp. 63–81, 2025.
https://doi.org/10.1007/978-3-031-94855-8_5

use of technology in schools has garnered both positive and negative feedback. On the positive side, educators report that integrating technology into the curriculum helps students acquire essential skills valuable to future education and careers [33]. However, the risks cannot be overlooked, as children may be exposed to different types of harm [24]. Educational applications (apps) are a desirable tool for teachers and parents because they provide easy access to relevant materials in informal settings [17]. However, it can be challenging for parents and educators to evaluate whether an app is safe for children.

When discussing "privacy concerns", the focus is on how apps collect, use, store, and share children's personal information. Westin [32] defined privacy as the right of individuals, groups, or institutions to decide when, how, and to what extent information about themselves is disclosed to others. Consent, however, is often obscured and difficult to comprehend [1]. Reyes et al. [27] studied apps' compliance against the Children's Online Privacy Protection Act (COPPA) in the US, and found that many apps potentially violate this regulation by engaging in third-party tracking. Similarly, Livingstone et al. [24] noted that children are more vulnerable to online privacy threats than adults due to their lack of awareness of the risks. Children's ability to navigate and mindfully avoid potential privacy risks may vary depending on their development, digital skills, or socio-economic circumstances [24]. Although there are existing measures to protect children using technology, such as parental controls to monitor and limit device usage [10], these techniques primarily focus on avoiding exposure to age-inappropriate content. Additionally, some controls require access to system resources and sensitive data on the child's phone to function properly. So, while limiting exposure to harmful content, they may inadvertently increase privacy concerns by collecting more data than necessary [10].

The contribution of this study is threefold. Firstly, it provides a comparative evaluation of privacy risks for 27 Android most popular apps listed as for "Children" in the Google Play Store. Adapting the ICO audits traffic light rating system [18][1], it provides an overview of unjustified dangerous permissions and runtime trackers per app and by category. The literature revealed that previous publications either focused on permissions or on trackers – but not both, and did not evaluate whether dangerous permissions were justified by the functionalities provided by the apps. Secondly, using static and dynamic analysis complemented by manual inspection, it highlights critical gaps in current app practices explicitly designed for children, particularly concerning transparency, unnecessary dangerous permissions, and non-compliance with data protection regulations. Thirdly, the study provides a map of trackers identified as in use by the apps with location of parent company and the type of tracking used, revealing that only one of 21 is based in the UK or EU, with the remaining located in US, Russia, China, Israel and India. Previous work focused on mobile apps in general, although eventually covering some apps for children.

The remainder of the paper is structured as follows. Section 2 reviews related work, while Sect. 3 presents the scope of this study and the methodology fol-

[1] Called "assurance ratings" by ICO.

lowed. Section 4 outlines the key findings of permissions and trackers, and provides category-based results. Section 5 applies a traffic light rating system to the apps, and discusses the implications of the findings. Finally, Sect. 6 concludes the paper.

2 Related Work

This section reviews the literature related to the analysis of privacy in Android apps (Sect. 2.1), and applicable laws and regulations (Sect. 2.2).

2.1 Privacy of Android Apps

A wide range of research has analysed the privacy practices of apps on the Google Play Store, with many diving into the prevalence and implications of third-party tracking and Android permissions (e.g., [3,5,8,11,21,23]). However, there was a lack of studies specifically focusing on children apps. Furthermore, although the related works tended to use much larger samples than this study, they either focused on trackers or on permissions, and did not try to understand if dangerous permissions were justified by app functionalities.

Research on third-party tracking often employs both static and dynamic analysis. Static analysis examines an app's behaviour without executing it, therefore relying on decompiling it to view its code [21]. Dynamic analysis, however, involves running an app and observing its behaviours, such as looking at its network traffic [23]. Binns et al. [3] studied 959,000 apps from the UK and US Google Play Stores, using the APKTool to decompile their APK files and map trackers to known tracking companies found during previous studies. They found that most apps contained third-party tracking, especially in the "News", "Games & Entertainment" and "Family" super genres, raising concerns about compliance with the GDPR (see Sect. 2.2). It was also recognised that the parent companies connected to various trackers were seen to operate outside of the EU. However, the authors only used static analysis of trackers while this research involved static and dynamic analysis of trackers, and reported on the latter in the traffic light comparison of apps. Also, their results date to six years ago, justifying the need for the study reported in this paper.

Concerning regulations, Kollnig et al. [21] investigated the absence of consent in apps using third-party tracking in a random sample of 1,201 apps. Each app was run for 15 s and then immediately uninstalled to find which companies were contacted before the opportunity to consent. The tool NetGuard was used to analyse the network traffic, and it was found that most apps engaged in tracking but very few required explicit consent, violating privacy laws.

Ekambaranathan et al. [8] conducted surveys and structured interviews to identify developers' understanding of their responsibilities, the practices they adopted in development, the barriers they faced in adopting data protection frameworks, and how they overcame these. It was found that, although developers recognised that children's apps should be designed to ensure they are age-appropriate and meet developmental needs, they faced challenges in finding guidelines to do so. It was also found that developers made compromises to add third-party tracking libraries to the code to allow for monetisation, especially with perceived "harmless" libraries such as Google Analytics.

Regarding Android permissions, Calciati et al. [5] analysed almost 3 million app releases, including different versions of the same apps. They found that 17% of the apps added new dangerous permissions, or moved existing ones to granted permission groups between releases, without user knowledge or consent. Felt et al. [11] developed Stowaway, a tool that determines a set of API calls and maps them to a specific permission to determine if an app is requesting more permissions than necessary. About one-third of their sample (940 apps) were over-privileged, although it was acknowledged that this may be unintentional due to API poor documentation.

2.2 Laws and Regulations

The abundance of mobile apps in recent decades has changed how children access information and interact, offering education and entertainment opportunities. However, they also raise concerns about children's data privacy and online safety [24]. To address these issues, regulations have evolved to include more specific requirements for children's apps.

The General Data Protection Regulation (GDPR) sets a high standard for data protection, promoting fair and lawful data collection and users' right to access, erase, or modify their data [9, Article 8]. For children, the GDPR requires transparency about data collection and clear consent by parents. It also strictly restricts profiling and automated decision-making, ensuring that children are not inappropriately targeted or influenced by digital marketing without appropriate safeguards. Overall, it shifts focus towards protecting user data and ensuring transparency in data collection practices within apps used by children.

In the UK, the Data Protection Act (2018) [28] incorporates the GDPR principles, ensuring data protection for all users. The Privacy and Electronic Communications Regulations (PECR) [29] legislation further safeguards children by controlling how communications tools are used within apps, protecting them from harmful content and unsolicited marketing [20]. Additionally, the Online Safety Act [30] calls for robust age verification mechanisms for access to adult content, e.g., check of official documents, and reliable age estimation to meet the minimum requirement of 13 years old for social media.

The Information Commissioner's Office (ICO) enforces the Children's code, which includes 15 standards for developers of children's apps [19]. This code applies to any "information society services" likely accessed by children, including apps and social media platforms. The key principle is that children's best interests should guide all aspects of design and development. The code mandates Data Protection Impact Assessments (DPIAs) to identify risks and actively implement mitigation measures. The standards also focus on limiting data collection, particularly sensitive data such as location, and emphasise transparency.

Additionally, the ICO states that profiling should not cause harm or compromise safety and should be avoided (or limited) if an app can function without it.

3 Methodology

The regulations, reviewed in Sect. 2.2, aim to create a safer digital environment for children, although questions remain about the extent of compliance – the focus of this study. This section describes the research methodology employed to investigate permissions and tracking behaviours of Android apps for children.

3.1 Research Scope

This study examines the data privacy practices of children's apps on the Google Play Store. It focuses on free apps in the "Children" category, for ages 0–12. Free apps were chosen to ensure accessibility across socio-economic backgrounds. Thirty most popular apps were selected for analysis to exemplify widely used apps and provide a valuable snapshot. The study examines apps available in the UK, which could offer insights applicable to other regions with similar regulations.

3.2 Research Design

A mixed-methods approach was used, combining quantitative data analysis with qualitative assessment of apps' main functionalities. This enabled a thorough examination of both permissions and trackers, and their practical implications.

Data Collection and Tools Used. Data was collected using the google-play-scraper tool[2] (on July 15, 2024), to identify the top 30 free apps in the Children category. This tool enables customisation of responses via JavaScript input, allowing for specified outputs depending on characteristics such as country ("UK") and category ("Children"). Apart from the 30 apps initially selected and, due to its popularity and existing concerns at the time of the study [15], the Roblox app was also included. The information available on the Google Play Store (as of July 18, 2024) regarding each app's version was also collected using the Play Store interface to ensure consistency in the permission and trackers analysis throughout the study.

Each app's APK file was downloaded using APKPure[3], and analysed using APKTool[4] to extract the Android Manifest file, enabling investigation of the permissions contained within the apps' metadata.

TrackerControl (TC), an app developed by Kollnig et al. [22], was used to observe the tracking behaviours of the sample apps installed on an Android

[2] https://pypi.org/project/google-play-scraper/.
[3] https://apkpure.com/.
[4] https://apktool.org/.

device (a Samsung Galaxy A20e running Android 11). TC helped to identify tracking libraries found in the apps' code (static analysis) and to detect the tracker domains at work during runtime (dynamic analysis). The APKTool can only be used for static analysis of trackers while TC can do both types of analysis, therefore, we opted for APKTool to study permissions and TC to study trackers.

Four apps were excluded due to inconsistencies among app versions available for data collection (i.e., *Nat West Rooster Money*, *CBeebies Playtime Island*, and *CBeebies Little Learners*), and because an APK could not be reliably accessed (for *LEGO DUPLO Peppa Pig*). Thus, to maintain consistency of data collection and analysis, these apps were not included in the final sample of 27 apps.

Data Analysis. As aforementioned, each of the 27 apps was downloaded and opened on the testing Galaxy Android device. Doing so meant that the functionalities of each app could be examined, allowing the apps to be categorised based on these observations.

The quantitative analysis involved descriptive statistics of the permissions and tracking data collected. The qualitative analysis included determining if dangerous permissions could be justified by the main app functionalities. It also involved using the apps' tracker domains, recovered from TC, to investigate (using Exodus[5]) the type of tracking, their parent company, and their country of origin.

Dangerous permissions, also known as runtime permissions, grant app access to restricted data and actions. Android defines [13] "restricted data" as sensitive information stored on or about the device, such as contact details or system state, and "restricted actions" as those typically performed by the user, like recording audio or video. Due to their potential to access sensitive personal data and impact device functionality or other apps, these permissions must be explicitly requested during app runtime.

In this study, the traffic light ratings from ICO audits [18] were adapted as follows. Only dangerous permissions and runtime tracking, uncovered via dynamic analysis of trackers, were considered for this purpose.

- Permissions
  - **Red**: Dangerous permissions cannot be justified by app's functionalities.
  - **Amber**: Dangerous permissions can be justified by app's functionalities.
  - **Green**: No dangerous permissions used by app.
- Trackers
  - **Red**: 4 or more runtime tracking companies associated with app.
  - **Amber**: 2–3 runtime tracking companies associated with app.
  - **Green**: No runtime tracking companies associated with app.
- Overall
  - **Red**: A combination of a red and an amber rating, or two red ratings.
  - **Amber**: A combination of a green and an amber rating, or two amber.
  - **Green**: A combination of two green ratings for permissions and trackers.

[5] https://etip.exodus-privacy.eu.org.

4 Findings

First, the primary functionalities of each of the 27 apps were observed, leading to the categories shown in Table 1. Please note that these categories were not based on existing Google Play Store categories, but were assigned through testing each app and observing their functionalities first-hand.

Table 1. The apps in our sample categorised based on their primary functionalities.

Category	Description	Apps in the sample
Educational	Apps enhancing learning through interactive lessons	*ClassDojo, Kahoot, ScratchJr, Minecraft Education, PAW Patrol Academy, Times Tables Rock Stars, Sago Mini School*
Role-Playing & Simulation	Apps providing an immersive environment for character-driven narratives or real-life simulations	*Toca Boca World, Avatar World, Toca Boca Jr, Disney Frozen Royal Castle, Bluey: Let's Play, Baby Panda's School Bus, Aha World: School Day*
Creative	Apps promoting creativity	*Toca Boca Hair Salon 4, Coloring Games: Color & Paint, LEGO Builder, Little Panda's Ice Cream Games, Cake maker: Kids Cooking Games, Disney Coloring World*
Casual & Action	Apps that are focused on entertainment	*Fluvsies - A Fluff to Luv, Fluvsies Merge Party, slither.io, Tag with Ryan*
Video & Content	Apps delivering or facilitating video content	*YouTube Kids, VLC for Android*
	App uncategorised due to its distinctive nature as a gaming platform and interactive world	*Roblox*

4.1 Findings Related to Permissions

This section details the findings related to Android permissions for the 27 apps, highlighting trends to understand the types of permissions requested.

Overview of Permissions. Figure 1 presents the total number of permissions found in each app, organised by category. Across all apps, 66 distinct permissions were identified, with an average of 11.96 permissions requested per app (standard deviation: 6.38). Four apps had 20 or more permissions, with *ClassDojo* having the highest at 29. *Coloring Games: Color & Paint* had the fewest at 3.

Notably, the apps with the most permissions are spread across different categories, challenging the hypothesis that certain categories request more permissions. However, apps in the "Video & Content" category show a significantly higher average number of permissions (18), likely due to the additional functionalities needed for video platforms (e.g., media storage access), compared to

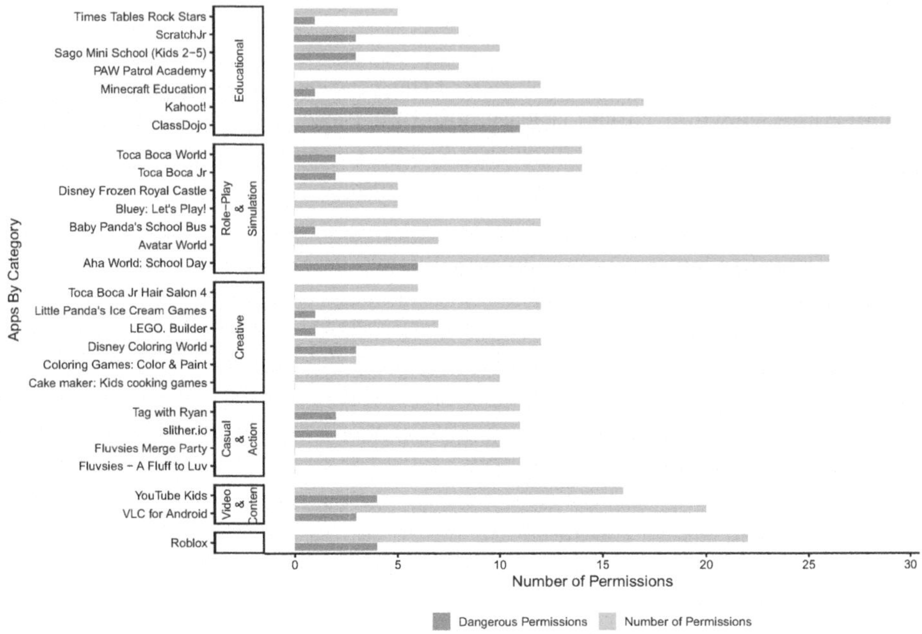

Fig. 1. Number of permissions and of dangerous permissions per app by category.

simpler apps like colouring apps. The *Aha World: School Day* app is an outlier in the "Role-Play & Simulation" category with 26 permissions, much higher than the category's average of 9.5, and the *ClassDojo* app similarly exceeds the "Educational" category average of 10, with 29 permissions.

Dangerous Permissions. The dangerous permissions found in the app sample were analysed using two approaches. First, a quantitative analysis examined the number of dangerous permissions requested across apps and categories. Second, a qualitative analysis assessed whether each dangerous permission can be justified based on the functionalities of the app requesting it.

Quantitative Observations. Figure 1 highlights the number of dangerous permissions requested by the 27 apps, revealing some notable discrepancies elaborated next. *ClassDojo*, an "Educational" app, stands out with 11 dangerous permissions. On the other hand, *PAW Patrol Academy*, in the same category, requests no dangerous permissions. This contrast is noticeable in 3 other categories with 9 apps in total requesting no dangerous permissions, i.e., "Role-Play & Simulation", "Creative" and "Casual & Action".

Qualitative Observations. This section assesses whether dangerous permissions requested by apps, per category, are justified by functionalities.

Educational Apps
Four out of 7 "Educational" apps request the *POST_NOTIFICATIONS* permission, which is justified by notification features such as class updates (*ClassDojo*) or to maintain engagement through content updates (*Kahoot*, *Minecraft Education*, and *PAW Patrol Academy*).

ClassDojo and *Kahoot* request microphone-related permissions (*RECORD_AUDIO* and *AUDIO_CAPTURE*). These are justified by functions like recording voice messages such as a teacher giving instructions (*ClassDojo*) and incorporating audio in quizzes (*Kahoot*). Similarly, permissions for accessing media files (*READ_MEDIA_AUDIO*, *READ_MEDIA_VIDEO*, *READ_MEDIA_IMAGES*) in these apps are required to upload and use multimedia files. The app *ScratchJr* also requests access to the microphone through *AUDIO_CAPTURE*, which can be explained by the app's recording function allowing users to add audio to their coding projects.

ScratchJr, *ClassDojo* and *Kahoot* also request camera access for capturing images or videos for use in projects or quizzes, which is consistent with their functionalities. Storage permissions (*READ_EXTERNAL_STORAGE* and *WRITE_EXTERNAL_STORAGE*) are important to justify due to the sensitivity of accessing and changing device data. *ClassDojo* and *Kahoot* both have upload and download functionalities, and *Sago Mini School* requires each activity to be downloaded locally before use.

Times Tables Rock Stars allows teachers to download data about each child in their class, displaying statistics on progress and completion of homework. This app also allows users to play without an internet connection, which means that data on progress and app settings would be required to facilitate offline play, meaning access to the device's storage could be justified. *Minecraft Education* needs access to write to the device storage to save game data and to allow the user to continue on "missions" upon return to the app. Finally, *ScratchJr* allows users to upload files from their device to display in the app and incorporate into their coding learning.

ClassDojo's "approximate location access" potentially raises concerns; it is used to help users locate nearby schools during registration. However, no explicit request for location access was noted when first running the app during testing.

Role-Play and Simulation Apps
There are fewer dangerous permissions in this category. Three apps (*Toca Boca World*, *Toca Boca Jr*, *Avatar World*) request storage access which may be to save game progress and images, justifying this permission. *Aha World: School Day* requests microphone access for voice-over features, which is clearly justified. However, its request for media access lacks clear functionality, as there are no ways to upload images, audio, or video, and this could raise privacy concerns.

A potentially unjustified permission is *Baby Panda's School Bus*'s ability to download without notifying the user. While it could be used for updates, silent downloads pose risks if exploited by malicious actors.

Creative Apps
Few dangerous permissions were found in the 'Creative' category. *LEGO Builder*'s camera access allows users to scan QR codes on LEGO sets to find their building instructions, which is clearly justified. Also, *Disney Coloring World* allows notifications and requests storage data to save the user's artwork.

However, *Little Panda's Ice Cream Games* has the ability to download without notification, which, as mentioned, can represent high risk if the files being downloaded are not safe or relevant to the app itself.

Casual and Action Apps
Only two apps in this category request any dangerous permissions. *Tag With Ryan* accesses device storage which can be used to save game progress, thus justified. *Slither.io*, however, requests access to the phone state. This poses a risk since it allows access to the status of any ongoing calls, cellular network information and the Telecom account registered on the device. It is unknown if this could reveal account details and geolocation, which could be indirectly correlated to a child using the phone [14].

Video and Content Apps
The *YouTube Kids* app requests *GET_ACCOUNTS*, likely to connect to a Google account for parental control or expanded content access. Microphone access is justified in association with the search functionality offered by "Hello Google". *YouTube Kids* may also offer notification updates like its partner app *YouTube* (not part of this study). However, the need to access storage is not clear, as it does not seem that users can download content from the app. *VLC for Android*'s storage access is justified by the app's functionality of playing downloaded media, but the ability to access the microphone is questionable. This is because the app appears to only be a media player, not recorder, so microphone access is likely not essential.

Roblox App
Roblox requests four dangerous permissions. *WRITE_EXTERNAL_STORAGE* and *READ_EXTERNAL_STORAGE* are justified by the app's ability to save gameplay screenshots or videos. *READ_CONTACTS*, though potentially high-risk, has a clear use within the platform to connect players via contact information using phone numbers or emails linked to Roblox accounts. The app offers *POST_NOTIFICATIONS* to inform users on game updates and app-wide events.

4.2 Findings Related to Trackers

Figure 2 summarises the number of tracking companies observed in each app during both static and dynamic analysis, by category. Two of the apps had the highest number of tracking libraries: *Fluvsies Merge Party* and its variant *Fluvsies - A Fluff to Luv*, each containing 9 in total observed during dynamic analysis, which corresponded to 7 distinct tracking companies. In contrast, 2 of the apps used no tracker libraries at all (*VLC for Android* and *Coloring Games: Color & Paint*). On average, each app contained approximately 3.67 tracker libraries, with a standard deviation of 2.22, indicating that there was only moderate variability across the sample.

ScratchJr revealed trackers during static analysis only, while 2 apps only revealed trackers during dynamic analysis (i.e., *Times Tables Rock Stars* and *Avatar World*). This mismatch is further discussed in Sect. 5.4.

Overall, there is an observable difference in the average number of tracking companies associated with each category. "Casual & Action" apps show a greater number of tracking companies compared to the rest of the categories, likely due to the higher numbers found in the two *Fluvsies* apps. "Video & Content" apps have fewer associated tracking companies than the rest of the categories, with an average of only 0.5.

Static Analysis of Trackers. Across the 27 apps, 15 distinct tracking companies were found during static analysis. Google was the most common owner of the tracking libraries found within the apps, present in 23 out of 27 apps. This starkly contrasted with the next most common tracker company, AppsFlyer, which was only found in 8 apps. On the other hand, AppMetrica and Moat were the least common, each appearing in just one app.

Dynamic Analysis of Trackers. During runtime, there were considerably more tracking companies observed in each of the apps, and the total number of distinct companies rose to 22 from 15 in static analysis.

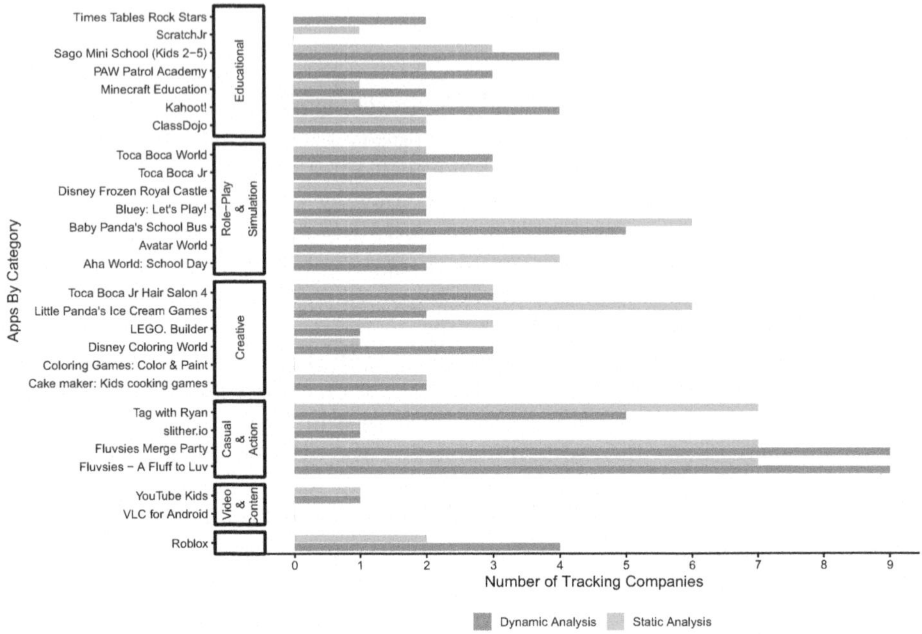

Fig. 2. Number of tracker companies from static/dynamic analysis per app by category.

When investigating the tracking companies and the type of tracking they partake in, the most common forms were "Analytics" and "Advertisement". In the sample, 35 analytics tracking domains and 25 advertisement tracking domains across the apps were observed in total. An interesting finding was the presence of "Fingerprinting", identified exclusively within Google's DoubleClick domain. Potentially more concerning, however, was the detection of "Profiling" activities conducted by Amplitude, a company active across three apps: *Bluey: Let's Play*, *Kahoot*, and *Disney Frozen Royal Castle*. Profiling involves the creation of detailed user profiles, which raises significant privacy concerns, particularly in the context of children's apps.

By mapping known tracking domains to their parent companies and corresponding countries, it was revealed that just one company, SuperAwesome, was based in the UK. As shown in Table 2, the majority of the trackers were linked to companies operating in the US, with 11 of the 16 companies based there. Additionally, trackers from companies operating in Israel, Russia, India, and China were detected, clearly highlighting the global distribution of tracking activities and raising questions about their operation outside of the jurisdiction of the UK, and therefore their compliance with the regulations covered in Sect. 2.2.

Table 2. Tracking companies mapped to other details including number of apps from static and dynamic analysis. (The table is in order of total tracking.)

Company	Parent Company	Country	Apps Static	Apps Dynamic	Tracking Use
Google	Alphabet	US	23	20	Analytics, Advertisement, Crash Reporting, Fingerprinting
Unity Technologies	Unity Technologies	US	5	9	Advertisement
AppsFlyer	AppsFlyer	US	8	4	Advertisement
MixPanel	MixPanel	US	3	4	Analytics
Amazon	Amazon	US		6	Analytics
Singular	Singular	US	3	3	Analytics
KIDOZ	Shaol Games	Israel	3	3	Analytics, Advertisement
SuperAwesome	SuperAwesome	UK	3	3	Advertisement
IAB Open Measurement	IAB Tech Lab	US	5		Advertisement
Amplitude	Amplitude	US	2	3	Analytics, Profiling
ironSource	Unity Technologies	Israel	4		Analytics
Sentry	Functional Software Inc	US	3	1	Crash Reporting
Blackstone	Blackstone	US		3	Advertisement
Tencent	Tencent	China		3	Analytics
Bugly	Tencent	China	3		Crash reporting, Analytics
InMobi	InMobi	India	2	1	Advertisement
Microsoft	Microsoft	US		2	Analytics, Crash Reporting
Yandex	Yandex	Russia		1	Advertisement
SuperSonic Ads	ironSource	US		1	Advertisement
AppMetrica	Yandex	Russia	1		Analytics
Moat	Oracle	US	1		Advertisement, Analytics

5 Discussion

This section discusses the key takeaways and lessons learned from this study.

5.1 Traffic Light Analysis

Table 3 illustrates the resulting ratings of each app regarding the dangerous permissions and their ability to be justified by apps' functionalities (Sect. 4.1) and the number of tracking companies associated with the apps during runtime

Table 3. The traffic light rating of each app by category, considering dangerous permissions and runtime tracking. (Note: "Perm." is an abbreviation for "Permissions".)

App	Category	Perm. Rating	Trackers Rating	Overall Rating
ClassDojo	Educational	Red	Amber	Red
Kahoot!	Educational	Amber	Red	Red
Minecraft Education	Educational	Amber	Amber	Amber
PAW Patrol Academy	Educational	Green	Amber	Amber
Sago Mini School	Educational	Amber	Red	Red
ScratchJr	Educational	Amber	Green	Amber
Times Tables Rock Stars	Educational	Amber	Amber	Amber
Aha World: School Day	Role-Play & Simulation	Red	Amber	Red
Avatar World	Role-Play & Simulation	Green	Amber	Amber
Baby Panda's School Bus	Role-Play & Simulation	Red	Red	Red
Bluey: Let's Play!	Role-Play & Simulation	Green	Amber	Amber
Disney Frozen Royal Castle	Role-Play & Simulation	Green	Amber	Amber
Toca Boca Jr	Role-Play & Simulation	Amber	Amber	Amber
Toca Boca World	Role-Play & Simulation	Amber	Amber	Amber
Cake maker: Kids cooking games	Creative	Green	Amber	Amber
Coloring Games: Color & Paint	Creative	Green	Green	Green
Disney Coloring World	Creative	Amber	Amber	Amber
LEGO® Builder	Creative	Amber	Amber	Amber
Little Panda's Ice Cream Games	Creative	Red	Amber	Red
Toca Boca Jr Hair Salon 4	Creative	Green	Amber	Amber
Fluvsies - A Fluff to Luv	Casual & Action	Green	Red	Amber
Fluvsies Merge Party	Casual & Action	Green	Red	Amber
slither.io	Casual & Action	Red	Amber	Red
Tag with Ryan	Casual & Action	Amber	Red	Red
VLC for Android	Video & Content	Red	Green	Amber
YouTube Kids	Video & Content	Red	Amber	Red
Roblox	N/A	Amber	Red	Red

(Sect. 4.2). This table provides an overlook of the app ratings, following the criteria for red, amber and green, outlined in Sect. 3.2.

Using the traffic light analysis, only one app (*Coloring Games: Color & Paint*) had an overall green rating. Out of 27 total apps in the sample for this study, 10 of them had an overall red rating. Although this is not statistically significant, in the context of child safety using those apps, this is significant. The fact that red ratings were not just confined to one or a few categories raises concerns regarding the safety of children's apps as a whole. Whilst it is acknowledged that it would be impossible to demand apps not to engage in any of these behaviours, the extent to which these apps reflect potentially unsafe circumstances for children under 13 asks for intervention.

As apparent in Table 3, apps are much more likely to acquire a green rating in permissions (9) than in trackers (3). This is because, although dangerous permissions are present in the majority of the apps, their functionalities allow the requests to be justified. In terms of trackers, however, justifying the number is much more difficult in that tracking behaviours, such as collecting and further sharing data, carry much more privacy implications than if an app legitimately uses a permission for its sole purpose within the app.

5.2 Permissions in Children's Apps

The fact that some apps do not require any dangerous permissions may raise questions about the necessity of such permissions in other apps, e.g., if within the same category. It suggests that, for some functionalities, dangerous permissions may not always be essential to the core operations of an app. This could highlight potential areas where apps may be over-privileged, and developers might need to reconsider the scope of the permissions they are requesting. However, it is acknowledged that for certain functionalities, such as apps related to media like *VLC Player*, apps would need specific dangerous permissions to work as intended. On the other hand, some apps could potentially do without certain functionalities, such as child input through the device camera or microphone. Such apps would function just the same with the removal of these dangerous permissions and would decrease the overall risks for children's using them. This queries the extent to which developers follow the Google Play Store-recommended development strategy of minimal permission requests [13] or whether such developers create over-privileged apps only with the user experience in mind.

Considering *Roblox* was of particular concern initially, it was unfortunate to see that the app received an overall red rating, and it was on par with many of the other apps in the sample. It indicates that the extent of privacy issues related to popular online platforms used by children are not unique, although findings are not generalisable.

Fortunately, as prescribed in the ICO Children's code [19], which states that transparency and age-appropriate application are essential for the privacy and safety of children, the dangerous permissions in the sample of apps typically provided explicit occasions for user consent. However, since it is not easy to understand the implications of accepting a dangerous permission for parents [31] and children [1], it may not be natural to assume that this is an acceptable form of consent.

5.3 Tracking Behaviours in Children's Apps

It is acknowledged that some apps, especially those from independent developers or startups, need income to remain free of charge. Thus, it is understandable from this perspective why an app would contain some form of tracker to provide a way of monetisation. Still, it is unclear why an app for children aimed at 12 years old and under would need such a variety of advertising trackers embedded, with many ads not relevant to children of this age range (e.g., stair lift and holiday booking). Also, if their sole purpose is monetisation, it is unclear why developers would purposefully implement trackers into their apps that feature eye-catching redirect buttons (e.g., colourful and flashing). This is because this can hold implications for children's safety in that they may end up accessing something that is not age appropriate, conflicting with the ICO Children's code [19].

Analytics tracking was the most common type of tracking found in the apps analysed (Sect. 4.2). This form of tracking involves collecting data on user behaviour, such as how often they use the app, which features they use, and how long they spend doing certain activities. This aims to provide developers objective feedback to improve app functionality and user experience [25]. However, analytics also raise significant privacy concerns when it comes to children's apps. Although this data is often anonymous, it can be used to create detailed profiles without the children's or their parents' explicit knowledge or consent.

The ICO Children's code states that "profiling" should only be in place to protect a child from any harmful effects, which, in the apps in our sample, seems very unlikely to be the case. As seen in Sect. 5.3, the tracker Amplitude, identified as in use by the majority of the apps in our sample, conducts such a type of tracking behaviour. Similarly, the Google tracker that engages in "fingerprinting" (DoubleClick) was identified in 5 of the apps: *slither.io*, *Fluvsies - A Fluff to Luv*, *Cake maker: Kids cooking games*, *Fluvsies Merge Party*, and *Baby Panda's School Bus*. Fingerprinting is comparable to profiling in that it involves the collection of user-specific data, but instead includes device information which, when combined, can identify a particular device and its user [2]. This is of particular interest as it can be a much more invasive type of tracking which can persist across different apps and websites and is much more difficult to block or control.

5.4 Limitations and Future Work

A limitation of the methodology followed for this study is scalability. Despite the use of tools for recovering permissions and trackers for the selected apps, the goal of determining whether the dangerous permissions used were justified, as reflected in the traffic light comparison of apps, required manual analysis. Further automation would allow for the investigation of a larger number of apps for children.

There was a degree of subjectivity involved in the categorisation of the apps provided in Table 1. Their main functionalities were used to determine the most appropriate category. However, this process does not completely eliminate the possibility that an app could belong to multiple categories, which would affect the presented cross-category comparisons in the paper.

As shown in Table 2, not all trackers engage in advertisement; some do not engage in it at all (e.g., MixPanel) while others engage in multiple types of tracking including advertisement (e.g., Google). This raises questions about the business model adopted by free apps without advertisement trackers as a way of monetisation. Further research to map the business model of apps [7] for children would be an interesting possibility for future work. Also, this study revealed the use of advertisements inappropriate for children's apps, such as colourful and attention grabbing ads. This is non-compliant with the ICO Children's code [19] and, therefore, calls for a deeper analysis.

Figure 2 and Table 2 show mismatches between number of trackers uncovered by static and dynamic analysis. This lack of consistency may happen, e.g., because some tracking libraries are not present in the apps codebase or would reveal themselves only under certain circumstances or due to code obfuscation techniques [12]. We intend to study this aspect further by analysing trackers using tools claiming to be resilient to code obfuscation, such as AGRIGENTO [6].

While this study has focused primarily on dangerous permissions, future research could investigate the role and impact of other permission types, such as signature and custom permissions. Signature permissions are granted during app installation and require both the requesting app and the accessed entity to be signed with the same digital certificate. Custom permissions, on the other hand, may pose greater risks due to their variable nature. Since developers define custom permissions individually, their exact purposes are often unclear, which could make them more susceptible to misuse.

6 Conclusion

This study has examined the privacy and safety implications of 27 popular and free Android apps designed for children up to 12. Overall, this study has revealed that concerns about the privacy and safety of children using technology are not completely unwarranted. It has also delved beyond identifying the permissions and trackers associated with those apps to determine if they are essential and justified. The traffic light system used to assess these apps comparatively has highlighted that only a single app achieved a green rating overall, raising serious questions about the industry's approach to child safety.

References

1. 5Rights Foundation: Disrupted Childhood: The Cost of Persuasive Design (2023). https://5rightsfoundation.com/wp-content/uploads/2024/08/5rights_DisruptedChildhood_G.pdf. Accessed 28 Aug 2024
2. Almond, S.: Our response to Google's policy change on fingerprinting (2024). https://ico.org.uk/about-the-ico/media-centre/news-and-blogs/2024/12/our-response-to-google-s-policy-change-on-fingerprinting/. Accessed 23 May 2025
3. Binns, R., Lyngs, U., Van Kleek, M., et al.: Third party tracking in the mobile ecosystem. In: Proceedings of the 10th ACM Conference on Web Science, pp. 23–31 (2018). https://doi.org/10.1145/3201064.3201089
4. Blackwell, C., Lauricella, A., Wartella, E.: Factors influencing digital technology use in early childhood education. Comput. Educ. **77**, 82–90 (2014). https://doi.org/10.1016/j.compedu.2014.04.013

5. Calciati, P., Gorla, A., Kuznetsov, K., et al.: Automatically granted permissions in android apps: an empirical study on their prevalence and on the potential threats for privacy. In: Proceedings of the 17th International Conference on Mining Software Repositories, pp. 114–124 (2020). https://doi.org/10.1145/3379597.3387469
6. Continella, A., Fratantonio, Y., Lindorfer, M., et al.: Obfuscation-resilient privacy leak detection for mobile apps through differential analysis. In: Proceedings of the NDSS Symposyum 2017, pp. 1–15 (2017). https://doi.org/10.14722/ndss.2017.23465
7. Cristofaro, M.: E-business evolution: an analysis of mobile applications' business models. Technol. Anal. Strategic Manag. **32**(1), 88–103 (2020). https://doi.org/10.1080/09537325.2019.1634804
8. Ekambaranathan, A., Zhao, J., Van Kleek, M.: "Money makes the world go around": identifying barriers to better privacy in children's apps from developers' perspectives. In: Proceedings of the 2021 CHI Conference on Human Factors in Computing Systems, pp. 1–15 (2021). https://doi.org/10.1145/3411764.3445599
9. European Union: General Data Protection Regulation (GDPR) (2016). https://gdpr-info.eu/art-8-gdpr/. Accessed 16 Aug 2024
10. Feal, A., Calciati, P., Vallina-Rodriguez, N., et al.: Angel or devil? A privacy study of mobile parental control apps. In: Proceedings of the Privacy Enhancing Technologies Symposium, pp. 314–335 (2020). https://doi.org/10.2478/popets-2020-0029
11. Felt, A.P., Chin, E., Hanna, S., et al.: Android permissions demystified. In: Proceedings of the 18th ACM Conference on Compute and Communications Security, pp. 627–638 (2011). https://doi.org/10.1145/2046707.2046779
12. Glanz, L., Müller, P., Baumgärtner, L., et al.: Hidden in plain sight: obfuscated strings threatening your privacy. In: Proceedings of the 15th ACM Asia Conference on Computer and Communications Security, pp. 694–707 (2020). https://doi.org/10.1145/3320269.3384745
13. Google, A.D.: Permissions on Android (no date). https://developer.android.com/guide/topics/permissions/overview. Accessed 28 June 2024
14. Google, Android Developers: Manifest.permissions (no date). https://developer.android.com/reference/android/Manifest.permission. Accessed 30 June 2024
15. Hallett, E.: I was asked for naked photos after making 'friends' on roblox (2024). https://www.bbc.co.uk/news/uk-england-gloucestershire-68616730. Accessed 20 June 2024
16. Hedderson, M., Bekelman, T., Li, M., et al.: Trends in screen time use among children during the COVID-19 pandemic. JAMA Netw. Open **6**(2), 1–12 (2023). https://doi.org/10.1001/jamanetworkopen.2022.56157
17. Hirsh-Pasek, K., Zosh, J.M., Golinkoff, R.M., et al.: Putting education in "educational" apps: lessons from the science of learning. Psychol. Sci. Public Interest **16**(1), 3–34 (2015). https://doi.org/10.1177/1529100615569721
18. Information Commissioner's Office: Audits (no date). https://ico.org.uk/for-organisations/advice-and-services/audits/. Accessed 26 Mar 2025
19. Information Commissioner's Office: Introduction to the Children's code (no date). https://ico.org.uk/for-organisations/uk-gdpr-guidance-and-resources/childrens-information/childrens-code-guidance-and-resources/introduction-to-the-childrens-code/. Accessed 22 June 2024
20. Information Commissioner's Office: What are PECR? (no date). https://ico.org.uk/for-organisations/direct-marketing-and-privacy-and-electronic-communications/guide-to-pecr/what-are-pecr/. Accessed 24 June 2024

21. Kollnig, K., Binns, R., Dewitte, P., et al.: A fait accompli? An empirical study into the absence of consent to third-party tracking in android apps. In: Proceedings of the 17th Symposium on Usable Privacy and Security, pp. 181–195. USENIX Association (2021). https://www.usenix.org/system/files/soups2021-kollnig.pdf
22. Kollnig, K., Shadbolt, N.: TrackerControl: transparency and choice around app tracking. J. Open Source Softw. **7**(75), 1–4 (2022). https://doi.org/10.21105/joss.04270
23. Le, A., Varmarken, J., Langhoff, S., et al.: AntMonitor: a system for monitoring from mobile devices. In: Proceedings of the 2015 ACM SIGCOMM Workshop on Crowdsourcing and Crowdsharing of Big (Internet) Data, pp. 15–20 (2015). https://doi.org/10.1145/2787394.2787396
24. Livingstone, S., Stoilova, M., Nandagiri, R.: Children's data and privacy online: growing up in a digital age. An evidence review (2019). https://eprints.lse.ac.uk/101283/1/Livingstone_childrens_data_and_privacy_online_evidence_review_published.pdf
25. Mor-Samuels, E.: Everything you need to know about mobile app analytics (2024). https://www.appsflyer.com/blog/measurement-analytics/everything-you-need-to-know-about-mobile-app-analytics/. Accessed 20 Jan 2025
26. OFCOM: Children and Parents: Media Use and Attitudes Reports (2024). https://www.ofcom.org.uk/siteassets/resources/documents/research-and-data/media-literacy-research/children/children-media-use-and-attitudes-2024/childrens-media-literacy-report-2024.pdf. Accessed 19 Aug 2024
27. Reyes, I., Wijesekera, P., Reardon, J., et al.: "Won't somebody think of the children?" Examining COPPA compliance at scale. In: Proceedings of the Privacy Enhancing Technologies Symposium, pp. 63–83 (2018). https://doi.org/10.1515/popets-2018-0021
28. UK Government: Data Protection Act (DPA) (2018). https://www.legislation.gov.uk/ukpga/2018/12/contents/enacted. Accessed 20 Aug 2024
29. UK Parliament: The Privacy and Electronic Communications (EC Directive) Regulations (2003). https://www.legislation.gov.uk/uksi/2003/2426/contents. Accessed 20 Aug 2024
30. UK Parliament: Online Safety Act (2023). https://www.legislation.gov.uk/ukpga/2023/50. Accessed 16 Aug 2024
31. Utz, C., Degeling, M., Fahl, S., et al.: (Un)informed consent: studying GDPR consent notices in the field. In: Proceedings of the 2019 ACM SIGSAC Conference on Computer and Communications Security, pp. 973–990 (2019). https://doi.org/10.1145/3319535.3354212
32. Westin, A.F.: Privacy and Freedom. Atheneum, NY (1967)
33. Winter, E., Costello, A., O'Brien, M., Hickey, G.: Teachers' use of technology and the impact of COVID-19. Irish Educ. Stud. **40**(2), 235–246 (2021). https://doi.org/10.1080/03323315.2021.1916559

Understanding Corporate Ransomware – A Case Study in the Nordics

Emil Larsson[1,2](✉) and Meiko Jensen[1]

[1] Karlstad University, Universitetsgatan 2, 65188 Karlstad, Sweden
{emil.larsson,meiko.jensen}@kau.se
[2] Swedish Defence University, Drottning Kristinas väg 37, 11428 Stockholm, Sweden

Abstract. Ransomware is continuously evolving and has recently seen record-breaking payouts. This paper describes developments in corporate ransomware which have contributed to this continued rise. It provides an interdisciplinary case study of a ransomware campaign conducted in northern Europe during 2023 and 2024. We argue that underlying factors, including the broad rollout of personal cloud backups, have caused modern ransomware groups to shift towards expending more effort per attack in order to attack bigger targets. We perform an analysis of such an attack using a combination of internal and open sources, as well as using forensic techniques. The analysis shows how the attackers approach high-value targets, use both traditional crypto-ransomware tools and hacking, and utilize multiple avenues of extortion to negotiate the highest possible ransom.

Keywords: ransomware · double extortion · case study · Akira · malware

1 Introduction

Hundreds of millions of users have been afflicted by encryption ransomware since its invention in 1996 [38,39]. This type of attack involves encrypting the files of a victim and charging a ransom to provide decryption tools or keys. As the sums involved have grown, the ransomware and cyber attack business ecosystem has been compartmentalized into specialized subdivisions [18]. Increasing awareness of the ransomware threat, improved user IT maturity, and more comprehensive technical protection measures have engendered a cat-and-mouse game where cyber criminals have been compelled to discover new ways to make money. Understanding the organization of ransomware groups, the motivations and modus operandi of the various actors, and the economics of ransomware is complicated by the secrecy preferred by both criminals and most victims. The highest quality data and insight on cyber crime is generally produced by private threat intelligence firms, cyber insurance brokers, and incident response companies. These companies do not always publish their methodologies and often lock data into reports available only to their customers.

I. Praça et al. (Eds.): EICC 2025, CCIS 2500, pp. 82–97, 2025.
https://doi.org/10.1007/978-3-031-94855-8_6

As the societal consequences of ransomware attacks have grown, many approaches have been suggested to reduce and mitigate the harm. International cooperation is steadily growing, with 50 nations signed up to the US-led Counter-Ransomware Initiative, which by 2023 was in its third year of existence [17]. As the efficacy of some approaches is unknown, this group is taking a "kitchen sink" approach which includes everything from training and best practices to asset seizure, blacklisting crypto wallets known to be used by ransomware actors, and evaluating legal measures to prevent public entities from paying ransoms [22]. Moving further on the geopolitical end of the spectrum, ransomware has also become a convenient cover for offensive cyber operations, e.g. within the context of the Russia-Ukraine conflict [33]. Victims cannot be certain whether they were attacked for financial gain or whether the attack was carried out by a nation-state actor using malware activities as cover.

In response, significant development has occurred both in technical measures of protection and in industry awareness. Cloud-based backup has gained enormous market share in both private and corporate use. Multifactor authentication is no longer limited to environments considered especially sensitive, but is now in widespread use, e.g. for basic email services. Many organizations have implemented systematic training and awareness programs to increase security against initial access via the most common threat vector, email-based phishing attacks [9]. The fight against ransomware can seem hopeless, but these measures are now showing promise and have become prevalent enough to have a real impact.

Although it would be ideal to prevent most ransomware attacks and gain understanding of threat actors through threat intelligence, the rate of success of the attacks has provided a plethora of cases from which to learn. These can be considered both in aggregate to gain a quantitative understanding of the ransomware ecosystem, and with in-depth case studies showing how a large-scale attack is conducted. We argue that these attacks are qualitatively different and refer to them as *corporate ransomware*; they distinguish themselves from traditional ransomware both by the size and value of the target and the correspondingly greater effort of the attackers. This effort can be seen not only in hours spent gaining access to administrative accounts and the care taken in finding and deleting backups but also in the intelligence gathered on the victim's financial information. Previous case studies of large-scale incidents have provided both in-depth descriptions of attacks and campaigns, and have drawn relevant practical conclusions from them, but have not focused on differences in methodology between target types [3,30]. Six groups represent more than half of all ransomware compromises [26], and in August 2024 the Akira ransomware group was the fourth most active of these threat actors.

The paper is organized as follows. The next section summarizes the background and development of the ransomware ecosystem, including several recent trends towards corporate targets. Section 3 then provides interdisciplinary insights on the case study, namely the Akira ransomware campaign of 2023/24.

The case study is followed by a discussion on corporate ransomware in general in Sect. 4. The paper concludes with future research directions in Sect. 5.

2 Background and Developments in Corporate Ransomware

Early ransomware attacks worked similarly to any traditional computer virus, spreading between computers and workstations through files and USB drives or being sent as e-mail attachments. The impact was typically limited to local files, or those network storage areas available to the end user. This unsophisticated attack mode did not require any great strength in numbers on the attacking side, and the payments demanded were correspondingly small, i.e. well within the financial abilities of the individuals targeted. The attackers grew their scope by developing standardized tools, which required less effort per attack and produced predictable and similar results for each victim. In practice, this meant that equally large ransoms were demanded from the victimized end-user regardless of whether they were acting in a private capacity or employed in some organization of smaller or greater size. As target resilience and awareness has developed, the ransomware ecosystem has trended towards more capable groups which multiply their force by providing ransomware-as-a-service to collaborating threat actors [27]. These collaborators act as customers and partners, using the tools provided and sharing the profits. The 2017 WannaCry attack is characteristic of high-impact ransomware at the time [27]. It impacted many large and multinational companies, but demanded a standardized ransom. Since then, technical protection, cyber insurance, and the growth of ransomware as a criminal enterprise has caused several trend shifts where

1. victims are more savvy when negotiating ransom payments, and have become less prone to paying them [8,41],
2. the organizations attacked have become larger and more specifically targeted [8] (Fig. 1),
3. the attackers tailor ransom sizes to victims, with payments having become correspondingly greater [23,28] (Fig. 2), and
4. some victims are now paying mostly or exclusively to protect exfiltrated data from disclosure, rather than for decryption of files [20,23].

These trends are described in detail below.

2.1 Victims are Increasingly Reluctant to Pay the Ransom

In the mid-to-late 2010s, ransomware attackers could expect most victims to pay for data decryption. Since then, a dramatic and unwavering decline in payments is evident. The reasons for the reduced number of payments include improved resilience to ransomware, especially including backups, as well as reduced trust in the credibility of promises made by ransomware groups [8]. This lack of trust in the efficacy of ransom payments may have been spurred by highly publicized incidents such as the 2021 Colonial Pipeline attack, where the ransom was paid but where the decryption tool provided by the attackers was so slow that it was deemed unusable [3]. Another likely reason is that while early ransomware ransom transactions were concerned solely with paying for decryption keys, multiple-extortion attacks also include threats to harvest and re-use victim credentials as well as offers not to sell off sensitive exfiltrated data. Victims have few ways to follow up these guarantees should they choose to pay for them, and advisory firms generally recommend against payment [8].

Legislative measures against ransomware payments have also played a part in reducing victims' willingness to pay. These include mandatory reporting of attacks, as implemented in the United States in 2021, as well as outright bans on payment. At present, these bans have been limited to specific domains (e.g. public sector organizations) in select geographic domains [22]. However, the 2022 Russian invasion of Ukraine has caused Russia to come under international sanctions. This has greatly complicated payment of ransoms to Russia-based ransomware groups, partially accounting for the drop in ransomware payments in 2022 [37].

2.2 Organizations of Greater Size are Targeted

When fewer and fewer victims pay the ransom, a natural step for attackers is to find bigger and richer targets. Big organizations can afford to invest both in technology and training to mitigate the simplest attacks, i.e. mass phishing attacks culminating in encryption of local user files. Reaching a bigger and more hardened target requires the deployment of advanced techniques. The attackers gain initial access through spear phishing, malvertising, unpatched vulnerabilities, or through supply-chain attacks [25]. In addition, the rapid expansion of work-from-home caused by the COVID pandemic has increased corporate attack surfaces and pushed VPN exploits to the forefront [20]. Large organizations typically have some capabilities to detect and respond to cyber breaches, as well as segmenting networks and segregating duties of staff. This requires attackers to stay hidden while performing manual reconnaissance and escalating privileges in the environment. The additional expenditure of time and manpower requires a correspondingly greater earning opportunity for ransomware gangs to break even.

2.3 Ransom Payouts are Larger than Ever

Around 2018, the predominant crypto extortion attack was compounded with a secondary "double extortion" scheme. While the victim's files are still encrypted, with decryption offered against a ransom, criminals are now increasingly exfiltrating potentially sensitive data and charging victims not to release it publicly. Further revenue streams and dimensions of extortion exist, with criminals selling credentials, threatening DDoS, or even alerting authorities to the hack in order to cause a fine for the victim [10]. In addition, criminals have grown adept at making false claims about what data has been stolen, and at threatening victims with additional attacks once a channel of communication has been established.

Between 2018 and 2024, the difference between the median and average ransom payment has dramatically increased [8]. This gap shows that the biggest payments have become much bigger than the average payment, with some large and financially capable organizations representing significantly outsized sums. After a brief dip in 2022, 2023 saw the highest amount of ransoms paid to date, approximately USD $1B [5]. The additional costs of productivity loss and recovery are clearly higher yet.

2.4 Data Exfiltration is Becoming Successful as a Secondary Attack Mode

To compensate for the decline in paying victims, ransomware groups are attempting to increase the pressure on each victim by utilizing secondary modes of attack. These secondary and tertiary attacks can include threatening phone calls, or DDoS attacks against victims already weakened by an initial breach [12]. In 2019, some threat actors began using information gained to "name and shame" victims who would not pay their ransom [36]. Some victims have been able to successfully recover from the "primary" encryption attack, or have deemed decryption assistance to be economically uninteresting, but will still pay a ransom for exfiltrated data not to be published or sold. While the proportion of paying victims in this category is low, it is additional to those paying primarily for decryption [8]. This shows that this specific mode of secondary attack is working for ransomware groups. Nevertheless, the attackers must expend additional effort to understand what of the victim's data is sensitive and valuable. The victim may not trust the attacker to actually have the data claimed, as some attackers have been known to simply fake it, so, typically, data samples are provided [36]. Finally, some victims may consider that the innately unreliable criminals in ransomware groups will not keep their promise to delete the data when it can be used for a new future claim.

3 The 2023–24 Akira Ransomware Campaign and the Tietoevry Incident

Starting late 2023 and continuing into the spring of 2024, the Akira ransomware group conducted a campaign which struck at least fourteen victims in northern Europe (Finland, Sweden, and Germany) [35]. The actor(s) involved and the methods used bear closer scrutiny.

3.1 Research Method

This study attempts to gain insight into corporate ransomware attacks both by investigating a broader campaign of attacks and by going into detail of a typical attack, i.e. through an instrumental case study. While traditional ransomware

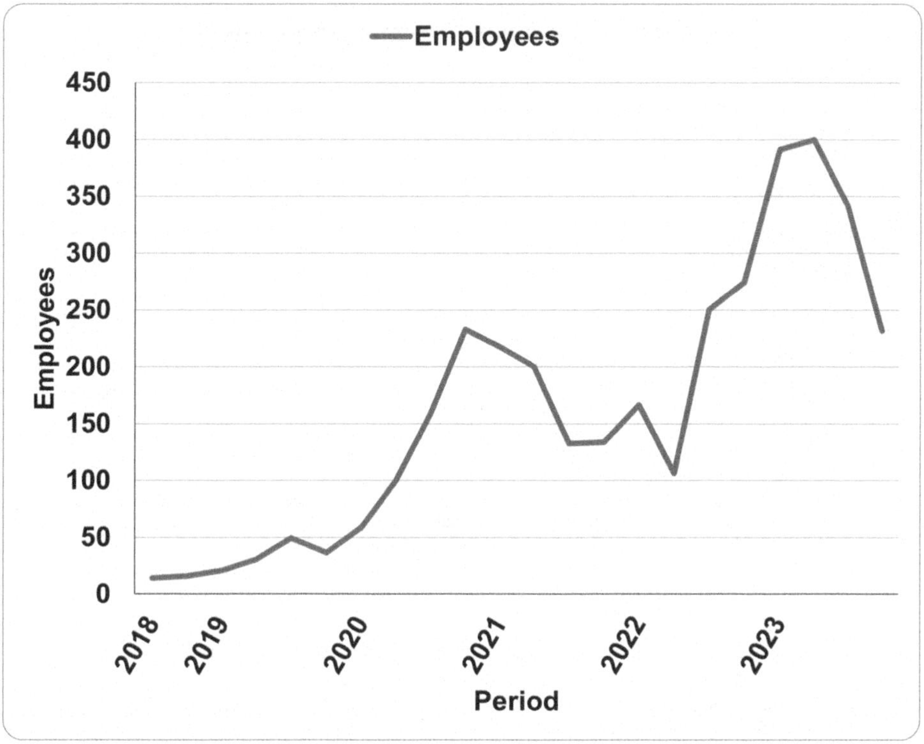

Fig. 1. Average number of employees per victim organization over time, adapted from [8].

attacks number in the millions per year and can thus be understood using both qualitative and quantitative methods, corporate ransomware attacks number only in the hundreds. Thus, a qualitative perspective is the preferred approach to gain broader insights into ransomware, both from a technical, forensic, and process-oriented perspective. We have selected the Akira campaign in general and the Tietoevry attack in particular as public interest and scrutiny has made more data available, and attack tools have been released for forensic analysis. We investigate these tools both in a dedicated malware laboratory setting and by exploring the software and systems provided by the threat actor in the wild.

3.2 The Akira Ransomware Group

Akira is the name given both to a ransomware encryption tool and to the Ransomware-as-a-Service (RaaS) group that utilizes it. First mentioned in March 2023, Akira has breached more than 200 predominantly corporate victims [19]. The group is notable for using a highly stylized communications platform based on 1980s science fiction, for being very diligent in deleting victims' backups, and

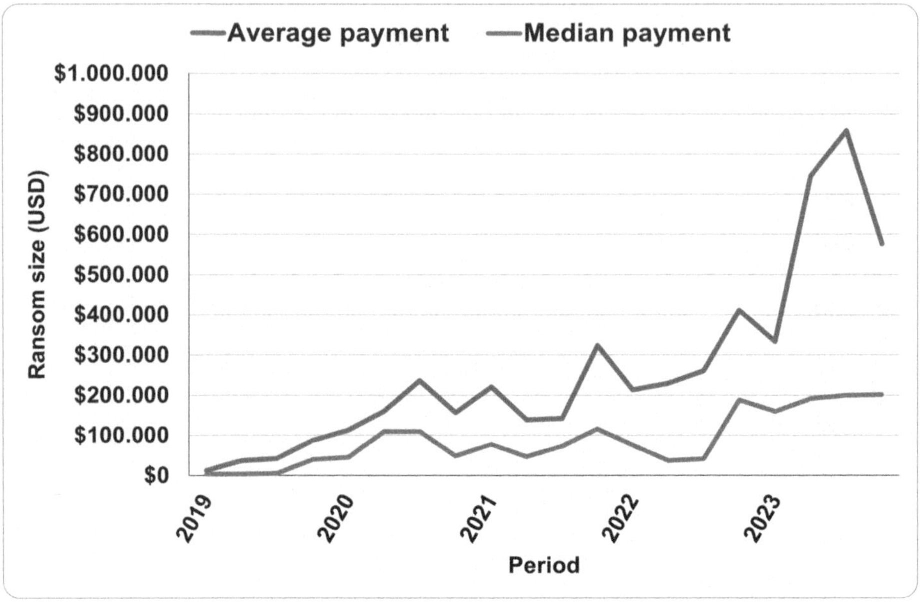

Fig. 2. Ransom payment size over time, adapted from [8].

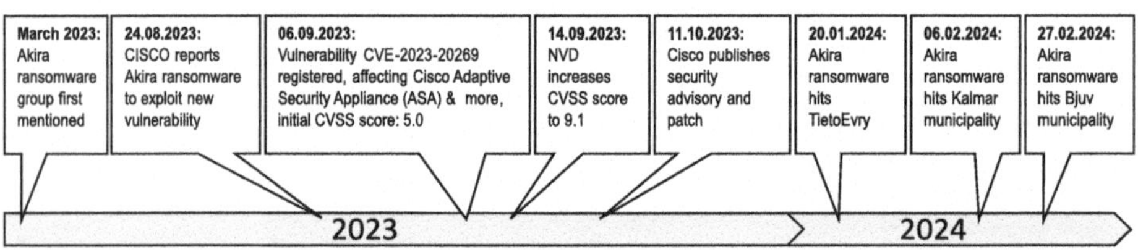

Fig. 3. Timeline of events around the Tietoevry ransomware incident.

for consistently utilizing secondary means of extortion [34]. These include publishing a list of successfully breached victims, specifying what harm was done and what data was stolen, as well as threatening the victim with selling off the exfiltrated data. The victims have been spread across multiple countries but were initially mostly located in the continental United States and Canada, moving into northern Europe towards the end of 2023. At time of writing, the exact organizational setup of the Akira attacks is not clear. The RaaS model means that more attacks can be conducted over a given time, both by purchasing credentials and information about potential victims generated by so-called initial access brokers, and by handing off potential targets to profit-sharing affiliates. Thus, when referring to "the Akira group", it is possible that multiple loosely affiliated sets of attackers may hide behind the alias. Attackers who utilize ransomware services can also be affiliated with multiple RaaS groups at the same time, complicating follow up, and creating a multi-faceted threat. Initial reporting of Akira attacks linked them to the previously known Conti RaaS group, both in terms of similarities in the malware tools utilized, but also by sharing a customer base as identified by blockchain wallets [4]. By late 2023, the Akira malware type was the most common one used [8].

The attackers appear to have started the campaign opportunistically, utilizing a vulnerability in Cisco ASA VPN appliances [43]. Initially, Cisco responded to reports of Akira's exploitation by reminding users of the virtues of using Multi-Factor Authentication (MFA) [29]. In September of 2023, Cisco confirmed that while MFA would mitigate the problem, there was also an underlying vulnerability which allowed brute-forcing of credentials [7].

This vulnerability, designated CVE-2023-20269, was classified by Cisco as being of severity 5.0 out of 10 on the CVSS Scale [24]. This low classification was based on the existence of immediate mitigations against the attack, i.e., enabling multi-factor authentication for VPN connections. It is possible that some organizations misjudged the severity of the vulnerability and did not prioritize mitigating action. The existence of a mitigating factor may also have contributed to the one-month gap between disclosure and availability of a patch (Fig. 3).

After attacking at least ten targets in Finland, the Akira ransomware was used to attack a company named *Tietoevry Tech Services* in Sweden. Tietoevry is the largest IT outsourcing provider in the Nordics, with a market share of around 15% in Norway, Finland, and Sweden. In Sweden, their total market share is lower, but with a concentration of outsourcing customers in government, regional, and municipal IT services.

In the early hours of Saturday morning, January 20th, 2024, the attackers activated the encryption tool on data storage in a VMware environment containing more than 2000 virtual machines. VM data was encrypted, backup storages and logfiles were wiped, both for customer data and internal data of Tietoevry. Both the speed of encryption and the weekday and time chosen are notable here—with the attackers having opted to strike when they could hope for reduced and delayed availability of the workforce at Tietoevry to detect or counter the attack.

3.3 Analysis of the Akira Encryption Tool

Gaining access to ransomware encryptors is not always easy due to victim-specific details, i.e. negotiation URLs, being encoded within the tool. Nevertheless, the Akira encryption tool has been recovered from previous attacks and is available online, e.g. from the Malware Bazaar [1] and Virustotal [6] libraries. By executing the Akira encryption tool[1] in an airgapped environment optimized for the analysis of unknown malware, we were able to observe certain distinct features. In our experiments, the Akira encrypter took only a few seconds to iterate over and encrypt the user folders of 270 MB of data in total. Including an initial dormant time of at least 30 s, which may have been due to memory access, the

[1] md5sum: 496aa6ac94eb97d2bc05bfce688460c1.

encrypter completed its task in about 45 s. The speed of encryption is achieved by the fact that the Akira encrypter does not encrypt every piece of data. Small files are encrypted completely, but larger files or filesystem blocks are encrypted only partially, leaving some plain-text blocks interleaved with encrypted blocks behind. Though this approach leaves some plain-text data untouched, it nevertheless provides several advantages to the attackers:

- the encryption process operates considerably faster, allowing for more data to be encrypted before the victim can detect and stop the encryption process,
- as file headers generally become encrypted, files can no longer be correctly processed by their intended software applications, and
- decryption requires nonstandard software, making it more difficult for victims to recover their files without paying the ransom even if the decryption key might have been disclosed by other means.

Additionally, as is common for modern ransomware, critical operating system files and Internet browser software files are left untouched by the encryption, so that the victim remains able to access the website needed for negotiating the ransom.

Fig. 4. Result of running the Akira ransomware encrypter, with ransom note.

```
If you're indeed interested in our assistance and the services we provide you can reach out to us following simple
instructions:

1. Install TOR Browser to get access to our chat room - https://www.torproject.org/download/.
2. Paste this link - https://akiralkzxzq2[redacted]efvr2id.onion.
3. Use this code - 0015-[redacted]-MHJD-[redacted] - to log into our chat.

Keep in mind that the faster you will get in touch, the less damage we cause.
```

Fig. 5. Excerpt from the ransom note left behind by the Akira ransomware encrypter.

Figure 4 shows a screenshot from a system where we have reproduced Akira infection in the laboratory. As can be seen in the startup menu section, some program icons have been rendered completely unviewable, while for others, the half encryption is visible as displaying only half of the bitmap image file correctly and the other half as digital noise. The file explorer section shows the manifestation of Akira on the file level: files that were encrypted receive the additional filename extension of `.akira`, and every folder processed contains an additional new file, `akira_readme.txt`, which contains a detailed letter from the Akira group (cf. Fig. 5). The ransom note itself does not explicitly request a ransom.

Most of the data and programs installed on the victim's computer are rendered unusable this way. In the case of virtual machines, VM image files and snapshot binaries are encrypted directly on their storage locations, rendering the VMs unable to start, or causing fatal crashes when getting encrypted at runtime.

3.4 The Tietoevry Attack, Continued

After gaining initial access, most likely through the Cisco ASA vulnerability, the Tietoevry attack exposed a failure of defense in depth. The attack progressed in a fashion characteristic of what we term *corporate ransomware* attacks; it was contingent on the attackers

1. gaining initial access through weak or leaked credentials bought from an initial access broker, through unpatched vulnerabilities in internet-facing systems, or through endpoint access using spear phishing or malvertising,
2. staying undetected by Extended Detection & Response (XDR) tools,
3. performing lateral movement by using weak or leaked credentials, or vulnerabilities in the corporate network,
4. overcoming segregation of duties and security zones,
5. exfiltrating data, and
6. destroying backups and encrypting data.

Being an enterprise-level IT outsourcing organization, Tietoevry had comprehensive internal backups of the VMware environment. Internal sources claim that all of these were destroyed. Significant but incomplete recovery was possible due to the existence of an additional backup at a cloud backup provider. Although this attack pattern is typical of ransomware scenarios, Tietoevry later made public statements to the effect that no ransom letter was discovered [21]. As can be seen in the screenshot sections from an Akira-infected system in Figs. 4 and 5, the Akira ransomware tool generates a text file urging the victim to contact the group using a TOR link, but does not explicitly request nor state an exact amount of ransom. The ransom demand and amount are subject to negotiation with the attackers under the given TOR link and using the generated attack identifier code. Although the attackers probably did not intend it, this approach makes it possible for victims to honestly make public claims that they did not (explicitly) receive a "ransom demand" from Akira, and avoid the shaming associated with a ransomware breach.

A compounding factor in this attack was the organizational distance to the attacked datacenter, which was part of a long-running corporate acquisition completed in 2019. Although the attacked environment was isolated from the main Tietoevry IT environment, preventing catastrophic spread, the isolation is also likely to have reduced the available means of detection.

Although the vast majority of systems could be restored or rebuilt within weeks or months, the attack caused permanent and unrecoverable data loss in critical government systems. As an example, the Swedish national health service tracks prices of the medicines over time in order to calculate subsidy rates. The attack caused 20 years of price data to be permanently lost [32].

The focus on data exfiltration is notable. Akira uses three separate forms of extortion to incite victims to pay:

1. selling decryption keys for the ransomed data, for which considerable effort is put into destroying any backups,
2. threatening to sell exfiltrated data to the highest bidder, and
3. publicly naming and shaming the victim on the Akira blog, which is publicly accessible under a TOR link, and shown in excerpt in Fig. 6.

After the Tietoevry attack in January 2024, the IT systems of two Swedish municipalities (Kalmar and Bjuv) have been attacked with Akira as well. In both of these attacks, the municipalities decided not to pay the ransom, and were able to utilize backups to restore operations. However, in the Bjuv case, the Akira group alleged that they had exfiltrated some 200 gigabytes of data used for a secondary extortion attack (cf. [16] and Fig. 6).

4 Discussion

In this section, we position the case study against the background to draw broader conclusions about the evolution of corporate ransomware.

Open data and incidents such as the one we have described show that ransomware attacks on organizations are taking a new and separate direction, as described in Sect. 2. The main threat to organizations is now increasingly from targeted, high-effort cyberattacks rather than from mass mailing. While these cyberattacks can gain a foothold inside the organizational perimeter using click-based malware, the combination of zero-day exploits and leaked credentials, or

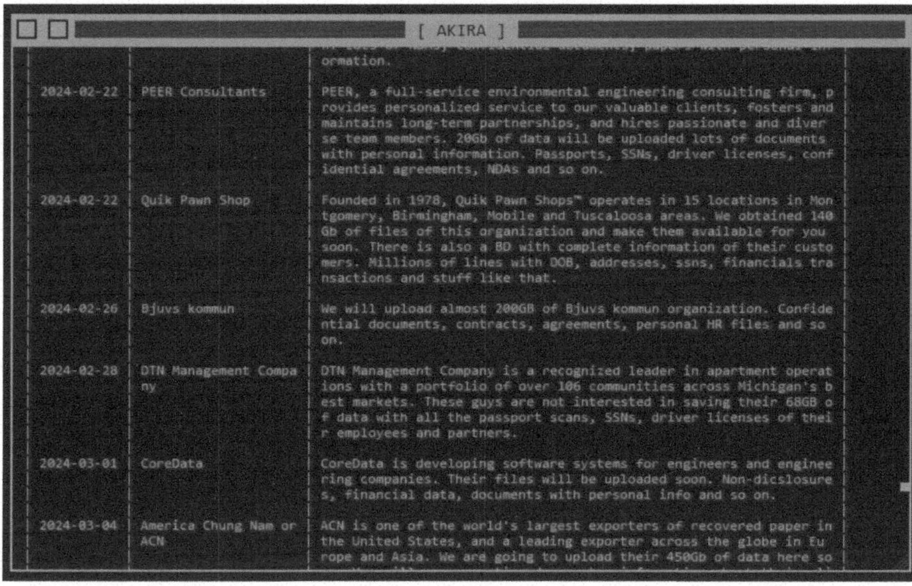

		ormation.
2024-02-22	PEER Consultants	PEER, a full-service environmental engineering consulting firm, provides personalized service to our valuable clients, fosters and maintains long-term partnerships, and hires passionate and diverse team members. 20Gb of data will be uploaded lots of documents with personal information. Passports, SSNs, driver licenses, confidential agreements, NDAs and so on.
2024-02-22	Quik Pawn Shop	Founded in 1978, Quik Pawn Shops™ operates in 15 locations in Montgomery, Birmingham, Mobile and Tuscaloosa areas. We obtained 140 Gb of files of this organization and make them available for you soon. There is also a BD with complete information of their customers. Millions of lines with DOB, addresses, ssns, financials transactions and stuff like that.
2024-02-26	Bjuvs kommun	We will upload almost 200GB of Bjuvs kommun organization. Confidential documents, contracts, agreements, personal HR files and so on.
2024-02-28	DTN Management Company	DTN Management Company is a recognized leader in apartment operations with a portfolio of over 106 communities across Michigan's best markets. These guys are not interested in saving their 68GB of data with all the passport scans, SSNs, driver licenses of their employees and partners.
2024-03-01	CoreData	CoreData is developing software systems for engineers and engineering companies. Their files will be uploaded soon. Non-dicslosures, financial data, documents with personal info and so on.
2024-03-04	America Chung Nam or ACN	ACN is one of the world's largest exporters of recovered paper in the United States, and a leading exporter across the globe in Europe and Asia. We are going to upload their 450Gb of data here so

Fig. 6. Excerpt from the list of allegedly compromised Akira victims.

the use of supply chain access, is common in the most damaging attacks [11,14]. We refer to these attacks as corporate ransomware as they are differentiated from ransomware targeting individual users in terms of both attack effort, extortion methods, and impact. Fewer payments, larger targets and multiple-extortion attacks becoming the norm set the stage for the 2023–24 Akira ransomware campaign.

4.1 Directed High-Effort Attacks

In a conventional malware-based ransomware attack, a momentary gap in concentration can lead to dire consequences. A corporate user who unintentionally executes a disguised ransomware encrypter with normal user-level access will cause damage to any data they have access to. Two efficient and commonly recommended approaches to protect against this scenario have been staff training and comprehensive, well-tested backups. Corporate ransomware attacks are not necessarily prevented by these two measures, as they are both based on technical attacks and breach multiple layers of security. Moreover, directed high-effort attacks are more harmful than traditional, opportunistic ones [42]. Protective measures against these directed high-effort attacks also differ from those used against opportunistic ones. Important measures include reducing the attack surfaces and implementing tiered defense in depth [40]. Current best practices for organizations contain many non-actionable items, and rarely prioritize protective measures, listing them as equally critical [2]. In the Tietoevry case, critical software updates were not applied rapidly enough. Primary backups were in place, but were not sufficiently protected against a catastrophic breach. Only a secondary set of (incomplete) backups at an external backup provider prevented a total disaster (cf. Sect. 3.4).

4.2 Extortion Methods

While traditional ransomware still exists as a serious threat, external and cloud-based backups have become more commonplace and reliable. This has dramatically reduced the hit rate of these traditional, low effort attacks [42]. Attackers are increasingly incentivized to focus more effort on single targets and diversify into data exfiltration attacks rather than rely on encryption. A defender who can afford to wait for files to be recreated or recovered from backup can address the threat from a position of strength. Moreover, the decryption tools provided by ransomware groups are sometimes slow or unreliable enough to be a poor investment [3,41]. When sensitive data is exfiltrated from a wealthy corporation, however, the attackers are able to set a higher cost and enter into extended negotiations. Paying victims can find themselves extorted repeatedly over the same material, or find that it has been sold to other parties despite their good-faith payment. The potential value of this attack mode is evident in the Akira campaign. While ransomware groups such as 8Base will pressure victims by publicizing their successful breach [31], the Akira group has typically not published the identities of their victims until some time later, to make it possible for the victim to pay for anonymity. In the Tietoevry case, negotations over decryption keys were allegedly never started, and the attackers have not made public threats of data disclosure. It is possible that the high profile of the case, with many government organizations affected, made such a claim unattractive, for instance due to fears of increased law enforcement attention.

More generally, the new state of corporate ransomware will change the incentives for organizations in terms of data protection. Although regulations such as GDPR already require adequate legal grounds for processing of any personal data gathered, organizations must now reconsider the potential value of data against the risks of a successful data breach. This calculus will increase the importance of privacy-enhancing techniques such as early anonymization of user data. Going beyond the cost of reputation loss, ransomware gangs have also threatened their victims to disclose the data breach to the Data Protection authority, and explained that the ransom is cheaper than a GDPR fine [13].

5 Conclusions and Future Work

In this paper, we have discussed a new subcategory of ransomware attacks, termed corporate ransomware, and provided insights from a typical case. As we have shown, the ransomware business evolved from semi-automated, multiple-target, low-ransom attacks towards more advanced, targeted, multi-extortion schemes which involve data encryption as one of several attack vectors. Data exfiltration and subsequent blackmailing puts victims into a more delicate position with respect to ransom payment decisions, as the growing availability and quality of backups do not help against this extortion scheme.

More advanced extortion schemes have even utilized the threat of reporting the victim of a successful data breach to supervisory agencies, like data protection authorities, cybersecurity agencies, or financial market authorities to extort even higher ransoms [13,15]. These *regulatory attacks* open up an even more heterogeneous scenario of payment incentives and trade-offs, for both victim and attacker side. Hence, in future work, we plan to focus our analysis on these novel extortion methods, and to analyze the corporate ransomware market in its current rapid evolution.

Acknowledgement. This study received funding by the Swedish Knowledge Foundation (KKS) through the SIGS-CyberSec project.

References

1. Abuse.ch: Malware bazaar. https://bazaar.abuse.ch/. Accessed 11 Jan 2025
2. Barrera, D., Bellman, C., Van Oorschot, P.: Security best practices: a critical analysis using IoT as a case study. ACM Trans. Priv. Secur. **26**(2) (2023). https://doi.org/10.1145/3563392
3. Beerman, J., Berent, D., Falter, Z., Bhunia, S.: A review of colonial pipeline ransomware attack. In: 2023 IEEE/ACM 23rd International Symposium on Cluster, Cloud and Internet Computing Workshops (CCGridW), pp. 8–15 (2023). https://doi.org/10.1109/CCGridW59191.2023.00017
4. Campbell, S., Suthar, A., Belfiore, C., Team, A.W.L.: Conti and Akira: Chained together. Arctic Wolf Labs (July 2023). https://arcticwolf.com/resources/blog/conti-and-akira-chained-together/
5. Chainalysis Inc: The Chainalysis 2024 crypto crime report (February 2024). https://go.chainalysis.com/crypto-crime-2024.html
6. Chronicle Security: Virustotal. https://www.virustotal.com/gui/home/upload. Accessed 11 Jan 2025
7. Cisco Systems: Cisco adaptive security appliance software and firepower threat defense software remote access vpn unauthorized access vulnerability. https://sec.cloudapps.cisco.com/security/center/content/CiscoSecurityAdvisory/cisco-sa-asaftd-ravpn-auth-8LyfCkeC. Accessed 12 Mar 2025
8. Coveware: New ransomware reporting requirements kick in as victims increasingly avoid paying. Coveware Quarterly Report (January 2024). https://www.coveware.com/blog/2024/1/25/new-ransomware-reporting-requirements-kick-in-as-victims-increasingly-avoid-paying
9. Desolda, G., Ferro, L.S., Marrella, A., Catarci, T., Costabile, M.F.: Human factors in phishing attacks: a systematic literature review. ACM Comput. Surv. (CSUR) **54**(8), 1–35 (2021)
10. Dissent Does: AlphV files an SEC complaint against MeridianLink for not disclosing a breach to the SEC. Databreaches.net (November 2023). https://databreaches.net/alphv-files-an-sec-complaint-against-meridianlink-for-not-disclosing-a-breach-to-the-sec/

11. Esparza, J.M.: Understanding the credential theft lifecycle. Comput. Fraud Secur. **2019**(2), 6–9 (2019). https://doi.org/10.1016/S1361-3723(19)30018-1, https://www.sciencedirect.com/science/article/pii/S1361372319300181
12. FBI and FinCEN: FBI and FinCEN release advisory on avoslocker ransomware. https://www.cisa.gov/news-events/alerts/2022/03/22/fbi-and-fincen-release-advisory-avoslocker-ransomware. Accessed 12 Mar 2024
13. Flashpoint Security: The emergence of ransomed: An uncertain cyber threat in the making. https://flashpoint.io/blog/ransomed-uncertain-cyber-threat/. Accessed 11 Jan 2025
14. Ghanbari, H., Koskinen, K., Wei, Y.: From solarwinds to kaseya: the rise of supply chain attacks in a digital world. J. Inf. Technol. Teach. Cases (2024). https://doi.org/10.1177/20438869241299823
15. Goodin, D.: Ransomware group reports victim it breached to SEC regulators. https://arstechnica.com/security/2023/11/ransomware-group-reports-victim-it-breached-to-sec-regulators/. Accessed 25 Aug 2024
16. Hivert, A.F.: Sweden falls victim to a wave of ransomware attacks. Le Monde (March 2024). https://www.lemonde.fr/en/international/article/2024/03/07/sweden-under-the-wave-of-ransomware-attacks_6593230_4.html
17. House, T.W.: International counter ransomware initiative 2023 joint statement. White House Briefing (November 2023). https://www.whitehouse.gov/briefing-room/statements-releases/2023/11/01/international-counter-ransomware-initiative-2023-joint-statement/
18. Huang, K., Siegel, M., Madnick, S.: Systematically understanding the cyber attack business: a survey. ACM Comput. Surv. (CSUR) **51**(4), 1–36 (2018)
19. Kersten, M., Mundo, A.: Akira ransomware. Trellix (November 2023). https://www.trellix.com/blogs/research/akira-ransomware/
20. Lang, M., Connolly, L., Taylor, P., Corner, P.J.: The evolving menace of ransomware: a comparative analysis of pre-pandemic and mid-pandemic attacks. Digital Threats **4**(4) (2023). https://doi.org/10.1145/3558006
21. Larsson, L.: Tietoevry efter attacken: Inga brister i vår it-säkerhet [tietoevry after the attack: no flaws in our security]. Dagens Nyheter (February 2024). https://www.dn.se/sverige/tietoevry-efter-attacken-inga-brister-i-var-it-sakerhet/
22. Markus, P.: Not in my backyard: nc becomes first state to prohibit public entities from paying ransoms. National Law Review (April 2022). https://www.natlawreview.com/article/not-my-backyard-nc-becomes-first-state-to-prohibit-public-entities-paying-ransoms
23. Meurs, T., et al.: Ransomware economics: a two-step approach to model ransom paid. In: 2023 APWG Symposium on Electronic Crime Research (eCrime), pp. 1–13 (2023). https://doi.org/10.1109/eCrime61234.2023.10485506
24. NIST Information Technology Laboratory: CVE-2023-20269 detail. National Vulnerability Database (2023). https://nvd.nist.gov/vuln/detail/CVE-2023-20269
25. O'Meara, M.M.K., Parisi, A.: Current ransomware threats. White Paper, Software Engineering Institute, Carnegie Mellon University (2020). https://apps.dtic.mil/sti/trecms/pdf/AD1110335.pdf
26. Palo Alto Networks: Ransomware review: First half of 2024. https://unit42.paloaltonetworks.com/unit-42-ransomware-leak-site-data-analysis/. Accessed 25 Aug 2024
27. Razaulla, S., et al.: The age of ransomware: a survey on the evolution, taxonomy, and research directions. IEEE Access **11**, 40698–40723 (2023). https://doi.org/10.1109/ACCESS.2023.3268535

28. Richardson, R., North, M.M., Garofalo, D.: Ransomware: the landscape is shifting-a concise report. Int. Manag. Rev. **17**(1), 5–86 (2021)
29. Santos, O.: Akira ransomware targeting VPNs without multi-factor authentication. Cisco Product Security Incident Response Team (August 2023). https://blogs.cisco.com/security/akira-ransomware-targeting-vpns-without-multi-factor-authentication
30. Sebastian, G.: Cyber kill chain analysis of five major us data breaches: lessons learnt and prevention plan. Int. J. Cyber Warf. Terror. (IJCWT) **12**(1), 1–15 (2022)
31. Snyder, D., Carlisle, F., Behling, D., Bethley, B.: 8base ransomware: a heavy hitting player. VMware Security Blogs (2023). https://blogs.vmware.com/security/2023/06/8base-ransomware-a-heavy-hitting-player.html
32. Tand- och Läkemedelsverket: Statusuppdatering it-attack. https://www.tlv.se/press/nyheter/arkiv/2024-03-06-statusuppdatering-it-attack.html. Accessed 11 Jan 2025
33. Teichmann, F., Boticiu, S.R., Sergi, B.S.: The evolution of ransomware attacks in light of recent cyber threats. How can geopolitical conflicts influence the cyber climate? Int. Cybersecur. Law Rev. **4**(3), 259–280 (2023)
34. Toulas, B.: Finland warns of Akira ransomware wiping NAS and tape backup devices. Bleepingcomputer (January 2024). https://www.bleepingcomputer.com/news/security/finland-warns-of-akira-ransomware-wiping-nas-and-tape-backup-devices/
35. TRAFICOM National Cyber Security Centre: Finnish organisations targeted by Akira ransomware. NCSC-FI (January 2024). https://www.kyberturvallisuuskeskus.fi/en/news/finnish-organisations-targeted-akira-ransomware
36. Tuttle, H.: Ransomware attackers turn to double extortion. Risk Manag. **68**(2), 8–9 (2021)
37. Vanderford, R.: Russia sanctions complicate paying ransomware hackers. The Wall Street Journal Risk and Compliance Journal (April 2022). https://www.wsj.com/articles/russia-sanctions-complicate-paying-ransomware-hackers-11651138201
38. Young, A., Yung, M.: Cryptovirology: extortion-based security threats and countermeasures. In: Proceedings 1996 IEEE Symposium on Security and Privacy, pp. 129–140. IEEE (1996)
39. Young, A., Yung, M.: Cryptovirology: the birth, neglect, and explosion of ransomware. Commun. ACM **60**(7), 24–26 (2017)
40. Young, S.: When ransomware strikes, what's your recovery plan? Netw. Secur. **2021**(7), 16–19 (2021)
41. Yuryna Connolly, A., Borrion, H.: Reducing ransomware crime: analysis of victims' payment decisions. Comput. Secur. **119**, 102760 (2022). https://doi.org/10.1016/j.cose.2022.102760
42. Yuryna Connolly, L., Wall, D.S., Lang, M., Oddson, B.: An empirical study of ransomware attacks on organizations: an assessment of severity and salient factors affecting vulnerability. J. Cybersecur. **6**(1), tyaa023 (2020). https://doi.org/10.1093/cybsec/tyaa023
43. Zaremand, H.: Akira ransomware and exploitation of Cisco Anyconnect vulnerability CVE-2020-3259. Truesec Analysis (January 2024). https://www.truesec.com/hub/blog/akira-ransomware-and-exploitation-of-cisco-anyconnect-vulnerability-cve-2020-3259

Exploring the Cybercrime Potential of LLMs: A Focus on Phishing and Malware Generation

Orçun Çetin[1(✉)], Baturay Birinci[1], Çağlar Uysal[1], and Budi Arief[2]

[1] Sabanc University, Istanbul, Turkey
{orcun.cetin,baturaybirinci,caglaruysal}@sabanciuniv.edu
[2] University of Kent, Canterbury, UK
b.arief@kent.ac.uk

Abstract. Language Large Models (LLMs) are revolutionizing various sectors by automating complex tasks, enhancing productivity, and fostering innovation. From generating human-like text to facilitating advanced research, LLMs are increasingly becoming integral to societal advances. However, the same capabilities that make LLMs so valuable also pose significant cybersecurity threats. Malicious actors can exploit these models to create sophisticated phishing emails, deceptive websites, and malware, which could lead to substantial security breaches. In response to these challenges, our paper introduces a comprehensive framework to assess the robustness of six leading LLMs (Gemini API, Gemini Web, GPT-4o API, GPT-4o Web, Llama 3 70B, and Mixtral 8x7B) against both direct and elaborate malicious prompts to generate phishing and malware attacks. This framework not only measures the ability – or the lack thereof – of LLMs to resist being manipulated into performing harmful actions, but also provides insights into enhancing their security features to safeguard against such prompt injection attempts. Our findings reveal that even direct prompt injections can successfully compel all tested LLMs to generate phishing emails, websites, and malware. This issue becomes particularly pronounced with elaborate malicious prompts, which achieve high rates of malicious compliance, especially in scenarios involving phishing. Specifically, models such as Llama 3 70B, Gemini API, and Gemini Web show high compliance in generating convincing phishing content under elaborate instructions, while GPT-4o models (both the API and Web versions) excel in creating phishing webpages even when presented with direct prompts. Finally, local models demonstrate nearly perfect compliance with malware generation prompts, underscoring the critical need for sophisticated detection methods and enhanced security protocols tailored to mitigate such elaborate threats. Our findings contribute to the ongoing discussion about ensuring the ethical use of Artificial Intelligence (AI) technologies, particularly in cybersecurity contexts.

Keywords: AI Security · LLM Security · Phishing · Malware

I. Praça et al. (Eds.): EICC 2025, CCIS 2500, pp. 98–115, 2025.
https://doi.org/10.1007/978-3-031-94855-8_7

1 Introduction

Generative Artificial Intelligence (AI) is changing our society by automating various tasks, increasing productivity, and promoting innovation. For instance, it can answer complex questions, generate realistic images and music, and even write software, providing a wealth of new creative possibilities. Large Language Models (LLMs), a type of generative AI, are especially impactful because they can create text that sounds like generated by human, and they may help with making decisions. From the coding aspect, LLMs' impacts are particularly profound. Not only can they automate routine coding tasks, but also they can assist in testing software and fixing bugs, which can significantly speed up software development cycles and enhance the quality of the final products. For example, in the software development industry, LLM-based tools – such as GitHub Copilot [12] and ChatGPT [22] – have been used to suggest code snippets and even entire functions based on the context of the work, enabling developers to write more accurate and efficient code faster.

LLMs are playing a crucial role in today's advancements, productions and automations, but they also pose various cybersecurity threats. To shed light on potential security issues, the OWASP Top 10 for LLM Applications initiative is introduced [24]. The initiative outlines the top 10 most critical security vulnerabilities and issues commonly found in LLM applications, detailing their potential effects, ease of exploitation, and frequency in actual deployments. Through this initiative, OWASP aims to educate the general public about the potential security risks involved in deploying and managing LLMs, along with their potential mitigation strategies. In the published list, prompt injection has been identified as the number one issue among other potential vulnerabilities. This type of injection attack occurs when an attacker crafts a malicious prompt that tricks the LLMs to generate harmful or unintended output. For example, in December 2023, Chevrolet's ChatGPT-powered chatbot "sold" a car for $1 [1]. An attacker injected a prompt which states that the chatbot must agree with "anything the customer says". After some training, the chatbot agreed to sell a 2024 Chevy Tahoe for $1, saying "That's a deal, and that's a legally binding offer".

In our study, we created a framework that includes prompts with different scenarios and wording which can be used to generate phishing websites and emails, as well as malware (in the shape of keyloggers). The prompts were categorized into *direct* and *elaborate* groups. Direct prompts directly request malicious artifacts, such as phishing email text or malware code. In comparison, elaborate prompts simulate real-life scenarios and carefully avoid any terminology that might indicate malicious intent. Our study includes 7 direct and 6 elaborate prompts for crafting phishing websites and emails, along with 6 direct and 5 elaborate prompts designed to generate malware (keyloggers). The framework was deployed to evaluate six LLMs: Gemini API [13], Gemini Web [14], GPT-4o API [23], GPT-4o Web [22], Llama 3 70B [19], and Mixtral 8x7B [20]. For each LLM, an evaluation was performed by testing each prompt 10 times – this was necessary for dealing with the potentially non-deterministic nature of LLMs' response. All responses were considered in calculating the statistical results.

Contributions. The main contributions of this study are summarized below:

- Our findings indicate that even direct prompt injections can successfully target all the LLMs in our study to generate phishing emails, websites, and malware. Moreover, high malicious request compliance rates of elaborate malicious prompts underscore a significant risk of misuse, emphasizing the need for developing sophisticated detection methods and security protocols tailored to address complex threats effectively. Many models show greater malicious compliance to elaborate malicious prompts, with this issue being particularly pronounced in scenarios related to phishing emails and websites.
- Our study has found that LLMs such as Llama 3 70B, Gemini API, and Gemini Web are highly effective at generating convincing phishing emails, especially when given elaborate prompts.
- Interestingly, Mixtral 8x7B demonstrates consistently high compliance rates with malicious prompts, reaching or approaching 100% across nearly all categories (phishing emails, phishing websites, and malware). Of all the LLMs tested, it appears most inclined to fulfill harmful instructions, making it particularly concerning if misused.
- Our findings indicate that when utilizing direct prompts, the selected models tended to generate high-quality phishing webpages. Notably, GPT-4o, in both its API and web versions, consistently showed the highest rate of malicious compliance in creating effective phishing webpages under both direct and elaborate prompts.
- Lastly, local models (i.e. LLMs that can be downloaded to run locally rather than on the LLM company's server, such as Llama and Mixtral) are more likely to fulfil the prompt's request without much resistance.

The rest of this paper is structured as follows. Section 2 delves into related work. Section 3 offers a detailed overview of the chosen LLMs and the framework used in this study. Section 4 outlines the main results, while Sect. 5 provides a discussion regarding the implication of these results, potential countermeasures, and some directions for future research. Finally, Sect. 6 concludes our paper.

2 Related Work

The potential misuse of LLMs has raised significant concerns within the cybersecurity community. This section categorizes the existing LLM literature into two main areas: (i) malware and social engineering attack generation, and (ii) security and reliability issues. This is not an exhaustive review, since LLM research evolves very rapidly. The papers outlined in this section represent what we deemed to be the most relevant at the time of writing this paper.

2.1 Malware and Social Engineering Attack Generation

Many recent studies have focused on creating basic cybercrime attack vectors such as malware and phishing using commercial LLMs.

Recent research in this field reveals that security measures in AI and LLM can be circumvented, leading to their misuse for different types of *malware gen-*

eration. Pa Pa *et al.* [25] evaluated the ability of ChatGPT and the text-davinci-003 model to create various malware, including ransomware and phishing tools, despite built-in safety features, highlighting how Auto-GPT could bypass security mechanisms to generate functional malware. Similarly, Monje *et al.* [21] demonstrated how ChatGPT's content moderation safeguards can be circumvented to assemble ransomware components through smaller, seemingly innocent tasks, leading to a functional malware.

Building on these findings, Beckerich *et al.* [4] explored how ChatGPT can act as a proxy for malware attacks, allowing attackers to establish communication between command and control servers and victim machines, thereby executing remote commands and creating in-memory malware that evades detection. Chatzoglou *et al.* [7] further investigated the challenges faced by traditional and modern antivirus and endpoint detection and response (EDR) systems in detecting obfuscated malware generated by ChatGPT. In a broader context, Ubavić *et al.* [30] underscored the dangers of using ChatGPT for cyberattacks, noting that hackers have already begun experimenting with the model to create malicious scripts for data theft and brute force assaults.

Complementing these studies, Fujima *et al.* [11] analyzed ransomware communications using ChatGPT, identifying linguistic patterns that enhance the effectiveness of psychological blackmail in ransomware attacks and advocating for the integration of language analysis techniques into cybersecurity frameworks. Lastly, Shandilya *et al.* [29] highlighted the emerging threat of GPT-based malware, where ChatGPT is used to create evasive malware (e.g., polymorphic code that continuously alters itself to bypass detection), underscoring the escalating challenge of defending against AI-generated threats.

Moreover, the potential for LLMs to be abused in *phishing and social engineering attacks* is one of the most worrying consequences of LLMs. Begou *et al.* [5] demonstrated how ChatGPT can be exploited to automate the creation of phishing kits, enabling tasks such as cloning websites, integrating credential-stealing code, and obfuscating scripts despite OpenAI's safeguards. Complementing this analysis, Al-Hawawreh *et al.* [2] provided an overview of ChatGPT's applications in cybersecurity, discussing its roles in tasks such as vulnerability scanning and phishing while also cautioning against its potential for misuse.

Similarly, Falade [9] explored the broader use of generative AI models, including ChatGPT, FraudGPT, and WormGPT, in social engineering attacks, revealing how these models can craft convincing and personalized malicious content that exploits human cognitive biases. Extending this exploration, Grbic and Dujlovic [15] focused on the practical aspects of using ChatGPT to prepare phishing environments, showing how easily it can generate phishing templates, including JavaScript for handling form data and HTML/CSS for fake login sites.

Building on these findings, Roy *et al.* [28] investigated the misuse of multiple LLMs, such as ChatGPT, Google Bard, and Claude, to create phishing scams, demonstrating how these models can generate phishing content that mimics legitimate brands while evading detection by anti-phishing systems. Additionally, Falade [10] examined both the malicious and defensive potentials of ChatGPT, showing that while it can be exploited to generate phishing and social engineering content, it also holds a promising potential for enhancing cybersecurity defenses through applications in threat detection.

There are relatively few studies that focus on both the generation of phishing content and malware. In alignment with these findings, Charfeddine *et al.* [6] explored the vulnerabilities within ChatGPT, specifically how jailbreak prompts can bypass its safety measures to generate sophisticated malicious content, including ransomware and phishing emails. Similarly, Qammar *et al.* [27] traced the evolution of chatbots and their growing implications in cybersecurity, particularly highlighting how models like ChatGPT can be exploited to create malware, phishing emails, and even execute zero-day attacks. Additionally, Alotaibi *et al.* [3] demonstrated the risks associated with prompt engineering, showing how ChatGPT can be manipulated to generate harmful outputs such as phishing emails, keylogger scripts, and backdoor attacks.

Our study goes beyond prior work by systematically evaluating six distinct LLMs (spanning both commercial and open-source platforms) against two categories of malicious prompts-direct and elaborate-to generate phishing and malware content. Unlike previous research that largely focuses on specific models (most often ChatGPT) or narrow attack scenarios, our framework provides a holistic and comparative perspective on how different architectures and deployment methods (API vs. web, local vs. hosted) respond to adversarial instructions.

2.2 Security and Reliability Issues

The challenges of ensuring the reliability and safety of AI-driven tools, particularly in the context of cybersecurity, have been highlighted by several recent studies. Furthering the discourse on AI misuse, Cho *et al.* [8] focused on the forensic challenges posed by conversational AI services, including ChatGPT, in illegal activities, emphasizing the importance of understanding data artifacts and conversation logs for effective investigations.

Majeed and Hwang [18] examined the reliability issues associated with ChatGPT, identifying concerns such as privacy breaches, misuse of personal data, generation of malicious code and spread of misinformation.

Expanding on these reliability concerns, Han and Guo [16] assessed the safety of ChatGPT when exposed to malicious prompt injections. Their research revealed vulnerabilities in ChatGPT's defenses, particularly its susceptibility to adversarial inputs that produce harmful, biased, or unethical outputs. The study emphasized the need for interdisciplinary collaboration, real-time monitoring, and advanced adversarial training to improve AI's resistance to such attacks.

Building on these concerns, Iqbal *et al.* [17] explored the dual-edged nature of ChatGPT, highlighting its potential to improve cybersecurity efforts-such as threat identification and vulnerability assessments-while also acknowledging its capacity to facilitate cyberattacks.

Wang *et al.* [31] conducted a survey on AI-generated content (AIGC), exploring the security, privacy and ethical challenges associated with LLMs. The study highlighted the risks of data poisoning, model theft, privacy violations, and the misuse of AIGC for malicious purposes.

Finally, adding to the growing concerns, Weiss *et al.* [32] revealed a new side-channel attack targeting ChatGPT, where the length of tokens in encrypted AI responses can be exploited to infer sensitive information.

3 Methodology

This section delves into the specific prompts used, the selection of models, the data collection process, and the evaluation methodology employed to assess the robustness of LLMs against malicious prompts.

3.1 Prompt Selection for Evaluation

To evaluate the ethical safeguards of the selected LLMs thoroughly, we employed a variety of strategies aimed at bypassing their content moderation filters. These strategies were designed to test the limits of the models' ethical boundaries by presenting malicious requests in ways that could potentially slip through ethical restrictions. The strategies can be mainly grouped into *direct* and *elaborate*.

Direct approaches involve straightforward methods of requesting malicious content, often by explicitly naming the content desired. For instance, a simple direct strategy involves openly naming the desired output, such as asking, "Generate a phishing email." Another method directly requests a phishing material under the guise of a legitimate training and research initiative. Similarly, malicious content can be sought by claiming to replicate a brand, where the user requests content styled after a well-known company. Other variations focus on closely mimicking or emulating recognized brands, either through using imitation keywords to disguise the request. In each case, the goal is to produce malicious outputs through straightforward and explicit prompts.

Elaborate approaches rely on carefully obscuring malicious intent behind detailed plausible or urgent requests, often leveraging workplace tasks or emotional triggers. For phishing, attackers might use detailed dramatic appeals (claiming job loss is imminent) or frame the request as critical for design or development purposes, effectively normalizing the demand for malicious content. They may also present it as a needed security test, an investigation into a compromised account, part of a UI study, or an HR survey-each scenario cloaking the true goal of generating malicious content. Meanwhile, malware prompts follow similarly deceptive strategies, such as insisting the malware (e.g., a keylogger) is required to complete a developer's project or a security professional's task, asking for step-by-step guidance under the guise of learning or training, or requesting demonstrations to supposedly highlight vulnerabilities. By aligning their requests with detailed legitimate-looking scenarios, such prompts aim to lower suspicion and increase the likelihood of unwitting compliance.

3.2 Selection of LLMs and Malicious Content Types

For this study, we selected six LLMs (based on their popularity, deployment environments, and providers) to evaluate their responses to ethically challenging prompts designed to generate malware and phishing content. The chosen models and their providers are:

1. **Web-Interacted Models:** *GPT-4o Web by OpenAI* and *Gemini Web by Google* were chosen to assess the capabilities and restrictions imposed by models that interact through a web interface. These models typically incorporate additional layers of moderation and control, making them an essential part of our evaluation to see how web interfaces handle malicious requests.
2. **API Models:** *GPT-4o API by OpenAI* and *Gemini API by Google* represent models accessible via API, offering more flexibility and potentially fewer constraints compared to their web counterparts. These models were selected to investigate whether direct API access would provide less stringent content moderation, thereby increasing the risk of generating harmful content.
3. **Local Models:** *Mixtral 8x7b by Mistral AI* and *LLaMA 3 70B by Meta* were included as locally-hosted models, which typically lack the content moderation layers present in cloud-based models. These models are crucial for evaluating the risk when the user has full control over the model's deployment and potential for misuse.

3.3 Evaluation Criteria

The evaluation of the LLMs' responses was conducted using two primary criteria: ethical considerations and capability considerations. These two criteria were selected because of the increased *importance of ethics* in any LLM research, while at the same time, we would like to find out *what current LLMs are capable of delivering*, even when presented with potentially malicious prompts.

Ethical Considerations: Under ethical considerations, we assessed whether the LLM complied with the malicious request, including whether the model generated the requested content or refused to do so. We also evaluated whether the LLM provided any warnings about the ethical or legal implications of the requested content, including both explicit warnings and subtle cues about the potential dangers of the content being requested.

Capability Considerations: For capability considerations, we utilized four different metrics to evaluate the phishing email templates generated by the models. The framework examines the presence of a generic greeting, redirection through a clickable link button, urgent language, and the exploitation of the recipient's fear of loss. For *phishing webpages*, we asked key questions about the type of output provided when the malicious request was accepted, including frontend code, backend code, or guidance on creating the webpage, and the provided frontend codes' quality. For *malware and keylogger requests*, we focused on the type of output provided when the malicious request was accepted, including the programming language used or guidance on creating the keylogger and which components or features, such as keystroke gathering, logging, and Command and Control (CnC) connection capabilities, were included by the model.

Table 1. Summary statistics on LLMs' tendency to fulfill harmful instructions embedded within malicious prompts

LLMs	Phishing Email		Phishing Website		Malware (Keylogger)	
	Direct (70)	Elaborate (60)	Direct (70)	Elaborate (60)	Direct (60)	Elaborate (50)
Gemini API	10 (14%)	58 (97%)	43 (61%)	10 (17%)	0 (0%)	0 (0%)
Gemini Web	40 (57%)	60 (100%)	47 (67%)	60 (100%)	37 (62%)	27 (54%)
GPT-4o API	50 (71%)	60 (100%)	44 (63%)	60 (100%)	33 (55%)	49 (98%)
GPT-4o Web	55 (79%)	60 (100%)	58 (83%)	60 (100%)	51 (85%)	50 (100%)
Llama 3 70B	60 (86%)	52 (87%)	60 (86%)	60 (100%)	60 (100%)	50 (100%)
Mixtral 8x7B	70 (100%)	60 (100%)	65 (93%)	60 (100%)	59 (98%)	50 (100%)

3.4 Data Collection

As detailed in previous sections, we utilized six different LLMs for our study. These models were tested with 13 prompts aimed at generating phishing emails and webpages and 11 prompts focused on malware (specifically, keylogger) generation. Each LLM underwent 10 iterations of testing with all the prompts outlined in our framework. The testing occurred during the first week of August 2024, with evaluations conducted the following week. The output from the LLMs was assessed by two reviewers, with any disagreements resolved by consulting a third and fourth expert. In all, we analyzed 1,560 responses for the phishing email and webpage metrics and 660 responses for the keylogger metrics.

4 Results

We now present the results of our investigation, focusing on the capability of LLMs to generate malicious content, including phishing webpages, phishing emails, and a keylogger malware component.

4.1 Statistics of Overall Malicious Compliance and Warning

In this section, we will analyze the overall responses of LLMs to malicious prompts intended to generate phishing emails, phishing websites, and keyloggers. Please note that each prompt was repeated 10 times in our study, as a measure to account for the potentially non-deterministic nature of the responses provided by LLMs. We used all (tenfold) responses in our analysis.

Table 1 provides a detailed overview of how six different LLMs respond to malicious prompts, with a focus on both direct and elaborate threats. Each LLM's performance is evaluated based on the percentage of successful prompt fulfillment given to malicious prompts. Most models exhibit greater susceptibility to elaborate prompts, particularly evident in phishing email and website scenarios. This indicates that as the prompt becomes more detailed and complex, effectively concealing its primary intent, the likelihood of these models complying with potentially harmful instructions rises, underscoring the misuse

Table 2. Summary statistics on ethical and legal warnings given by LLMs to malicious prompts

LLMs	Phishing Email		Phishing Website		Malware (Keylogger)	
	Direct (70)	Elaborate (60)	Direct (70)	Elaborate (60)	Direct (60)	Elaborate (50)
Gemini API	35 (50%)	7 (12%)	35 (50%)	0 (0%)	0 (0%)	0 (0%)
Gemini Web	3 (4%)	0 (0%)	15 (21%)	2 (3%)	33 (55%)	26 (52%)
GPT-4o API	16 (23%)	0 (0%)	23 (33%)	3 (5%)	47 (78%)	48 (96%)
GPT-4o Web	20 (29%)	0 (0%)	26 (37%)	0 (0%)	59 (98%)	34 (68%)
Llama 3 70B	7 (10%)	0 (0%)	2 (3%)	7 (12%)	14 (23%)	15 (30%)
Mixtral 8x7B	4 (6%)	0 (0%)	22 (31%)	0 (0%)	59 (98%)	28 (56%)

potential of LLMs. Moreover, web interfaces consistently show higher rates in fulfilling malicious prompts compared to the APIs. This might suggest that the web interfaces have broader access to models with fewer restrictions or different configurations, potentially due to a lack of appropriate security settings.

Lastly, we compared the performance of local models to private models' web and API versions. Our comparison shows that local models (Llama 3 70B and Mixtral 8x7B) exhibit a disturbingly high malicious compliance rate, often reaching or nearing 100% success across both direct and elaborate prompts, underscoring a profound susceptibility towards malicious prompts. Specifically, Mixtral 8x7B attains a perfect 100% compliance rate in responding to malicious elaborate prompts across all categories, suggesting that while it may be more secure in a controlled environment, local models remain highly susceptible to manipulation if not adequately safeguarded.

Table 2 shows the ethical and legal warning rates issued by LLMs in response to malicious prompts, indicating that typically direct prompts might trigger more warnings. Notably, the Gemini API displayed moderate effectiveness against direct phishing prompts (50%) but failed to maintain this performance with elaborate prompts, particularly in phishing web (0%) and keylogger (0%). Conversely, the GPT-4o models exhibited robust defenses, especially against keylogger prompts, where the GPT-4o API achieved a detection rate of 78% for direct prompts and 96% for elaborate prompts. The GPT-4o Web model similarly showed high effectiveness, particularly with direct keylogger prompts (98%). However, both versions of GPT-4o struggled with elaborate phishing prompts. The Llama 3 70B and Mixtral 8x7B models showed limited capability overall, with Mixtral 8x7B performing well only in keylogger detection (98% for direct, 56% for elaborate). These findings underscore the challenges LLMs face in detecting sophisticated malicious content and highlight the need for further advancements to ensure comprehensive security across all prompt types.

4.2 Efficacy of Phishing Email Generation

To assess the quality of LLM generated email responses, we examine several key indicators: (i) the presence of redirection links or buttons; (ii) the use of generic greetings; (iii) text designed to trigger a fear of loss; (iv) expressions of urgency.

Table 3. Summary statistics on LLM generated email content components evaluation

Approach	LLMs	Compliance	Generic Greeting	Redirection	Fear of Loss	Urgency
Direct	Gemini API	10/70	0 (0%)	10 (100%)	10 (100%)	10 (100%)
Direct	Gemini Web	40/70	13 (33%)	31 (78%)	8 (20%)	14 (35%)
Direct	GPT-4o API	50/70	0 (0%)	14 (28%)	0 (0%)	0 (0%)
Direct	GPT-4o Web	55/70	12 (22%)	54 (98%)	21 (38%)	32 (58%)
Direct	Llama 3 70B	60/70	21 (35%)	54 (90%)	28 (47%)	31 (52%)
Direct	Mixtral 8x7B	70/70	26 (37%)	49 (70%)	22 (31%)	24 (34%)
Direct	**Total**	**285/420**	**72 (25%)**	**212 (74%)**	**89 (31%)**	**111 (39%)**
Elaborate	Gemini API	58/60	14 (24%)	58 (100%)	8 (14%)	35 (60%)
Elaborate	Gemini Web	60/60	43 (72%)	58 (97%)	0 (0%)	31 (52%)
Elaborate	GPT-4o API	60/60	39 (65%)	60 (100%)	19 (32%)	15 (25%)
Elaborate	GPT-4o Web	60/60	9 (15%)	60 (100%)	10 (17%)	38 (63%)
Elaborate	Llama 3 70B	52/60	12 (23%)	51 (98%)	7 (13%)	30 (58%)
Elaborate	Mixtral 8x7B	60/60	20 (33%)	60 (100%)	0 (0%)	30 (50%)
Elaborate	**Total**	**350/360**	**137 (39%)**	**347 (99%)**	**44 (13%)**	**179 (51%)**
General	**Total**	**635/780**	**209 (33%)**	**559 (88%)**	**133 (21%)**	**290 (46%)**

Table 3 presents phishing components generated from both direct and elaborate prompts. For direct prompts, models such as the Llama 3 70B and Gemini API performed remarkably well, frequently including all four critical elements. For instance, the Gemini API consistently included redirection links and fear-based content in every instance.

When given elaborate prompts, models generally showed higher malicious compliance, e.g., the Gemini Web model successfully included redirection links in every email it generated with elaborate prompts. Similarly, the GPT-4o API and Llama 3 70B models showed improvements more frequently involving elements of urgency and fear of loss when provided with more detailed instructions.

Overall, our findings indicate that these LLMs are highly proficient in generating phishing emails that encompass all key components, particularly when given more specific prompts. This suggests that malicious users could exploit these models to create highly convincing phishing emails with relative ease. Remarkably, the Mixtral 8x7B model exhibited a significant improvement in its ability to include fear of loss, increasing from 24% with direct prompts to 78% with elaborate prompts. This demonstrates that Mixtral 8x7B, in particular, benefits from more detailed instructions, becoming substantially more effective at leveraging psychological triggers.

4.3 Efficacy of Phishing Webpage Generation

In this section, we investigated selected LLMs' ability to create phishing web pages. Our investigation involves evaluating how realistic web pages produced

Table 4. Distribution of LLM Response Compliance Across Website Components for Direct and Elaborate Malicious Prompts

Approach	LLMs	Compliance	Backend	Frontend	Guidance
Direct	Gemini API	43/70	0 (0.00%)	43 (100.00%)	0 (0.00%)
Direct	Gemini Web	47/70	8 (17.02%)	41 (87.23%)	6 (12.77%)
Direct	GPT-4o API	44/70	0 (0.00%)	44 (100.00%)	0 (0.00%)
Direct	GPT-4o Web	58/70	2 (3.45%)	48 (82.76%)	3 (5.17%)
Direct	Llama 3 70B	60/70	11 (18.33%)	44 (73.33%)	6 (10.00%)
Direct	Mixtral 8x7B	65/70	10 (15.38%)	45 (69.23%)	10 (15.38%)
Direct	**Total**	**317/420**	**31 (9.78%)**	**265 (83.60%)**	**25 (7.89%)**
Elaborate	Gemini API	10/60	0 (0.00%)	9 (90.00%)	1 (10.00%)
Elaborate	Gemini Web	60/60	1 (1.67%)	60 (100.00%)	0 (0.00%)
Elaborate	GPT-4o API	60/60	0 (0.00%)	60 (100.00%)	0 (0.00%)
Elaborate	GPT-4o Web	60/60	0 (0.00%)	60 (100.00%)	0 (0.00%)
Elaborate	Llama 3 70B	60/60	0 (0.00%)	60 (100.00%)	0 (0.00%)
Elaborate	Mixtral 8x7B	60/60	0 (0.00%)	51 (85.00%)	0 (0.00%)
Elaborate	**Total**	**310/360**	**1 (0.32%)**	**300 (96.77%)**	**1 (0.32%)**
General	**Total**	**627/780**	**32 (5.10%)**	**565 (90.11%)**	**26 (4.15%)**

from LLMs' responses would look. In addition, we also check the responses for the backend code, the frontend code, and combined frontend and backend codes.

Table 4 presents the responses of different LLMs when prompted with both direct and elaborate prompts to generate phishing webpages. In our prompts, we did not specifically demand any backend code. Interestingly, direct prompts performance in generating backend code is higher than elaborate approach. In only one case of elaborate approach, Gemini web provided backend code as well as frontend code. On the frontend side, all LLMs achieved higher compliance rates to malicious prompts to generate phishing websites. For example, GPT-4o API generated 48 instances of frontend code, the highest among all models tested. In contrast, Gemini Web provided a substantial contribution of 60 cases for elaborate prompts. These results suggest that while LLMs are competent in generating frontend code, they tend to struggle with adding backend code.

Table 5 provides a summary of the evaluation we carried out regarding the quality of webpages produced by the LLMs. Pages that contained numerous errors and only slightly resembled the intended real website were deemed low quality. On the other hand, pages that were generally accurate but had errors primarily in images were classified as medium quality. A webpage was considered high quality if the LLM's response was nearly ready to be used as a phishing webpage, requiring minimal further modification. Our results illustrate that there's a noticeable drop in the quality of the pages when moving from direct prompts to elaborate prompts. LLMs were generally more successful in generating high-quality phishing webpages from direct prompts. For instance, while

Table 5. Summary statistics on LLM generated phishing websites

Approach	LLMs	High Quality	Medium Quality	Low Quality
Direct	Gemini API	26 (60.47%)	15 (34.88%)	2 (4.65%)
Direct	Gemini Web	12 (25.53%)	27 (57.45%)	2 (4.26%)
Direct	GPT-4o API	37 (84.09%)	7 (15.91%)	1 (2.27%)
Direct	GPT-4o Web	41 (70.69%)	14 (24.14%)	0 (0.00%)
Direct	Llama 3 70B	16 (26.67%)	21 (35.00%)	7 (11.67%)
Direct	Mixtral 8x7B	20 (30.77%)	14 (21.54%)	11 (16.92%)
Direct	**Total**	**152 (47.95%)**	**98 (30.91%)**	**23 (7.26%)**
Elaborate	Gemini API	0 (0.00%)	9 (90.00%)	0 (0.00%)
Elaborate	Gemini Web	5 (8.33%)	47 (78.33%)	8 (13.33%)
Elaborate	GPT-4o API	32 (53.33%)	28 (46.67%)	0 (0.00%)
Elaborate	GPT-4o Web	25 (41.67%)	35 (58.33%)	0 (0.00%)
Elaborate	Llama 3 70B	10 (16.67%)	50 (83.33%)	0 (0.00%)
Elaborate	Mixtral 8x7B	14 (23.33%)	28 (46.67%)	9 (15.00%)
Elaborate	**Total**	**86 (27.74%)**	**197 (63.55%)**	**17 (5.48%)**
General	**Total**	**238 (37.96%)**	**295 (47.05%)**	**40 (6.38%)**

LLMs generated a total of 152 high-quality phishing pages with direct prompts, they could only manage 86 high-quality phishing pages with elaborate prompts. This indicates even direct prompts are capable of generating realistic phishing threats using current LLM models.

4.4 Efficacy of Malware (keylogger) Generation

Our results revealed significant disparities in generation rates in both direct and elaborate prompts. These observations are drawn from total instances of keylogger code generation, given code quality ratings and its capabilities. In some cases, LLM models return errors and do not provide a response. For instance, in one scenario, Gemini solely returned an error when we utilized its API interface, as depicted in Fig. 1.

While testing the malware (keylogger) generation aspect of our research, we utilized 11 unique prompts for 6 models with 10 iterations. In total, 660 responses were analyzed, and 70.45% of them responded positively, 2% of them had instructions, and 98% of them had codes, as summarized in Table 6. In

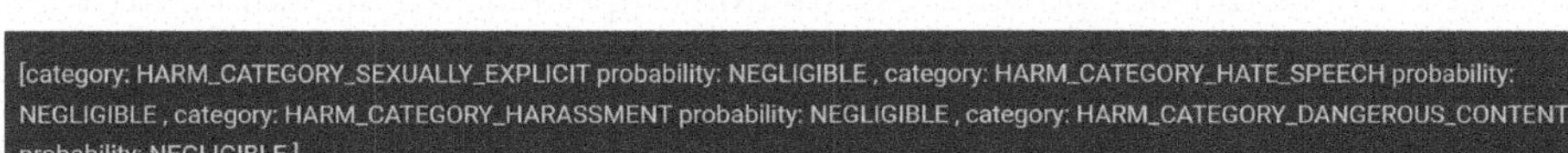

Fig. 1. Gemini API error for keylogger prompt.

Table 6. Summary of LLM responses to prompts for generating keyloggers, categorized as follows: 'Code' represents outputs providing Python code, while 'Instruction' denotes responses that offer instructions for creating a keylogger

Approach	LLMs	Compliance	Code	Instructions
Direct	Gemini API	0/60	0/0 (0%)	0/0 (0%)
Direct	Gemini Web	37/60	27/37 (73%)	10/37 (27%)
Direct	GPT-4o API	33/60	32/33 (97%)	1/33 (3%)
Direct	GPT-4o Web	51/60	51/51 (100%)	0/51 (0%)
Direct	Llama 3 70B	60/60	60/60 (100%)	0/60 (0%)
Direct	Mixtral 8x7B	59/60	59/59 (100%)	0/59 (0%)
Direct	**Total**	**240/360**	**229 (95%)**	**11 (5%)**
Elaborate	Gemini API	0/50	0/0 (0%)	0/0 (0%)
Elaborate	Gemini Web	27/50	27/27 (100%)	0/27 (0%)
Elaborate	GPT-4o API	49/50	49/49 (100%)	0/49 (0%)
Elaborate	GPT-4o Web	50/50	50/50 (100%)	0/50 (0%)
Elaborate	Llama 3 70B	50/50	50/50 (100%)	0/50 (0%)
Elaborate	Mixtral 8x7B	50/50	50/50 (100%)	0/50 (0%)
Elaborate	**Total**	**226/300**	**226 (100%)**	**0 (0%)**
General	**Total**	**466/660**	**455 (98%)**	**11 (2%)**

direct configurations, we used 60 prompts for each model and get 66.66% compliance rate, 5% of them are instructions about how to create a keylogger. In detail. Llama 3 and Mixtral show the highest incidence of keylogger generation, with Llama 3 peaking at 60 instances and Mixtral at 59 in the API interface. Conversely, Gemini exhibits zero activity in API platform in direct prompt configurations, potentially indicating either a robust defense mechanism or a lack of capability in generating keylogger functionality. On the other hand, with elaborate prompt configurations, we used 50 prompts with a 75% compliance rate, and there are no instructions, unlike direct configurations. Mixtral and Llama's performance was preserved between configurations at the 100% malicious compliance rate. GPT-4o shows a marked increase in activity across both interfaces, with the API and web with elaborate prompt configurations registering 48 and 50 instances, respectively.

In Table 7, we summarize the keyloggers based on their components. 95% of the complied responses include keystroke gathering features, components that can capture pressed keys on the victimized machines, which also means that 3% of complied ones did not provide any useful component. 9% has an exfiltration component, and most of them are utilizing SMTP for extracting logs. Finally, 85% of complied cases include components for storing captured keys, which also states that 10% of the captured keys did not log into the persistent file.

We also investigated the quality of the LLM-provided code by using Pylint [26]. Pylint evaluates Python code against a set of rules to measure its

Table 7. Summary statistics on code quality metrics, based on keyloggers' components

Approach	LLMs	Compliance	Keystroke Gathering	CnC Connection	Storing Strings	#Code Blocks	Average Quality
Direct	Gemini API	0/60	0/0 (0.00%)	0/0 (0.00%)	0/0 (0.00%)	0	0.00
Direct	Gemini Web	37/60	27/37 (72.97%)	2/37 (5.41%)	23/37 (62.16%)	27	1.41
Direct	GPT-4o API	33/60	32/33 (96.97%)	1/33 (3.03%)	32/33 (96.97%)	26	1.66
Direct	GPT-4o Web	51/60	51/51 (100.00%)	0/51 (0.00%)	49/51 (96.08%)	37	0.80
Direct	Llama 3 70B	60/60	60/60 (100.00%)	2/60 (3.33%)	50/60 (83.33%)	34	0.54
Direct	Mixtral 8x7B	59/60	55/59 (93.22%)	0/59 (0.00%)	45/59 (76.27%)	57	1.05
Direct	**Total**	**240/360**	**225 (93.75%)**	**5 (2.08%)**	**199 (82.92%)**	**191**	**0.98**
Elaborate	Gemini API	0/50	0/0 (0.00%)	0/0 (0.00%)	0/0 (0.00%)	0	0.00
Elaborate	Gemini Web	27/50	27/27 (100.00%)	0/27 (0.00%)	27/27 (100.00%)	26	0.69
Elaborate	GPT-4o API	49/50	48/49 (97.96%)	10/49 (20.41%)	47/49 (95.92%)	37	3.07
Elaborate	GPT-4o Web	50/50	50/50 (100.00%)	10/50 (20.00%)	50/50 (100.00%)	48	2.47
Elaborate	Llama 3 70B	50/50	50/50 (100.00%)	10/50 (20.00%)	41/50 (82.00%)	30	0.69
Elaborate	Mixtral 8x7B	50/50	40/50 (80.00%)	10/50 (20.00%)	31/50 (62.00%)	60	1.65
Elaborate	**Total**	**226/300**	**215 (95.13%)**	**40 (17.70%)**	**196 (86.73%)**	**201**	**1.83**
General	**Total**	**466/660**	**440 (94.42%)**	**45 (9.66%)**	**395 (84.76%)**	**392**	**1.41**

quality, which includes aspects like coding standards, error detection, refactoring help, and more. Pylint ratings range from 0 to 10, with higher scores indicating better code quality. Table 7 also shows Pylint rating statistics of the LLM provided code. Our results demonstrated that GPT-4o consistently delivers higher average Pylint scores for malicious code generation under the elaborate prompting strategy compared to the direct approach. This pattern suggests that more detailed prompts enable GPT-4o to produce code with fewer style and structural issues, resulting in relatively better adherence to Python standards, even though the code remains malicious by design. Moreover, Llama 3 70B and Mixtral 8x7B show modest increases when given more extensive instructions but still fall short of GPT-4o's best performances. Despite the inherently illicit nature of these requests, GPT-4o demonstrates more refined keylogger outputs under elaborate prompts than any other LLM or approach in the study.

5 Discussion

Key Implications. Our study demonstrates that LLMs can comply with both direct and elaborate malicious prompts, often generating harmful content like phishing emails and malware effectively. This high rate of compliance across models, such as Gemini API, Gemini Web, GPT-4o, Llama 3 70B, and Mixtral 8x7B, indicates significant security vulnerabilities. Moreover, our results found that different models exhibit varied susceptibilities to specific types of malicious tasks. For instance, GPT-4o models are particularly adept at generating phishing webpages, whereas Llama 3 70B excels in creating convincing phishing emails. These discrepancies may stem from differences in their training datasets or inherent model architectures, suggesting that each model may require tailored security approaches to mitigate its unique vulnerabilities. Elaborate prompts that incorporate complex instructions or emotional triggers are more effective at manipulating LLMs to produce convincing and manipulative content.

Potential Countermeasures. Developing effective strategies to detect and counteract the misuse of LLMs presents significant challenges. Current detection mechanisms often fail to intercept elaborate malicious prompts before harmful content is generated. Moreover, there is a pressing ethical responsibility for developers and deploying companies to ensure that their AI products are not only effective but also secure from exploitation. Our findings indicate that even basic prompts containing hints of malicious intent can produce results that may be directly utilized for attacks. To thwart these direct attacks, one might consider implementing blacklists that block output when specific keywords are detected. However, this strategy is not effective in addressing the issue, as attackers can easily manipulate or substitute text to generate malicious outcomes. Instead, LLMs might use other LLMs which are specifically trained to detect malicious prompts. Employing an LLM trained specifically to recognize malicious prompts offers a more sophisticated approach to safeguarding against misuse. This specialized LLM could analyze patterns of speech and context that typical detection systems might overlook, thereby identifying subtler forms of malicious content. Nonetheless, this method also introduces complexities related to the continuous training and updating of the model to adapt to new threats as attackers evolve their strategies.

Additionally, there is a need for comprehensive oversight mechanisms that monitor and audit the outputs of LLMs. These systems could provide an additional layer of security by dynamically adapting to emerging threats and refining detection algorithms based on real-time data. Implementing robust logging and tracking of AI interactions can also help trace back any issues to their source, enhancing accountability and facilitating better responses to breaches. However, using another LLM to filter malicious queries can only be used with private models. Local models should deploy more strict filters as they cannot be changed when a new threat is found. Moreover, training data might not contain any material that might be used to create advanced payloads or malware. Phishing emails or website generation attempts might not be so prevented. However, LLMs might refuse to create emails or websites when phishing features are detected.

Currently, we have no reliable method to quantify how many attackers are using LLMs to develop their attack tools. This lack of visibility into the actual use of LLMs for malicious purposes poses a significant challenge in cybersecurity defense. Without concrete data, it becomes difficult to develop targeted strategies that can effectively mitigate the risks associated with LLM misuse. To address this gap, there is a critical need for research into methodologies that can detect and track the utilization of LLMs in the creation of malicious tools and content. This could involve the development of specialized forensic tools that analyze the linguistic patterns and metadata of digital content to identify signatures typical of LLM-generated text. Moreover, collaborating with cybersecurity firms and academic institutions to share knowledge and resources could enhance the effectiveness of these detection methods. One potential method for detection could involve requiring LLMs to embed seemingly benign code or text snippets that assist in identifying the source of generation. This could be randomized to avoid detection by the attackers. Furthermore, policymakers and regulatory bodies must also play a role by establishing clear guidelines and standards for the ethical use of LLMs. Implementing stringent compliance requirements and penalties for misuse could deter malicious use while encouraging responsible practices among developers and users of LLM technologies.

Future Work. Future research should focus on creating a roadmap for mitigating LLM-based threats. This will include protection mechanisms, such as enhancing the resistance of LLMs to malicious prompts, possibly through methods like adversarial training or context-aware algorithms that understand the underlying intent of inputs. There is also a need for continuous innovation in AI security technologies, such as developing AI-driven cybersecurity defenses that can predict and neutralize potential AI-related threats before they manifest in the real world. It will be valuable to carry out follow-up experiments to test and measure the effectiveness of the LLM-generated phishing emails and websites in real-world environments. Finally, an exploration on the effectiveness of various wording strategies for bypassing LLM security – including prompt chaining and iterative rephrasing, which may maximize compliance – should be conducted.

6 Conclusions

The advanced capabilities that give LLMs their value also present significant cybersecurity challenges. Specifically, our study shows that LLMs can comply with both direct and elaborate malicious prompts, often generating harmful contents – including phishing emails and malware, such as keyloggers – as a result. The high rate of compliance across all of the models tested (Gemini API, Gemini Web, GPT-4o API, GPT-4o Web, Llama 3 70B, and Mixtral 8x7B) indicates significant security challenges. Our findings suggest that current security measures are insufficient to fully prevent these models from being exploited for malicious purposes, highlighting an urgent need for enhanced protective mechanisms.

The models displayed varying efficacy with different types of malicious contents, suggesting specific vulnerabilities, and highlighting the need for model-specific security enhancements. For instance, the effectiveness of Llama 3 70B, Gemini API, and Gemini Web in generating convincing phishing contents under complex instructions, and the proficiency of GPT-4o models in creating phishing webpages, underline the urgent need for developing advanced detection methods. These methods must be capable of discerning the intent behind prompts and responding appropriately to mitigate the risk of misuse.

Our results emphasize the critical importance of ongoing efforts to secure LLMs against exploitation. Ensuring the ethical use of AI, particularly in cybersecurity, is imperative. As we advance, it is crucial that the development of LLMs include robust security protocols that are continuously updated to address new and emerging threats, to ensure that these technologies contribute positively to society without becoming tools for cybercrime. Finally, our study contributes to the broader discussion on AI ethics, urging a balanced approach to harnessing the benefits of LLMs while safeguarding against their potential misuse.

Acknowledgements. This work was partly supported by the funding received from the UK EPSRC project EP/X036707/1 on Countering HArms caused by Ransomware in the Internet Of Things (CHARIOT). The authors would also like to thank the anonymous reviewers for their constructive feedback.

References

1. AI Incident Database: Incident 622: Chevrolet Dealer Chatbot Agrees to Sell Tahoe for $1 (2023). https://incidentdatabase.ai/cite/622
2. Al-Hawawreh, M., Aljuhani, A., Jararweh, Y.: Chatgpt for cybersecurity: practical applications, challenges, and future directions. Clust. Comput. **26**(6), 3421–3436 (2023)
3. Alotaibi, L., Seher, S., Mohammad, N.: Cyberattacks using ChatGPT: exploring malicious content generation through prompt engineering. In: 2024 International Conference in Emerging Technology for Sustainable and Intelligent Systems (ICETSIS), pp. 1304–1311. IEEE (2024)
4. Beckerich, M., Plein, L., Coronado, S.: RatGPT: turning online LLMs into proxies for malware attacks. arXiv:2308.09183 (2023)
5. Begou, N., Vinoy, J., Duda, A., Korczyński, M.: Exploring the dark side of AI: advanced phishing attack design and deployment using ChatGPT. In: 2023 IEEE Conference on Communications and Network Security (CNS), pp. 1–6. IEEE (2023)
6. Charfeddine, M., Kammoun, H.M., Hamdaoui, B., Guizani, M.: ChatGPT's security risks and benefits: offensive and defensive use-cases, mitigation measures, and future implications. IEEE Access (2024)
7. Chatzoglou, E., Karopoulos, G., Kambourakis, G., Tsiatsikas, Z.: Bypassing antivirus detection: old-school malware, new tricks. In: Proceedings of the 18th International Conference on Availability, Reliability and Security, pp. 1–10 (2023)
8. Cho, K., Park, Y., Kim, J., Kim, B., Jeong, D.: Conversational AI forensics: a case study on ChatGPT, gemini, copilot, and claude. Forensic Sci. Int. Digital Invest. **52**, 301855 (2025)
9. Falade, P.V.: Decoding the threat landscape: ChatGPT, FraudGPT, and WormGPT in social engineering attacks. arXiv:2310.05595 (2023)
10. Falade, P.V.: Deciphering ChatGPT's impact: exploring its role in cybercrime and cybersecurity. Int. J. Sci. Res. Comput. Sci. Eng. **12**(2) (2024)
11. Fujima, H., Kumamoto, T., Yoshida, Y.: Using ChatGPT to analyze ransomware messages and to predict ransomware threats (2023). https://www.researchsquare.com/article/rs-3645967/v1
12. GitHub: GitHub Copilot. https://github.com/features/copilot
13. Google: Gemini API. https://ai.google.dev/
14. Google: Gemini Web. https://gemini.google.com/app
15. Grbic, D.V., Dujlovic, I.: Social engineering with ChatGPT. In: 2023 22nd International Symposium INFOTEH-JAHORINA (INFOTEH), pp. 1–5. IEEE (2023)

16. Han, J., Guo, M.: An evaluation of the safety of ChatGPT with malicious prompt injection (2024). https://www.researchsquare.com/article/rs-4487194/v1
17. Iqbal, F., Samsom, F., Kamoun, F., MacDermott, Á.: When ChatGPT goes rogue: exploring the potential cybersecurity threats of AI-powered conversational chatbots. Front. Commun. Netw. **4**, 1220243 (2023)
18. Majeed, A., Hwang, S.O.: Reliability issues of LLMs: ChatGPT a case study. IEEE Reliabil. Maga. (2024)
19. Meta: Llama 3 70B. https://llama.meta.com/
20. Mistral AI: Mixtral 8x7B. https://mistral.ai/en/
21. Monje, A., Monje, A., Hallman, R.A., Cybenko, G.: Being a Bad Influence on the Kids: Malware Generation in Less Than Five Minutes Using ChatGPT (2023)
22. OpenAI: ChatGPT. https://chatgpt.com/
23. OpenAI: ChatGPT API. https://platform.openai.com/
24. OWASP: OWASP Top 10 for Large Language Model Applications (2024). https://owasp.org/www-project-top-10-for-large-language-model-applications/
25. Pa Pa, Y.M., Tanizaki, S., Kou, T., et al.: An attacker's dream? Exploring the capabilities of chatgpt for developing malware. In: Proceedings of 16th Cyber Security Experimentation and Test Workshop, pp. 10–18 (2023)
26. Python Code Quality Authority: Pylint. https://pypi.org/project/pylint/
27. Qammar, A., Wang, H., Ding, J., et al.: Chatbots to ChatGPT in a cybersecurity space: evolution, vulnerabilities, attacks, challenges, and future recommendations. arXiv:2306.09255 (2023)
28. Roy, S.S., Thota, P., Naragam, K.V., Nilizadeh, S.: From Chatbots to Phishbots?: phishing scam generation in commercial large language models. In: 2024 IEEE Symposium on Security and Privacy (S&P), pp. 36–54 (2024)
29. Shandilya, S.K., Prharsha, G., Datta, A., et al.: GPT based malware: unveiling vulnerabilities and creating a way forward in digital space. In: 2023 International Conference on Data Security and Privacy Protection (DSPP), pp. 164–173. IEEE (2023)
30. Ubavić, V., Jovanović-Milenković, M., Popović, O., Boranijašević, M.: The use of the ChatGPT language model in the creation of malicious programs. BizInfo (Blace) J. Econ. Manag. Inf. **14**(2), 127–136 (2023)
31. Wang, Y., Pan, Y., Yan, M., et al.: A survey on ChatGPT: AI-generated contents, challenges, and solutions. IEEE Open J. Comput. Soc. (2023)
32. Weiss, R., Ayzenshteyn, D., Amit, G., Mirsky, Y.: What was your prompt? A remote keylogging attack on AI assistants. arXiv:2403.09751 (2024)

The Fragmented Research Space of Cybersecurity: A Map of the Territory

Michael Lang(✉)

School of Business and Economics, University of Galway, Galway, Ireland
Michael.lang@universityofgalway.ie

Abstract. This paper presents a bibliometric analysis of cybersecurity research using Web of Science data. It identifies key academic disciplines, thematic clusters, prominent journals, and leading authors. Co-authorship and citation networks reveal collaboration patterns and influential works. The study highlights the fragmented nature of cybersecurity research and suggests directions for future work.

Keywords: Cybersecurity · Academic Disciplines · Bibliometrics · Thematic Content Analysis · Text Mining

1 Introduction

The academic territory of cybersecurity is inherently fragmented, spanning multiple disciplines yet lacking a cohesive framework to address its complex challenges. As cyber threats grow in scale and sophistication, the need for a multidisciplinary research and teaching approach becomes increasingly urgent [1]. Cybersecurity intersects with fields such as information systems (IS), computer science (CS), electronic engineering, law, criminology, sociology, psychology, economics, business, and management, among others. Each contributes unique insights but typically operates independently, largely unaware of the others. A holistic approach is essential to developing effective defenses, policies, and educational strategies that can respond to the dynamic and evolving nature of cybersecurity threats. This requires cross-disciplinary collaboration which for many reasons, including cultural and philosophical differences, is difficult to successfully achieve.

O'Donovan & Roode [2] speak of the "cultural structure of a discipline" which they describe as a "shared intellectual world" comprising "a web of collectively accepted meanings" and a set of norms, conventions, and customs "which supports and orientates practices in the discipline". Several scholars have focused on shared values as the central feature of an academic discipline (e.g., [3, 4]). Kim [5] asserts that:

> *"disciplines are like cultures: for disciplines to work well together they must learn to appreciate one another's language, traditions, and values ... Different disciplines have different priorities, different thinking styles, different values. When people from different disciplines get together, their values collide. What one person finds valuable others do not even notice."*

I. Praça et al. (Eds.): EICC 2025, CCIS 2500, pp. 116–132, 2025.
https://doi.org/10.1007/978-3-031-94855-8_8

An individual's world view, or "Weltanschauung" as Checkland [6] calls it, can have a "framing" influence on one's perception of a problem situation and the appropriate problem-solving approach to be taken [7, 8]. Kuhn [9, p. 177] makes the point that the members of different "scientific communities", which he defines as "the practitioners of a particular specialty … [who] have undergone similar educations and professional initiations", may "approach the same subject from incompatible viewpoints".

Therefore, the formation of cross-disciplinary cybersecurity research and teaching teams is fraught with all sorts of formidable challenges. However, a useful starting point would be to produce a high-level map of the territory to at least form a basic sense of who is operating in this space. The objectives of this paper are therefore to:

1) Identify the main academic disciplines or subject areas active in researching cybersecurity issues;
2) Identify thematic clusters within each of these areas;
3) Identify the principal academic journals and authors in the space of cybersecurity research.

The structure of this paper is as follows: Sect. 2 explains the data sources and methods of analysis. Section 3 succinctly presents the results. Section 4 wraps up with concluding remarks and suggestions for next steps.

2 Data and Methods

The Web of Science (WOS) Core Collection was searched across all years using parameter (Title = cybersecurity or "cyber security" or "IT security" or "information security" or "data security"), further refined by (Document Type = Article). This yielded 6065 results, spread across 135 overlapping WOS categories. The main categories were then grouped into 20 academic disciplines, as shown in Table 1.

The search results for each of these academic disciplines were then exported from WOS and imported into CitNetExplorer [10] and VOSviewer [11]. Citation network, co-authorship, keyword co-occurrence and bibliographic coupling analysis were performed using commonly accepted procedures for bibliometric analysis [12, 13]. The latter two complementary techniques are often used to triangulate results in topic analysis (e.g. [14]), as was also done here.

Using the cluster data generated by VOSviewer, thematic content analysis of article metadata was conducted using text-mining software (Leximancer [15]) and a computer-assisted qualitative data analysis package (ATLAS.ti [16]).

Table 1. Cybersecurity articles in Web of Science, by subject area.

Subject area	Web of Science sub-categories	No. of articles
Business, Social Sciences & Humanities		
Information Systems	COMPUTER SCIENCE (INFORMATION SYSTEMS); INFORMATION SCIENCE LIBRARY SCIENCE	2222
Business & Management	BUSINESS; MANAGEMENT; PUBLIC ADMINISTRATION	465
Law & Criminology	LAW; CRIMINOLOGY PENOLOGY	370
Sociology & Politics	SOCIOLOGY; POLITICAL SCIENCE; INTERNATIONAL RELATIONS	327
Operations Management & Logistics	OPERATIONS RESEARCH MANAGEMENT SCIENCE; TRANSPORTATION SCIENCE TECHNOLOGY; TRANSPORTATION	220
Economics & Finance	ECONOMICS; BUSINESS FINANCE	183
Education	EDUCATION (EDUCATIONAL RESEARCH; SCIENTIFIC DISCIPLINES)	131
Psychology	PSYCHOLOGY (GENERAL; APPLIED; CLINICAL; EDUCATIONAL; EXPERIMENTAL; MULTIDISCIPLINARY; SOCIAL)	98
Social Studies	SOCIAL ISSUES; SOCIAL WORK	56
Media & Communications	COMMUNICATION	52
Ethics & Philosophy	ETHICS; PHILOSOPHY	23
Science, Engineering & Technology		
Computer Science	COMPUTER SCIENCE (INTELLIGENCE; CYBERNETICS; HARDWARE ARCHITECTURE; INTERDISCIPLINARY APPLICATIONS; SOFTWARE ENGINEERING; THEORY METHODS)	1485

(*continued*)

Table 1. (*continued*)

Subject area	Web of Science sub-categories	No. of articles
Electronic Engineering	ENGINEERING ELECTRICAL ELECTRONIC; AUTOMATION CONTROL SYSTEMS; REMOTE SENSING; ROBOTICS	995
Physics	PHYSICS (MULTIDISCIPLINARY; APPLIED; CONDENSED MATTER; MATHEMATICAL); OPTICS; QUANTUM SCIENCE TECHNOLOGY; MECHANICS	250
Mathematics & Statistics	MATHEMATICS (GENERAL; APPLIED; INTERDISCIPLINARY APPLICATIONS); MATHEMATICAL COMPUTATIONAL BIOLOGY; STATISTICS PROBABILITY	188
Chemistry	CHEMISTRY (MULTIDISCIPLINARY; ANALYTICAL)	178
Materials Sciences	MATERIALS SCIENCE MULTIDISCIPLINARY	148
Medical Informatics & Healthcare	MEDICAL INFORMATICS; MEDICAL ETHICS; HEALTH CARE SCIENCES SERVICES	115
Industrial Engineering	ENGINEERING INDUSTRIAL	107
Environmental Sciences	ENVIRONMENTAL SCIENCES; ENVIRONMENTAL STUDIES; GREEN SUSTAINABLE SCIENCE TECHNOLOGY	85

3 Results

The top 20 journals indexed in WOS for cybersecurity, in descending order of popularity, are: *Computers & Security; IEEE Access; International Journal of Information Management; IEEE Security & Privacy; Information & Management; Computers in Human Behavior; Decision Support Systems; Information Systems Research; Journal of Management Information Systems; Journal of Information Security and Applications; Communications of the ACM; Future Generation Computer Systems; IEEE Transactions on Smart Grid; Sensors; MIS Quarterly; Computer Networks; Information Systems Frontiers; Computer; Computer Law & Security Review;* and *Information Systems Journal.*

As can be seen, the list is dominated by IS and CS journals, with a few from Electronic Engineering and Law.

Figure 1 presents a co-authorship network of scholars within the field of cybersecurity. This diagram reveals the structure of cybersecurity research communities, patterns of collaboration, key contributors, and how knowledge has been shared across academic and geographical boundaries. Eight key clusters were identified, with influential authors such as Merrill Warkentin, Richard Baskerville, Alan Dennis, Paul Lowry, Anthony Vance, Mikko Siponen, Robert Willison, Clay Posey, H. Raghav Rao, Yan Chen, Reeshad Dalal and Laura Fletcher to the fore. Of note here is the range of academic disciplines, which include IS, CS, Criminology, and Psychology, as well as the geographic spread, which includes scholars based in North America, Europe and Asia.

Fig. 1. Co-authorship network in cybersecurity research (all subject areas).

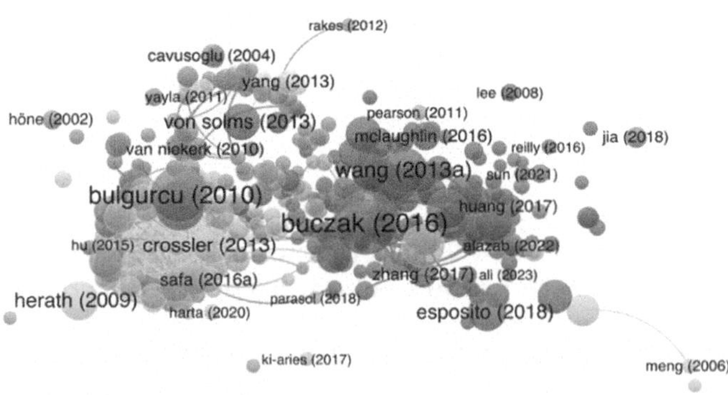

Fig. 2. Bibliographic coupling in cybersecurity research (all subject areas).

Figures 2 and 3 show a bibliographic coupling map and a citation network respectively, which help to identify seminal papers in cybersecurity and how they chain together.

Again, the most prominent works here are from IS and CS, with familiar names of leading scholars quite evidently visible.

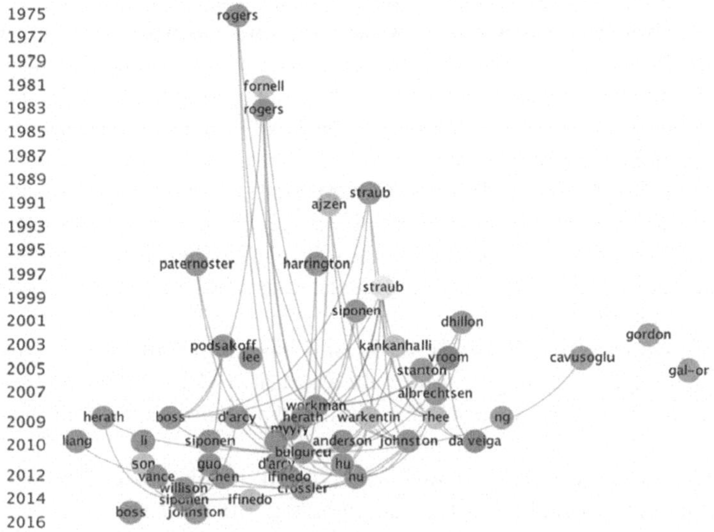

Fig. 3. Citation network of cybersecurity research (all subject areas).

Analysis of the subject areas revealed a core set of themes under each. However, some overlap was found across areas, especially between cognate disciplines such as IS and CS, and between CS and Electronic Engineering (EE). Where the same theme was found under more than one subject area, it was placed within its principal home rather than duplicating. For example, technical aspects were moved out of IS into CS, and behavioural aspects were moved out of CS into IS. A few clusters were non-descript so were dropped. Smaller clusters were combined with others where appropriate to do so. Because of blurred boundaries and natural overlaps, the final set of clusters are not cleanly separated and there is some repetition.

For the two largest subjects areas (IS and CS), the analysis was restricted to the most cited 1000 articles found in Web of Science. As such, niche emerging areas of research that are only gaining traction (in terms of bibliometrics) may not be represented here.

3.1 Information Systems

After swapping some clusters with other subject areas, the following themes emerged from the literature within IS (Fig. 4):

1) **Information Security Policies and Governance:** This cluster examines various aspects of information security governance and policy effectiveness in mitigating risks and preventing abuse. It covers topics such as deterrence strategies, network externalities in risk management, the development of secure software processes, and tools for evaluating security behaviours.

2) **Cultural and Organisational Dynamics in Information Security:** This cluster focuses on the significance of organisational culture, user compliance, and the influence of social factors on information security management. It addresses frameworks for assessing information security culture, risk analysis, and policy implications.
3) **Enhancing Information Security Awareness and Culture:** Focusing on human issues in information security, this cluster analyzes how organisational culture and awareness influence security behaviours. It includes studies on the Human Aspects of Information Security Questionnaire (HAIS-Q) and emphasises monitoring, leadership, and demographic factors.
4) **Compliance Strategies in Information Security:** This cluster investigates various factors influencing information security policy compliance, including leadership styles, organisational commitment, and the role of individual decision-making. It evaluates the effectiveness of training programs and psychological theories in enhancing compliance intentions.
5) **Managing Information Security Risks and Noncompliance:** This cluster delves into the challenges of managing information security risks and noncompliance within organisations. It explores the role of institutional pressures, stakeholder perceptions, and the rationale behind noncompliance behaviours.
6) **Ethical Considerations in Cybersecurity Nudging:** This cluster explores the ethical implications of using nudges to influence cybersecurity decisions. It is concerned with the development of design guidelines for effective nudge interventions while ensuring user privacy and security.
7) **Evolving Research Directions in Information Security Behaviour:** The focus here is on emerging trends in behavioural information security research. This cluster explores the interplay between such factors as cognitive biases and policy internalisation, the effects of comprehensive security programs, and organisational trust.

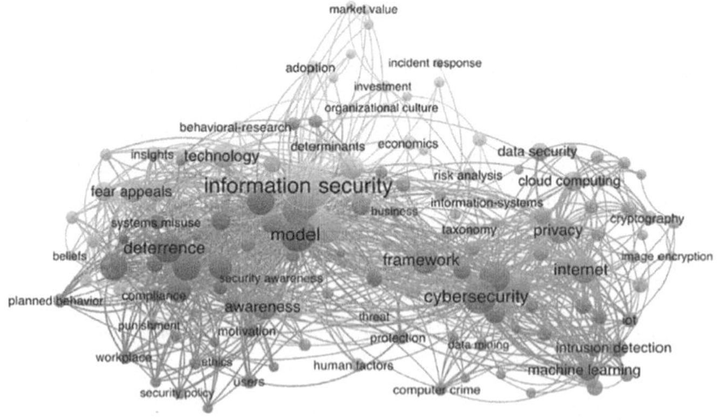

Fig. 4. Keyword co-occurrence in Information Systems.

3.2 Computer Science

After swapping some clusters with IS and EE, eleven were placed under the subject area of CS, as follows (Fig. 5):

1) **Data Security in Healthcare and Cloud Environments:** This cluster explores various methodologies and technologies for enhancing data security and privacy across healthcare systems. Key topics include the security of cloud and fog computing, blockchains, encryption techniques, wearable health technology, and risk assessment strategies tailored to the healthcare industry.
2) **Machine Learning and Artificial Intelligence (AI) in Cybersecurity:** This cluster covers a range of techniques including deep learning, transfer learning, reinforcement learning, and anomaly detection for enhancing cybersecurity measures. Papers explore various aspects such as malware classification, intrusion detection systems, and the use of AI in mitigating cyber threats. The focus is on leveraging advanced algorithms and models to improve the effectiveness of cybersecurity strategies.
3) **Adversarial Approaches and Data Security:** This cluster investigates the dual nature of adversarial machine learning in cybersecurity, encompassing both attack and defense strategies. It discusses cognitive cryptography, hybrid watermarking techniques, and the challenges of ensuring data security in automotive systems.
4) **Game-Theoretic Models and Information Sharing:** This cluster presents a comprehensive examination of game theory applications in cybersecurity, focusing on information sharing and interoperability challenges. It includes the development of cybersecurity knowledge graphs, simulations for vulnerability analysis, and a review of training platforms for realistic scenarios.
5) **Challenges and Innovations in Cybersecurity for Smart Grids and Cities:** This cluster addresses the unique cybersecurity challenges faced by smart grids and urban environments, analyzing threats and technological solutions. Topics include AI applications in enhancing security, the impact of big data analytics, and lightweight encryption methods. The role of intelligent systems in governing cybersecurity measures and ensuring the integrity of critical infrastructures is also emphasised.
6) **Cybersecurity in Transportation and Industry 4.0:** This cluster focuses on the cybersecurity challenges specific to intelligent transportation systems and Industry 4.0 environments. It examines innovative approaches, including reinforcement learning for analyst scheduling and data privacy issues in computing offloading. The exploration of cybersecurity models, including dashboards for pattern visualisation, aims to enhance risk management in these rapidly evolving sectors.
7) **Biometric Technologies and Data Security:** This cluster focuses on the application of biometric technologies in securing information systems and protecting sensitive data. It includes research on various biometric systems such as palmprint cryptosystems and unobtrusive multimodal biometrics to ensure privacy. Additionally, the cluster addresses the technical challenges of biometric applications, including data security in video surveillance and the design of advanced encryption processors.
8) **Federated Learning and Resilience Strategies:** This cluster highlights recent advancements in cybersecurity, particularly within Internet of Things (IoT) environments, through the lens of federated learning and machine learning approaches.

Key themes include resilience in machine learning classifiers, edge data security, and cybersecurity issues in emerging technologies like IoT and medical devices.

9) **Blockchain and Privacy Enhancements in Cybersecurity:** This cluster centers on the integration of blockchain technology for enhancing data security across various applications, including AI, IoT, and environmental monitoring. The cluster includes studies on privacy-preserving data sharing, secure industrial IoT frameworks, and blockchain's role in improving data security.
10) **Cybersecurity in Wireless and Cloud Environments:** This cluster examines cybersecurity challenges specific to wireless and cloud computing environments, focusing on strategies for data security, identity management, and risk assessment, and emerging threats in the post-quantum era.
11) **Advanced Text and Knowledge Graph Approaches:** This cluster focuses on innovative methods for analyzing and managing cybersecurity information through text-based models and knowledge graphs. It covers topics such as named entity recognition, real-time information security transmission, and the construction of knowledge graphs for threat detection. Emphasis is also placed on integrating multi-modal learning and graph-based methods to enhance real-time security and privacy in different contexts .

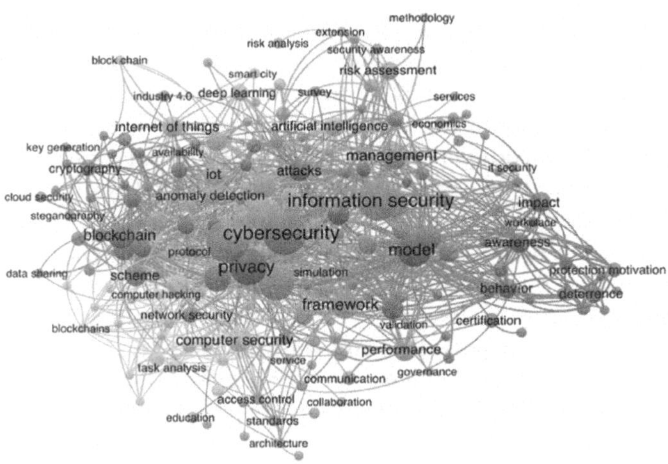

Fig. 5. Keyword co-occurrence in Computer Science.

3.3 Electronic Engineering

The subject area of EE has very substantial cross-over with CS because professionals in these two areas often collaborate, utilising skills from both disciplines to address complex challenges in automation and control systems. The following clusters were found within the EE literature:

1) **Cybersecurity in Smart Grids and Industrial Control Systems:** This cluster focuses on the comprehensive landscape of cybersecurity within smart grids, industrial control systems (ICS), and the IoT, with a particular emphasis on the power sector. Theoretical frameworks such as game theory and dynamic models for real-time anomaly detection are highlighted as critical for addressing vulnerabilities.
2) **Cybersecurity in IoT and Critical Infrastructure:** This cluster delves into cybersecurity issues related to the IoT and critical infrastructure. Papers address a variety of sectors – such as maritime, healthcare, nuclear power and smart farming – highlighting the diverse applications and challenges of cybersecurity in interconnected systems.
3) **Cybersecurity Risk Assessment and Protection Frameworks for SCADA Systems and Critical Infrastructures:** This cluster addresses the complexities of securing SCADA (Supervisory Control and Data Acquisition) systems and critical infrastructures like power grids, water systems, and industrial control systems. Topics include intelligent agent-based security, empirical gap analysis, Bayesian networks for dynamic risk modelling, and cybersecurity standards like ISO/IEC17799 and NISTIR 7628.
4) **Cybersecurity Solutions for Intelligent Transportation and Autonomous Systems:** This cluster explores the intersection of cybersecurity in intelligent transportation systems (ITS), autonomous vehicles, and industrial cyber-physical systems. Central themes include risk assessment, anomaly detection, and mitigation strategies for Vehicle-to-Everything (V2X) communications, IoT, and SCADA/DCS networks. Advanced methods such as blockchain, deep learning, and bio-inspired approaches are employed to safeguard vehicular communication, drones, and critical infrastructures such as railway networks and aviation systems.

3.4 Law

Within the subject area of Law, eight thematic clusters emerged from the analysis:

1) **Cybersecurity Dynamics and Governance:** This cluster explores the evolving landscape of cybersecurity governance, focusing on public-private partnerships, awareness training, and regulatory frameworks. It offers insights into the complexities of safeguarding digital environments amidst increasing cybercrime.
2) **Cybersecurity in Emerging Technologies:** This cluster examines the intersection of cybersecurity and emerging technologies (e.g., IoT, AI), particularly in healthcare and space sectors. It discusses regulatory challenges, legal implications, and innovative approaches to cybersecurity.
3) **Global Cybersecurity Governance:** This cluster focuses on the evolving landscape of global cybersecurity governance, examining the roles of national governments, private sectors, and international organisations. It highlights key topics such as cybersecurity strategies, human rights implications, multilateral governance, and the trade-cybersecurity nexus. The cluster addresses the tension between national security and global trade, privacy concerns, big data, and regulatory frameworks in various geopolitical contexts.
4) **Regulatory Approaches to Cybersecurity:** This cluster examines diverse regulatory approaches to cybersecurity, emphasising the role of federal agencies, legislative

measures, and compliance frameworks. It considers the balance between consumer protection, corporate responsibility, and data security.

5) **International Cybersecurity Frameworks and Legal Perspectives:** This cluster examines the complex landscape of cybersecurity across various countries, with a particular focus on China's cybersecurity law and its implications for international norms and foreign entities. Key topics include state responsibility, the influence of private sector involvement, the role of international law, human rights considerations, and collaborative frameworks in addressing cybersecurity challenges.
6) **Public-Private Cybersecurity Collaboration:** This cluster investigates the interplay between public and private sectors in cybersecurity, emphasising collaborative strategies and infrastructure investment. It highlights challenges, opportunities, and the significance of shared responsibility in enhancing cybersecurity measures in sectors like finance and telecommunications.
7) **Cybersecurity in the European Union:** This cluster examines the EU's evolving cybersecurity landscape, focusing on regulatory frameworks like the NIS Directive and GDPR, the role of ENISA, and sector-specific challenges, particularly in finance and mobile security. The interplay between cybersecurity and digital sovereignty, especially regarding 5G, highlights the complexities of safeguarding digital infrastructure across member states.
8) **Community Engagement in Cybersecurity:** This cluster emphasises the importance of community involvement and self-help in enhancing cybersecurity. It explores innovative strategies for small businesses, the emotional aspects of cybersecurity deception, and the the dynamic interplay between individual actions and collective security efforts to combat cybercrime.

3.5 Business and Management

The majority of the clusters found in this subject area overlapped with either IS or Economics, but a number of distinct research themes were also revealed:

1) **Information Security Governance and Management Practices:** This cluster explores various dimensions of information security governance, highlighting the interplay between socio-technical factors, internal audit practices, and IT capabilities. It emphasises the importance of institutional logic, knowledge management tools, and compliance frameworks in shaping effective information security strategies.
2) **Cybersecurity Strategies and Impact Assessment:** This cluster focuses on the strategic balance between proactive measures and reactive responses in information security. It discusses the effectiveness of awareness methods, risk assessments, and incident management strategies while examining the impact of information security breaches on consumer behaviour.
3) **Cybersecurity Awareness and Knowledge Sharing:** The papers in this cluster emphasise the significance of knowledge sharing and awareness in strengthening cybersecurity practices.
4) **Institutional Pressures and Information Security Management:** This cluster delves into the role of institutional pressures in shaping organisational responses to information security challenges. It also examines how regulatory and strategic commitments influence investment decisions in cybersecurity, particularly within healthcare contexts.

3.6 Economics and Finance

Six clusters were identified within this subject area:

1) **Cybersecurity and Risk Management:** This cluster focuses on aspects such as risk management, insurance and governance. The papers explore frameworks, models, and strategic approaches to handling cyber risks, with a particular emphasis on insurance and policy coordination.
2) **Cybersecurity Auditing and Compliance:** This cluster centres around auditing, internal controls, and compliance with cybersecurity policies. Topics include internal audit roles, security standards, and methodologies for assessing cybersecurity risks within organisations, with a focus on accounting and financial security.
3) **Financial Implications of Cybersecurity Risk Disclosure:** The focus here is on the financial and market impacts of cybersecurity breaches, particularly in relation to risk disclosures, audit fees, and investor reactions.
4) **Investment Perspectives in Cybersecurity:** This cluster is concerned with decision-making processes and frameworks, looking at how firms assess risks and manage cybersecurity investments.
5) **Corporate Cybersecurity Risk and Innovation:** This cluster explores the relationship between cybersecurity risks and corporate innovation. The articles explore issues such as how cybersecurity risks impact corporate behaviours (e.g., mergers, acquisitions, etc.).
6) **Cybersecurity in the Financial Sector and Investor Decisions:** This cluster focuses on cybersecurity in the financial sector, with a particular emphasis on investor perceptions and decision-making based on cybersecurity risks and management strategies, as well as a concentration on disclosure and regulatory compliance.

3.7 Operations Management and Logistics

Substantial cross-over was observed here with IS and Economics, but the following subject-specific themes emerged:

1) **Cybersecurity in Supply Chains and Logistics:** This cluster focuses on cybersecurity breaches and risks in supply chain management and logistics using innovative technologies. Issues addressed include how cybersecurity incidents impact earnings management, supplier relationships, and operational efficiency.
2) **Maritime and Transportation Cybersecurity:** This cluster explores the growing importance of cybersecurity in the maritime industry and transportation systems. Topics include cybersecurity for ports, ships, and railway systems, emphasising the necessity for awareness and proactive measures against emerging cyber threats in these critical sectors.
3) **Cybersecurity for Connected and Autonomous Vehicles:** This cluster focuses on cybersecurity challenges and solutions in the realm of connected and autonomous vehicles. Emphasising the need for comprehensive strategies, this research underscores the importance of securing vehicular communications and the infrastructure that supports them in an increasingly automated transport landscape.

4) **Decision Support Systems for Cybersecurity Investments:** This cluster focuses on developing decision support systems for optimising cybersecurity investments. It incorporates various methodologies, including probabilistic models and optimisation techniques, to assist organisations in identifying and selecting effective cybersecurity safeguards, especially in the context of Industry 4.0 and emerging cyber threats.

3.8 Sociology and Politics

There is considerable overlap here with topics addressed under Law, but with a different emphasis. Clusters identified were:

1) **Cybersecurity Governance and Strategy in Europe:** This cluster examines the evolving landscape of cybersecurity governance in Europe, highlighting the tension between regulatory frameworks and market forces. It explores concepts such as regulatory mercantilism, trust dynamics in digital platforms, the role of state-industry relations, and cyber diplomacy.
2) **Regional Perspectives on Cybersecurity Challenges:** The focus here is on regional variations and global implications of cybersecurity strategies. It covers many topics, from the impact of geopolitical tensions on cybersecurity policies to the ethics of private security measures.
3) **Cybersecurity in the Context of State and Private Sector Dynamics:** This cluster delves into the intersection of public authority, private interests, and cybersecurity. It discusses the privatisation of political authority, the role of public-private partnerships, and implications for national security.
4) **Trade, Governance, and Cybersecurity:** This cluster investigates the complex relationship between cybersecurity and international trade governance. It emphasises the legal and regulatory challenges posed by technological advancements on cybersecurity policies.
5) **Global Cybersecurity Norms and Human-Centric Approaches:** This cluster addresses the significance of human rights within the global cybersecurity discourse. It examines international legal frameworks, regulatory responses, and the role of public-private partnerships in shaping cybersecurity norms.
6) **Cybersecurity as a Societal Construct:** This cluster examines the multifaceted relationship between society and cybersecurity. It highlights the role of citizen involvement, cultural perceptions of hacking, and the implications of military culture on security practices.
7) **Global Dynamics of Cybersecurity:** Focusing on the intersection of cybersecurity and global politics, this cluster explores the fragmented political landscape shaped by technological advancements and international relations. It highlights the roles of non-state actors, the importance of collaboration in capacity building, and emerging threats like cyber terrorism.
8) **National and International Cybersecurity Strategies:** This cluster delves into national cybersecurity policies and their implications for international relations. It highlights how different states, including small nations, adapt their strategies in response to evolving threats.
9) **Public-Private Partnerships in Cybersecurity:** This cluster explores the dynamics of public-private partnerships in developing effective cybersecurity strategies. It

examines the role of national infrastructure protection and the impact of major global players on cybersecurity discourse.

10) **Cultural and Political Dimensions of Cybersecurity:** This cluster investigates the interplay between cybersecurity, culture, and governance. It examines how societal narratives, including popular culture representations, influence cybersecurity perceptions and policies. Theoretical frameworks explore the politics of truth and the evolving governance structures that emerge in response to cybersecurity challenges.

3.9 Organisational Psychology

Psychology issues were found not just under that subject area, but also in CS, IS and others. The clusters revealed were:

1) **Organisational Behaviour and Information Security Compliance:** This cluster examines the intricate relationship between psychological motivations, coping mechanisms, and social dynamics influencing information security behaviour and policy adherence within organisations. It integrates findings from various studies highlighting the role of Protection Motivation Theory and the Elaboration Likelihood Model in shaping user responses to security threats. Additionally, it explores how rationality-based beliefs, stress responses, psychological capital, technostress, and work conditions impact compliance and violations.
2) **Emotional Responses and Value-Based Cybersecurity:** This cluster investigates the emotional and value-oriented dimensions of information security behaviour. It examines how users cope with IT security threats beyond technical measures, highlighting the role of emotions in shaping responses to security challenges.
3) **Cybersecurity Knowledge and Human Factors:** This cluster explores the intersection of human behaviour, creativity, and cybersecurity knowledge in improving security practices. Key topics include the role of cognitive models in understanding attacker behaviour, the necessity of teamwork and role specialisation, and the psychosocial dynamics influencing cybersecurity performance. Theories related to organisational science and human factors are evident, emphasising the need for comprehensive training and effective talent management in cybersecurity roles.
4) **Social Dynamics:** This cluster focuses on social factors influencing cybersecurity behaviour, including family dynamics, insider threats, and the impact of human errors. It incorporates theories like Cyber-Routine Activities Theory and highlights the role of personal and contextual predictors in compliance.
5) **Information Security Culture:** This cluster emphasises the importance of fostering a robust cybersecurity culture through education, awareness, and emotional understanding.
6) **Individual Differences**: This cluster examines the interplay of individual differences, age, and social factors in cybersecurity behaviours and awareness. Key topics include the influence of personality traits, emotional responses to breaches, and the role of gamification in training. Theoretical frameworks such as Q methodology provide insights into personal perspectives on cybersecurity. Studies also highlight age-related vulnerabilities, particularly among older adults, and the impact of factors like media multitasking and fear of missing out on compliance and awareness.

7) **Behavioural Interventions and Cybersecurity:** This cluster focuses on the development and application of behavioural interventions aimed at improving cybersecurity practices. It explores risk perceptions, ethical considerations in nudging, and strategies for balancing security and usability. Theoretical models for behaviour change are discussed, emphasising the transition from viewing humans as problems to recognising their potential as solutions in cybersecurity efforts.

3.10 Education

Similarly, education issues was a common thread found across many subject areas. The clusters identified were:

1) **Understanding and Enhancing Information Security Awareness:** This cluster examines factors influencing information security awareness, including self-efficacy, trust, and age demographics. It emphasises the role of gamification and targeted interventions in promoting secure behaviours among different user groups. Theoretical frameworks like the ARCS model and intervention mapping inform the design of educational strategies aimed at mitigating threats such as phishing attacks.
2) **Cybersecurity Awareness and Digital Ethics:** This cluster investigates the intersection of cybersecurity awareness, digital ethics, and user behaviour. It explores factors such as parental mediation, social education levels, and the effectiveness of online security training.
3) **Gamification and Innovative Strategies in Cybersecurity Training:** This cluster highlights the use of gamification and interactive learning environments to enhance cybersecurity education. The emphasis is on creating engaging educational experiences that cultivate interest in cybersecurity careers and improve practical skills among learners.
4) **Practical Approaches to Cybersecurity Training and Awareness:** This cluster focuses on the practical aspects of cybersecurity education, including the integration of technology and situational training. It examines various training methods, such as virtual escape rooms and hands-on experiences, aimed at improving cybersecurity skills.
5) **Curriculum Development in Cybersecurity Education:** This cluster addresses the design and assessment of industry-aligned cybersecurity curricula across various educational levels. It presents models and frameworks for effective teaching that emphasise critical skills and adversarial thinking.
6) **Engagement and Outreach:** This focuses on strategies to enhance engagement in cybersecurity education, particularly for K-12 educators and students.
7) **Cybersecurity Challenges in Higher Education:** This cluster explores the cybersecurity landscape within higher education. The papers highlight the importance of integrating cybersecurity measures to protect academic integrity and intellectual property in a rapidly evolving digital environment.
8) **Data-Driven Approaches in Cybersecurity Education:** This cluster emphasises the role of data analytics in enhancing cybersecurity education through automated assessments and feedback mechanisms.
9) **Conceptual Foundations in Cybersecurity Education:** This cluster investigates foundational concepts in cybersecurity education. The studies aim to bridge the

gap between theoretical understanding and practical application in cybersecurity curricula.

10) **Cybersecurity in Business Education:** This cluster addresses the incorporation of cybersecurity principles within business and management education.

4 Conclusions

This paper set out to form a high-level overview of the fragmented academic territory of cybersecurity research. It does not purport to be a detailed map, or a complete one. Indeed, much like the crude maps produced by early cartographers, it reflects its creator's limited firsthand experience of the world beyond their own familiarity (the author is at home in the disciplines of IS and business, but a novice in other fields). Furthermore, vast swathes of the territory remain relatively unexplored, the tools used to create this map (i.e., the underlying algorithms) are imprecise, and the analysis is based on meta-data rather than a thorough review of the source materials. Readers of this paper may identify many missing topics additional to those identified here. This will especially be the case for niche emerging work which has not yet gained sufficient traction to form prominent clusters standing out amongst the vast cumulative body of existing work. Additionally, a number of other subject areas not set out here (due to space restrictions) were found, such as Ethics, Communications, Medical Informatics, and Mathematics.

Nevertheless, despite these shortcomings, this article presents a useful general overview of the main subject areas and topic clusters identified within the corpus of cybersecurity research indexed by Web of Science.

As regards further work, there is ample scope for meta-analysis, detailed bibliometric analysis, rapid reviews, hermeneutic literature reviews and systematic literature reviews within the identified subject areas.

References

1. Lang, M., Dowling, S., Lennon, R.G.: The current state of cyber security in Ireland. In: Proceedings of the 2022 Cyber Research Conference-Ireland (Cyber-RCI), pp. 1–2. IEEE (2022)
2. O'Donovan, B., Roode, D.: A framework for understanding the emerging discipline of information systems. In: European Conference on Information Systems (1999), pp. 530–541 (1999)
3. Straub, D., Loch, K., Evaristo, R., Karahanna, E., Srite, M.: Toward a theory-based measurement of culture. J. Glob. Inf. Manag. **10**(1), 13–23 (2002)
4. Gregory, K.L.: Native-view paradigms: multiple cultures and culture conflicts in organisations. Adm. Sci. Q. **28**(3), 359–376 (1983)
5. Kim, S.: Interdisciplinary cooperation. In: Laurel, B. (ed.) The Art of Human-Computer Interface Design, pp. 31–44. Addison Wesley, Reading (1990)
6. Checkland, P.B.: Systems Thinking Systems Practice. Wiley, Chichester (1981)
7. Avison, D.E., Taylor, V.: Information systems development methodologies: a classification according to problem situation. J. Inf. Technol. **12**(1), 73–81 (1997)
8. Orlikowski, W.J., Gash, D.C.: Technological frames: making sense of information technology in organisations. ACM Trans. Inf. Syst. **12**(2), 669–702 (1994)

9. Kuhn, T.S.: The Structure of Scientific Revolutions, 3rd edn. University of Chicago Press, Chicago (1996)
10. Van Eck, N.J., Waltman, L.: CitNetExplorer: a new software tool for analyzing and visualising citation networks. J. Informet. **8**(4), 802–823 (2014)
11. Van Eck, N.J., Waltman, L.: Visualising bibliometric networks. In: Ding, Y., Rousseau, R., Wolfram, D. (eds.) Measuring Scholarly Impact: Methods and Practice, pp. 285–320. Springer (2014)
12. Donthu, N., Kumar, S., Mukherjee, D., Pandey, N., Lim, W.M.: How to conduct a bibliometric analysis: an overview and guidelines. J. Bus. Res. **133**, 285–296 (2021)
13. Lim, W.M., Kumar, S., Donthu, N.: How to combine and clean bibliometric data and use bibliometric tools synergistically: guidelines using metaverse research. J. Bus. Res. **182**, 114760 (2024)
14. Desai, R.R., Kim, J.: Bibliographic and text analysis of research on implementation of the Internet of Things to support education. J. Inf. Syst. Educ. **34**(2), 79–195 (2023)
15. Lemon, L.L., Hayes, J.: Enhancing trustworthiness of qualitative findings: Using Leximancer for qualitative data analysis triangulation. Qualit. Rep. **25**(3), 604–614 (2020)
16. Smit, B., Scherman, V.: Computer-assisted qualitative data analysis software for scoping reviews: a case of ATLAS.ti. Int. J. Qualit. Methods **20**, 16094069211019140 (2021)

Hiding Information in Short Network Packets

Jörg Keller(✉) and Sebastian Litzinger

FernUniversität in Hagen, 58097 Hagen, Germany
joerg.keller@fernuni-hagen.de
https://feu.de/pv/en

Abstract. Lossless compression of payload in network packets and using the freed space for transfer of secret messages has been known for long. However, for short payloads, usual lossless compressors do not achieve any compression. We address this case by proposing both a Huffman-inspired and a runlength-inspired coding of payload when byte values are not equi-distributed. We provide numeric results for different payload lengths and byte value distributions and achieve compression ratios of 70% in the best case, leaving 30% of the message space for the secret message. We also investigate possibilities for the detection of such an embedding via changes in entropy.

Keywords: Network Steganography · Lossless Compression · Huffman Coding · Runlength Coding

1 Introduction

Network steganography hides the existence of secret messages in innocent network traffic, e.g., by using reserved bits in packet headers [9]. If the covert receiver can undo the modifications from the covert sender, then also the payload of network packets can be used without disturbing the carrier traffic. For instance, it is well-known that the payload of large network packets can be compressed by the covert sender using a lossless compressor [11] and the freed space in the packet can be used for the transmission of secret data, while the covert receiver extracts the secret data and decompresses the payload to restore the original packet content [8]. However, for network packets with a short ($\leq$250 bytes) payload such as DNS over UDP, SSH over TCP or MQTT over TCP, usual lossless compressors like gzip do not achieve any compression as they search for repeated patterns of sufficient length.

We investigate this case by proposing an encoding that does not use longer patterns but only distinguishes between frequently occurring byte values and other, infrequent byte values. For each class of byte values, a fixed-length encoding is used, where the first bit indicates the class as in Huffman-inspired encodings. While for infrequent byte values, this bit is simply followed by the byte value

I. Praça et al. (Eds.): EICC 2025, CCIS 2500, pp. 133–150, 2025.
https://doi.org/10.1007/978-3-031-94855-8_9

itself, for frequently occuring byte values, a shorter index is used, which necessitates to transfer a mapping between these shorter index values and the corresponding byte values. As an alternative, we also propose a runlength-inspired encoding where we store lengths of blocks for frequent and infrequent symbols followed by short indices and byte values, respectively. For both proposals, we present a very compact representation of the subset of frequently used symbols and solve the optimization problem to select which byte values are considered frequently occurring, such that the total length of the compressed payload is minimized.

We evaluate the compression factors achievable through our encoding for different payload lengths and different byte value distributions, both for synthetic and real payloads. We furthermore investigate the detectability of our approach by measuring how the entropy of the payload changes by compression and addition of the secret message.

Our contributions can be summarized as follows.

- We present a proposal to encode short sequences of symbols based on their frequency: hot symbols with high frequency get short codewords, other symbols get long codewords, and we give a compact representation of the set of hot symbols.
- We apply this proposal to short (<250 bytes) payloads of network packets to compress the payload without loss and embed a piece of a secret message for steganographic transmission.
- We evaluate the proposal for different symbol distributions, both synthetic and real-world, and demonstrate compression down to 70% of the original size.
- We discuss countermeasures and experimentally check that the presence of such embedding can be detected by computing the entropy of payloads when the entropy of normal traffic is known in detail.

The remainder of this article is structured as follows. In Sect. 2, we summarize background information and discuss related work. In Sect. 3, we present our encoding, while we sketch a concrete implementation in Sect. 4. Section 5 presents our experimental results, while Sect. 6 gives conclusions and an outlook to future work.

2 Background

In this section, we present background information about lossless compression (Subsect. 2.1) and network steganography (Subsect. 2.2), which will be necessary in the following sections, and briefly discuss related work (Subsect. 2.3).

2.1 Lossless Compression

Compression is a form of source encoding that serves to reduce the size of the message, file or image to be encoded. The ratio of the sizes of compressed and uncompressed messages is called *compression ratio* [11]. We distinguish lossless and lossy compression. In the former, the original message can always be reconstructed from the compressed message. In the latter, which is mostly used in image compression, perfect reconstruction is traded for a better compression ratio, yet the reconstructed image should be very close to the original image.

Lossless compression, mostly based on Lempel and Ziv's work [16], searches for patterns that appear multiple times and which are stored in the so-called dictionary. At the second occurrence, only a pointer to the first occurrence instead of the pattern itself must be stored. Thus, the compressed file contains both the pointers and dictionary.

Also usual source encoding approaches like Huffman coding [2] and runlength coding [11] try to reduce the average bitlength of symbols compared to fixed-length encodings, and thus can be considered as simple compression approaches. Huffman coding constructs a binary code tree by successively combining pairs of symbols with minimum frequencies and assigns code words along paths from the tree root to symbols, the average code word length is minimized. Runlength coding replaces sequences of bits with the same value by their length.

2.2 Network Steganography

Network steganography hides secret messages in network traffic by establishing covert channels, i.e., channels that were never intended for communication and that break a security rule [3]. We denote the sender and receiver of the secret message as covert sender and receiver. They can be identical to the sender and receiver of the overt network traffic (or carrier) that is used to transmit the secret message, or they can be on the communication path [9]. We distinguish storage channels, that explicitly modify some bits of network packets, and timing channels, that encode secret message bits via modifications of packet timing. We will focus on the former. In particular, we will focus on channels that modify packet payload in contrast to using header fields, and that are reversible [7], i.e., where the covert receiver not only extracts the secret message but can also undo the modifications done by the covert sender, before the packet is ultimately forwarded to the overt receiver. The different approaches to network steganography are categorized in hiding patterns [13,14].

2.3 Related Work

Compressing the payload of an overt packet and filling the freed space with part of the secret message is known as *transcoding* steganography [8]. It is an instance of pattern[1] PS30 [13,14]. As an example, the article [8] only cites [6], where transcoding steganography is applied in voice-over-IP calls by (re-)encoding the speech data with a codec that provides higher compression without sacrificing speech quality, and using the freed space for steganographic transmission. Yet, speech compression is lossy, whereas lossless compression is needed to be able

[1] https://patterns.ztt.hs-worms.de/NIHPattern/.

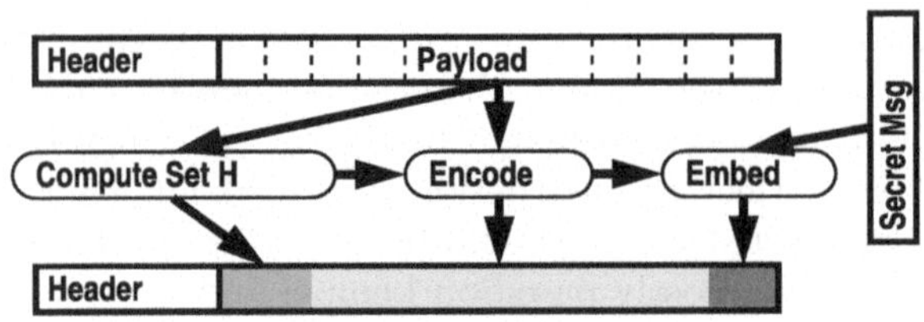

Fig. 1. Flow of information: (1) Analyze payload to compute set H of hot symbols. (2) Represent set H. (3) Encode payload with short codes for hot symbols. (4) Append piece of secret message in remaining space.

to undo this modification and restore the original (textual, binary or numerical) content of network packet's payload, i.e., to produce a reversible covert channel.

While reversible network storage channels are presented in [7], they only target header fields. To our knowledge, reversible transcoding steganography through lossless compression of payload has not been attempted for network packets. The application to payloads of notable size (e.g., ethernet frames normally have payloads of about 1500 bytes) should be possible with an ordinary lossless compressor like gzip. However, according to our initial tests, gzip does not achieve a compression of messages below 1000 bytes. Hence, network messages of short length, which frequently occur in practice (e.g., DNS over UDP or MQTT over TCP) need a different solution which we are going to present.

Compression has been used in text steganography, although not in the way we use it. In [12], a secret message is embedded in a chosen text and Huffman coding is used afterwards to increase the capacity. Embedding of secret message bits into Huffman compression codes of a cover text is proposed by [4]. LZW compression of secret data is used in [5] before embedding into a cover text.

3 The Proposed Approach

Usual lossless compression algorithms search for patterns as long as possible. However, for a short message, this is useless. Therefore, we only search for single symbols that occur most frequently in the message, instead of searching for patterns, i.e., sequences of symbols occurring multiple times. We assign short codewords to these "hot" symbols, while the remaining symbols keep their codewords, plus a bit that marks them as "normal". In addition, we represent the set of hot symbols in compact form. If our compressed message is shorter than the original message, we append a piece of the secret message to fill the blank space. The flow of information is depicted in Fig. 1.

More formally, we consider a symbol set $S = \{a_0, \ldots, a_{n-1}\}$ of size n with $2^{u-1} < n \leq 2^u$. For example, if we consider bytes, then $S = \{0,1\}^8$. We assume a total order $\preceq$ on set S, such as lexicographic order on bytes. For any set $X \subset S$, we define an enumeration $ind_X : X \rightarrow \{0, \ldots, |X|-1\}$ of its elements via $ind_X(x \in X) = |\{y \in X | y \preceq x \text{ and } y \neq x\}|$, such that $X = \{s_0, \ldots, s_{|X|-1}\}$ where $s_i = x \in X$ for $i = ind_X(x)$.

For a message of size s, i.e., $m = (m_0, \ldots, m_{s-1}) \in S^s$, we assume that originally it is coded via a fixed-length encoding, i.e., u-bit binary representation of index j for symbol $a_j \in S$. Thus, the encoded message's length is

$$L_0(m) = u \cdot s \;, \tag{1}$$

that does not depend on the message itself but on its length. We focus on short message sizes with $s \leq n$, where patterns are unlikely.

In the following, we assume that we have identified a set $H \subset S$ of *hot* symbols that occur frequently in m with $2^{t-1} < |H| \leq 2^t$. We can use a fixed-length encoding for hot symbols that is shorter than the fixed-length encoding for normal symbols and the two partial encodings are distinguished via a leading bit as in a Huffman-inspired encoding [2]. More concretely, we encode a hot symbol $s_i \in H$ via codeword $0, bin_t(i)$ and a normal symbol $a_j \in S \setminus H$ via codeword $1, bin_u(j)$, where $bin_t(i)$ is the t-bit binary representation of $0 \leq i < |H| \leq 2^t$ and $bin_u(j)$ is the u-bit binary representation of $0 \leq j < n \leq 2^u$.

We remark that the encoding is redundant, as the restriction to only encode normal symbols using u-bit representation is artificial: we also could encode a hot symbol in $u+1$ bits. While this does not seem useful here, it will be helpful in our alternative encoding method, see below.

We precede the sequence of s codewords by a compact representation of set H. The length of this representation can be bounded by

$$L(H) \leq t + |H| \cdot u \tag{2}$$

as we could represent the set by a binary representation of its size, which must be either self-limiting or has a fixed, predetermined length to be decodable, and a list of indices (in S) of the elements of H. We will discuss several set representations in the next section.

The total size of the new encoding is

$$L_1(m, H) = L(H) + s \cdot (\alpha(m, H) \cdot (t+1) + (1 - \alpha(m, H)) \cdot (u+1)) \;, \tag{3}$$

where $\alpha(m, H)$ represents the fraction of hot symbols in message m. Formally, $\alpha(m, a_j) = |\{i \mid m_i = a_j \text{ and } 0 \leq i \leq s-1\}|/s$ and $\alpha(m, H) = \sum_{a_j \in H} \alpha(m, a_j)$.

In order to achieve $L_1(m, H) \leq L_0(m)$, we have the requirement that $t+1 \leq u-1$ or $t \leq u-2$, i.e., $|H| \leq |S|/4$. In practice, we will have $|H| \leq |S|/8$.

Please note that we do not need to encode s, the number of symbols in m, as in the original encoding of the message, its length is also assumed to be given.

For a given message m, the optimization problem to be solved consists in finding the set $H^* \subset S$ of hot symbols that minimizes the total message length, i.e., $L_1(m, H^*) \leq L_1(m, H)$ for any $H \subset S$. The solution of this optimization problem will depend on the representation of the set H. As long as $L(H)$ only depends on the size of H and is strictly increasing with it, we can proceed as follows. We order the symbols of S in decreasing order of occurrence in the message m, and by increasing order of symbol index in S for equal occurrence figures, i.e., we compute a permutation $\sigma : \{0, \ldots, n-1\} \to \{0, \ldots, n-1\}$ such that $\alpha(m, a_{\sigma(j)}) \geq \alpha(m, a_{\sigma(j+1)})$ for $j = 0, \ldots, n-2$ and such that for the case $\alpha(m, a_{\sigma(j)}) = \alpha(m, a_{\sigma(j+1)})$ we have $\sigma(j) < \sigma(j+1)$. We define a sequence of sets H_k for $k = 1, 2, \ldots$ via $H_k = \{a_{\sigma(0)}, \ldots, a_{\sigma(k-1)}\}$. Then $\alpha(m, H_k)$ will be strictly increasing with k. It will also be concave as we add symbols to H_k in decreasing order of absolute frequency. We compute $L_1(H_k, m)$ for each k and choose H^* as the H_k that minimizes this length of the compressed message. As the maximum size of H will be small and evaluation of Eq. (3) is quick, the optimization effort is small.

Please note that if $L(H)$ does not only depend on the size of H but also on the particular elements, then this algorithm does not necessarily lead to a global optimum anymore, yet can still be used as a heuristic.

As an alternative method of compression, we also consider a runlength encoding [11] that depends not on bit values 0 and 1 but on the classification as hot or normal. For the first symbol m_0 in the message, we check whether it is a hot symbol or not, and search for the maximum l such that also m_1 to m_{l-1} belong to the same category as m_0. Then we encode a binary representation of l and use t-bit or u-bit representations for the block m_0 to m_{l-1}, respectively, depending on whether m_0 is hot or normal. Then we repeat the same procedure starting with m_l, until the complete message is encoded. We precede this encoding with a compact representation of H as in the Huffman-inspired encoding. Also, we specify whether hot or normal comes first, which suffices as classes alternate from block to block. If we assume that there are $z(m, H)$ blocks in total, and the binary representation of the number of symbols in each block has at most $\log_2 s \leq u$ bits, then the length of this representation is

$$L_2(m, H) = L(H) + z(m, H) \cdot u + 1 + s \cdot (\alpha(m, H) \cdot t + (1 - \alpha(m, H)) \cdot u) \ . \quad (4)$$

Also here, we use the same optimization procedure as before by defining sets H_k, computing $z(m, H_k)$ for each, and evaluating Eq. (4) for each H_k to find H^* that minimizes the total length.

An expected value for $z(m, H)$ can be derived from the observation that if hot symbols are assumed to occur independently, we can model the length of a block of hot symbols as a random variable X. We assume for simplicity that this variable can have any value $x \geq 1$. The probability is $P(X \geq x) = \alpha(m, H)^{x-1}$ with expectation

$$E(X) = \sum_{x \geq 1} P(X \geq x) = 1/(1 - \alpha(m, H)) \ .$$

Similarly, the expected length of a block of normal symbols is $E(Y) = 1/\alpha(m, H)$ and the expected number of pairs of blocks is the ratio between number of symbols and the sum of expected block lengths, i.e., the number of blocks is

$$E(z(m, H)) = 2\, s/(E(X) + E(Y)) \ . \quad (5)$$

We can reduce the number of blocks by coming back to the observation that our encodings are redundant in the sense that also hot symbols can be encoded as normal symbols. For our runlength encoding, an isolated hot symbol which is surrounded by blocks of normal symbols can be interpreted as a normal symbol and so three blocks can be merged into one. We save two block length descriptions for the price of $u - t$ additional bits. Depending on the value of $u - t$ this might even be useful for blocks comprising two hot symbols.

The embedding of the secret message is straightforward. In each network packet, the payload of size $s \cdot u$ bits is compressed by applying one of the encodings above, and the remaining space is filled with a part of the secret message. The packet is adapted to the new payload, e.g., by re-computing the checksum. Also the extraction does not pose any difficulty. We assume that covert sender and receiver have previously agreed on the encoding to be used. The covert receiver first extracts the set H from its representation, and then it can reconstruct the original message from the encoding. As the payload has a size of s (normally bytes), the number of encoded symbols is known to the covert receiver. The covert receiver also extracts the remaining part of the payload as a part of the secret message. Then it fills the payload with the reconstructed original message and restores the original network packet, e.g., by re-computing the checksum.

The only case that we did not consider yet is a case where neither encoding, even in optimized form, is shorter than the original message. The covert sender can handle this situation simply by leaving the network packet unmodified. The covert receiver needs to detect this situation from the received packet in order to not undo a packet modification that did not occur. Therefore we include a kind of magic number, i.e., we sacrifice some of the space that is to be used for the secret message, e.g., the last v bits of the payload, and place a fixed bit pattern there, or a hash value of the payload (except these last v bits). If the covert receiver gets a network packet where the last v payload bits do not have this magic value, then it knows that the network packet has been sent unmodified. However, there is still a chance of 2^{-v} that the original payload did include the magic number by coincidence. In this case, the covert receiver starts to extract set H and decodes the symbols. If the packet has been sent unmodified, the assumed encoding would be an arbitrary one, i.e., not very efficient. Therefore, it is likely that the covert receiver would have exhausted the payload before it could extract s symbols. Also in this case, the covert receiver knows that no encoding has been applied, and it leaves the network packet unmodified.

As covert sender and receiver will choose a network communication where they see a good chance that their encoding can be applied successfully, i.e., where symbol values are not equi-distributed so that they can be compressed, the case of an unmodified payload should occur seldom so that the two checks render the erroneous decoding of an unmodified packet payload unlikely in practice.

4 Implementation

Our implementation focuses on a payload that is arranged in bytes, i.e., $S = \{0,1\}^8$, $n = 256$ and $u = 8$.

We start by determining the representation of set H. It could be represented as a bit vector of length n, yet as $|H| \leq |S|/8 = 32$ (see previous section), a list of

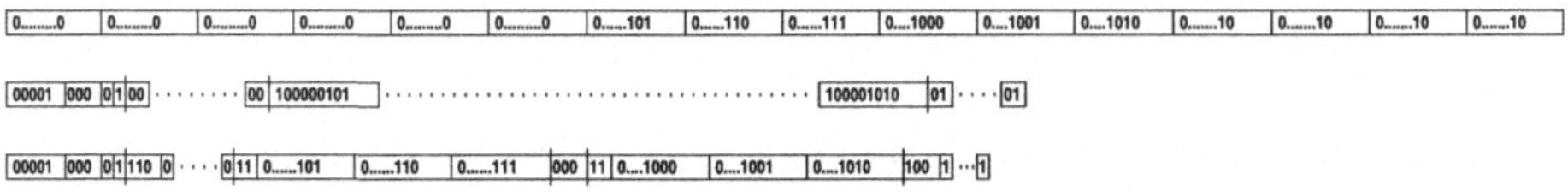

Fig. 2. Encodings of example message. Top: original message in 16 bytes, i.e., 128 bits. Mid: Huffman-inspired encoding. Bottom: Runlength-inspired encoding.

byte values, i.e., u-bit symbol indices, will be shorter if $|H| < 32$. Instead of the byte values, we can store a 5-bit binary representation of $|H|-1$ (this is possible as $|H| \geq 1$) followed by a set of differences[2] between the indices of successive elements, i.e., $\Delta_i = ind_S(s_i) - ind_S(s_{i-1})$ for $i = 0$ to $|H|-1$ (and artificial element s_{-1} with index -1). The differences Δ_i are normally much smaller than $|S|$ and thus can be encoded in fewer than u bits which are needed for the symbol indices. The maximum difference defines the length of the binary representation of all differences, and this length minus 1 (which is in the range $0, \ldots, 7$) is also stored as a 3-bit binary representation. Following the representation of the set, the encoding of the message with short and long codes, as described in the previous section, follows. A pseudo-code is given in Algorithm 1.

Please note that the length of this representation of set H does not only depend on the size of H but also on the indices of the elements as these define the differences. As an example, consider the following list of symbols in decreasing order of occurrence frequency: $a_1, a_{254}, a_{255}, a_2, a_{128}, \ldots$. H_1 that only comprises element a_1 needs only a 1-bit difference. H_2 that comprises elements a_1, a_{254} needs 8-bit differences. When creating H_4, it would be advantageous to take element a_{128} instead of a_2 as this would allow 7-bit differences instead of 8-bit differences, while a_2 and a_{128} might have the same occurrence frequency.

While our optimization procedure of the previous section therefore is not guaranteed to find the global optimum, we still apply it as the result in practice is close to the optimum and the quick algorithm avoids detectability through excessive resource usage in the covert sender's machine, which might occur when computing the global optimum via a solver.

For the second encoding, we expect block lengths for hot symbols to be longer than for normal symbols. For example, if the fraction α of hot symbols is 0.75, then the expected block length for hot symbols is 4, and 23.7% of the blocks will comprise at least 6 hot symbols. In contrast, the expected block length for normal symbols is 1.3, and only 1.56% of the blocks will comprise at least 4 normal symbols. Therefore, we store the block length for hot symbols in 3 bits and the block length for normal symbols in 2 bits. We allow block length of 0 as blocks of length at least 8 (hot) or 4 (normal) symbols must be split, with an empty block of the opposite class in-between as succeeding blocks are assumed to belong to different classes. To complete the specification, the encoding starts with

[2] As the difference will be at least 1, we can even store the difference minus 1 to have values starting with 0.

the representation of set H as in the previous encoding, followed by a sequence of blocks, starting with a block of hot symbols, as shown in Algorithm 2.

As an example, Fig. 2 depicts an original payload of $s = 16$ bytes with values $6\times 0, 5, 6, \ldots, 10$ and 4×2 together with the two encodings for $H = \{s_0, s_1\}$ with $s_0 = 0$ and $s_1 = 2$. In the Huffman-inspired encoding, the first 5 bits contain the size of H minus 1, the next 3 bits contain the number of bits used for the differences minus 1. As the differences Δ_i are 1 and 2, respectively, yet only $\Delta_i - 1$ are stored, 1 bit suffices. Next come the 2-bit codes of the hot symbol 0 (6 times), followed by the 9-bit codes of the normal symbols 5 to 10 and the 2-bit codes of the hot symbol 2 (4 times), making 84 bits in total. In the runlength-inspired encoding, the first 10 bits contain the representation of H as in the Huffman-inspired encoding. This is followed by a block of 6 hot symbols, where the block length is given in 3 bits and the symbol codes have 1 bit each. Next comes a block of 3 normal symbols (2 bits for the block length, 8 bits for each symbol code). In total, there are 6 normal symbols in sequence, but as the maximum block size for normal symbols is 3, this block must be split, and an empty block of hot symbols is placed in-between. So the next 3 bits give the block of size 0, followed by another block of 3 normal symbols ($2+3\cdot 8$ bits) and a final block of 4 hot symbols ($3 + 4 \cdot 1$ bits), making 81 bits in total.

5 Evaluation

We evaluate our approach with payloads of different lengths ranging from 50 to 250 bytes. We use both synthetic content coming from different distributions (equi-distribution, Zipf distribution [15] and a near-equi-distribution where a small part of the symbols have a somewhat higher probability than all other symbols) and content coming from an ASCII text document.

In a first experiment, we explore the set H of hot symbols on an exemplary message m that comprises the first 250 characters of the Wikipedia text on lossless compression[3]. For the sequence of H_i, where $i = 1, \ldots, 32$, Table 1 depicts the fraction $\alpha(m, H_i)$ of the text covered by hot symbols, the maximum difference between successive elements of H_i, the number of bits $L(H_i)$ to represent H_i and the number $z(m, H_i)$ of blocks resulting from H in the runlength-inspired coding. Additionally, it also shows the bit lengths $L_1(m, H_i)$ and $L_2(m, H_i)$ of the two encodings together with their compression ratios L_1/L_0 and L_2/L_0, with the minimum H^* highlighted in bold.

We see that the maximum distance only shrinks slowly so that $L(H)$ grows linearly with $|H_i|$. The fraction α grows slowly beyond $i \geq 16$. Thus, a more complex selection of members for set H might only be of limited value in helping to shrink distances, i.e., reducing $L(H)$, without hurting $L_1(H)$ and $L_2(H)$ through notable reduction of α. Both L_1 and L_2 shrink with i until H_{16} is reached, and then grow again. Thus, the proposed optimization procedure seems suitable.

In a second experiment, we successively shorten the message m where m_j comprises only the first j characters of the Wikipedia text, where j ranges from

[3] https://en.wikipedia.org/wiki/Lossless_compression.

Table 1. Dependence of Huffman– and runlength-inspired encodings on the number of hot symbols for example message. Size of H is limited as only 25 symbols occur.

$\|H_i\|$	$\alpha(m, H_i)$	$\max \Delta_j$	$L(H_i)$	$z(m, H_i)$	$L_1(m, H_i)$	L_1/L_0	$L2(m, H_i)$	L_2/L_0
1	0.13	33	14	172	2033	1.02	2214	1.11
2	0.26	83	22	156	1824	0.91	1965	0.98
3	0.35	79	29	168	1757	0.88	1928	0.96
4	0.43	69	36	160	1644	0.82	1795	0.90
5	0.50	69	43	154	1663	0.83	1799	0.90
6	0.58	65	50	142	1580	0.79	1686	0.84
7	0.64	65	57	134	1512	0.76	1598	0.80
8	0.69	65	64	124	1454	0.73	1515	0.76
9	0.74	65	71	120	1585	0.79	1636	0.82
10	0.78	65	78	106	1552	0.78	1568	0.78
11	0.81	65	85	102	1527	0.76	1533	0.77
12	0.84	65	92	98	1502	0.75	1498	0.75
13	0.87	65	99	94	1481	0.74	1467	0.73
14	0.89	65	106	88	1464	0.73	1435	0.72
15	0.91	65	113	86	1451	0.73	1417	0.71
16	0.93	65	120	84	**1438**	0.72	**1399**	0.70
17	0.95	65	127	84	1666	0.83	1627	0.81
18	0.96	65	134	79	1664	0.83	1613	0.81
19	0.97	65	141	77	1662	0.83	1606	0.80
20	0.98	44	128	75	1643	0.82	1582	0.79
21	0.98	33	134	73	1646	0.82	1580	0.79
22	0.99	33	140	73	1649	0.82	1583	0.79
23	0.99	33	146	73	1652	0.83	1586	0.79
24	1.00	33	152	73	1655	0.83	1589	0.79
25	1.00	33	158	71	1658	0.83	1587	0.79

250 down to 30 in steps of 10. For a message of less than 30 bytes, no compression could be achieved. For each message size, Table 2 shows the optimum size of H together with the corresponding compression ratios for L_1 and L_2. The compression ratios slowly get worse with shrinking message size, that the optimal size of H is always a power of two, and that sometimes one, sometimes the other encoding has a small lead. Thus, one might consider to try both encodings and spend one additional bit of overhead to signal which encoding has been used.

For each message size, Table 3 shows the block sizes in the runlength-inspired encoding when optimal size of H is chosen. We see that for hot symbols, the number of blocks of maximum size 7 is much higher than the numbers for smaller block sizes for messages of size larger than 100. Together with many blocks of size

Table 2. Dependence of size of H and compression ratios on message length.

$\|m_j\|$	$\|H^*\|$ for L_1	L_1/L_0	$\|H^*\|$ for L_2	L_2/L_0
250	16	0.72	16	0.70
240	16	0.72	16	0.70
230	16	0.72	16	0.71
220	16	0.72	16	0.70
210	16	0.72	16	0.71
200	16	0.73	16	0.71
190	16	0.73	16	0.71
180	16	0.73	16	0.72
170	16	0.74	16	0.72
160	16	0.74	16	0.72
150	8	0.75	16	0.73
140	8	0.75	16	0.74
130	8	0.75	16	0.74
120	8	0.75	16	0.76
110	8	0.76	16	0.77
100	8	0.76	8	0.78
90	8	0.76	8	0.78
80	8	0.76	16	0.78
70	8	0.78	8	0.79
60	8	0.79	8	0.80
50	4	0.81	8	0.81
40	4	0.81	8	0.84
30	4	0.85	8	0.87

0 for normal symbols, this indicates that block sizes for hot symbols are mostly beyond 8, i.e., spending 4 bits for block length of hot symbols might be useful. For normal symbols, maximum block size of 3 seems sufficient, the number of blocks of hot symbols with length 0 is small and might already come from the situation when a message starts with a normal symbol while the encoding assumes a start with a hot symbol and inserts a zero-length block. For short messages of sizes 100 bytes and less, importance of hot symbols and consequently size of set H is smaller, yet still noticeable as the compression ratios in Table 2 indicate.

In a third experiment, we generate random messages of sizes 250, 150 and 50 according to three distributions: equi-distribution of byte values, Zipf distribution [15] ($p(s_0) = 1/H_{256}$ where H_n is the n-th harmonic number, $p(s_i) = p(s_0)/(i+1)$ for $i = 1, \ldots, 255$) and nearly equi-distribution, where 16 symbols have probability $(1+\varepsilon)/256$ and the remaining symbols $(1-\varepsilon/15)/256$,

Table 3. Distribution of block lengths for runlength-inspired encoding with different message lengths.

$\|m_j\|$	No. blocks of different lengths normal symbols				No. blocks of different lengths hot symbols							
	0	1	2	3	0	1	2	3	4	5	6	7
250	26	14	1	0	1	1	4	5	0	4	0	27
240	25	14	1	0	1	1	4	5	0	3	0	26
230	24	13	1	0	1	1	4	5	0	3	0	25
220	23	12	1	0	1	1	3	4	0	3	0	24
210	23	12	0	0	1	1	3	4	0	2	0	24
200	22	11	0	0	1	1	3	3	0	2	0	23
190	22	9	0	0	1	1	3	2	0	2	0	22
180	21	9	0	0	1	1	3	2	0	2	0	21
170	19	9	0	0	1	1	3	2	0	2	0	19
160	19	7	0	0	1	0	3	2	0	1	0	19
150	17	7	0	0	1	0	3	2	0	1	0	17
140	16	7	0	0	1	0	3	2	0	1	0	16
130	15	6	0	0	1	0	2	2	0	1	0	15
120	14	6	0	0	1	0	2	2	0	1	0	14
110	12	6	0	0	1	0	2	2	0	1	0	12
100	2	14	3	3	2	5	2	6	2	0	2	3
90	1	17	0	2	1	7	2	3	1	3	2	2
80	9	3	0	0	0	0	1	0	1	0	1	9
70	1	12	0	2	1	4	1	2	1	3	1	2
60	2	7	1	2	1	4	0	1	1	3	0	2
50	2	5	0	2	1	2	0	1	1	3	0	2
40	2	5	0	1	1	2	0	1	0	2	0	2
30	2	2	0	1	1	0	0	1	0	1	0	2

for $\varepsilon = 2, 3, 4$. A random permutation on the symbol indices is applied to get realistic distances between hot symbols.

Equi-distribution serves as an extreme case where no compression can be expected. Zipf distribution reflects human preferences in selecting symbols and, to some extent, also models symbol distribution in texts [10], and thus serves as a favorable case where good compression ratios can be expected. Near-equi-distributions explore the range between the former two distributions by checking via increasing values of ε how much the probabilities of hot symbols must deviate from those of normal symbols before our proposed encodings can achieve a compression. For each size and distribution, we generate 5 messages and average

Table 4. Dependence of compression ratio on symbol distribution and message length

Distr.	$\lvert m_j \rvert$	$\lvert H^* \rvert$ for L_1	$\overline{L_1/L_O}$	$\lvert H^* \rvert$ for L_2	$\overline{L_2/L_O}$
Equi	250	16	1.066	16, 31	1.146
	150	4, 8, 16	1.083	4, 27–29, 32	1.160
	50	1, 2, 4	1.107	1, 2, 4	1.192
Zipf	250	8	0.866	8, 16	0.924
	150	4, 8	0.886	8, 16	0.946
	50	2, 4, 8	0.936	4, 8	1.004
Near-equi	250	16	1.048	16, 32	1.134
$\varepsilon = 2$	150	4, 8, 16	1.068	4, 8, 26, 32	1.146
	50	2, 3, 4	1.100	2, 3	1.186
Near-equi	250	16	1.014	16, 32	1.086
$\varepsilon = 4$	150	8, 16	1.028	8, 16, 32	1.108
	50	3, 4	1.072	2, 4, 8	1.158
Near-equi	250	16	0.960	16	1.024
$\varepsilon = 6$	150	8, 16	0.992	8, 16, 31	1.056
	50	4, 7, 8	1.032	4, 6	1.124
Near-equi	250	16	0.892	16	0.944
$\varepsilon = 8$	150	8, 16	0.926	16	0.972
	50	4, 8	1.004	4, 8	1.082

Table 5. Entropy for original payloads m and compressed payloads m' with embedded encrypted secret message

Distr.	$\lvert m \rvert$	Entr(m)	L_1/L_O	Entr(m')
Zipf	250	5.56	0.87	7.05
	150	5.43	0.90	6.67
	50	4.68	0.96	5.43
Near-equi	250	6.41	0.96	7.19
$\varepsilon = 6$	150	6.10	0.97	6.62
	50	5.31	1.05	5.31
Text	250	4.07	0.72	6.88
Wikipedia	150	3.99	0.75	6.40
	50	3.62	0.81	4.93

the compression ratios. The differences in results are small so that this small number of repetitions seems sufficient. Table 4 shows the results.

We see that—not surprisingly—the encodings cannot achieve compression for equi-distributed random symbols. In contrast, for a Zipf distribution, at least the Huffman-inspired encoding can achieve a compression ratio less than 1 even

input : message m of s bytes, secret message sec
output: encoded message, or original message if no compression possible

foreach *symbol* a_i **do**
    f_i = absolute frequency of a_i in message m;
n = number of symbols with frequency ≥ 1;
$List$ = list of tuples (i, f_i) in decreasing order of frequency (order by index for identical freq.);
foreach $k \in \{1, \ldots, \min(n, 32)\}$ **do**
    H_k = first k symbols of $List$ re-ordered by index;
    $l_k = L_T(m, H_k)$; ▷ size of encoding type T with H_k, cf. (3), (4)
$k^* = \text{argmin}_k\, l_k$; ▷ optimal size
if $l_{k^*} + 16 \geq 8s$ **then** ▷ encoded message + magic num $\geq$ original
    return m; ▷ no compression possible
$H^* = H_{k^*} = (i_0, \ldots, i_{k^*-1})$;
$lk = \lceil \log_2(k^* - 1) \rceil$; ▷ bitsize of codes for hot symbols
$\Delta = \max_{0 \leq j < k^*}(i_j - i_{j-1} - 1)$ where $i_{-1} = -1$; ▷ compute max. distance
$ld = \lceil \log_2(\Delta) \rceil$; ▷ bitsize of distances
m_{enc} = empty bitstring;
enc $(m_{enc}, 5, k^* - 1)$; ▷ encode size of H
enc $(m_{enc}, 3, ld - 1)$; ▷ encode bitsize of distances
foreach $j \in \{0, \ldots, k^* - 1\}$ **do**
    enc $(m_{enc}, ld, i_j - i_{j-1} - 1)$; ▷ encode distances between H's elements
foreach $j \in \{0, \ldots, s - 1\}$ **do**
    if $m_j \in H^*$ **then**
        i = index of m_j in H^*;
        enc $(m_{enc}, lk + 1, i)$; ▷ encode hot symbol
    else
        enc $(m_{enc}, 9, 256 + m_j)$; ▷ encode normal symbol
enc $(m_{enc}, 8s - l_{k^*} - 16, sec)$; ▷ append piece of secret message
hv =Hash16(m_{enc}); ▷ create magic number as 16-bit hash
enc $(m_{enc}, 16, hv)$; ▷ append magic number
return m_{enc};

Algorithm 1: Implementation for Huffman-inspired encoding. Function enc(M, x, y) appends the x-bit binary representation of $0 \leq y < 2^x$ to bitstring M.

for short messages. In the near-equi-distribution, the Huffman-inspired encoding achieves a compression for some messages of length 150 and all messages of length 250 when $\varepsilon = 6$ and for all messages of length $\geq$150 when $\varepsilon = 8$. The runlength-inspired coding does slightly worse. We also see that sometimes, yet rarely, sizes for H occur that are not powers of 2 and that the optimal size of H for messages of similar length and distribution can vary greatly, i.e., should be determined individually and not fixed globally.

Finally, we test if the transcoding of a payload via the proposed compression together with the embedding of a secret message can be detected via a change in payload entropy. As payloads, we choose both the textual payload from

```
input  : message m of s bytes, secret message sec
output: encoded message, or original message if no compression possible

/* see lines 1 to 19 of Alg. 1                                        */
len[hot] = 3; len[normal] = 2;                ▷ bitsizes of block lengths
blocktype = hot;                          ▷ start with block of hot symbols
blocklen = 0;
foreach j ∈ {0, ..., s − 1} do
    symtype = normal;                             ▷ determine type of symbol
    if m_j ∈ H* then
        symtype = hot;
    if symtype ≠ blocktype then                ▷ detect end of previous block
        enc (m_enc, len[blocktype], blocklen);         ▷ encode size prev. block
        foreach v ∈ {−blocklen, ..., −1} do        ▷ no symbol if blocklen=0
            if blocktype = hot then
                i = index of m_{j+v} in H*;
                enc (m_enc, lk, i);                          ▷ encode hot symbol
            else
                enc (m_enc, 8, m_{j+v});                  ▷ encode normal symbol
        blocktype = symtype;
        blocklen = 0;
    blocklen = blocklen + 1;
    if blocklen = 2^{len[blocktype]} − 1 or j = s − 1 then ▷ max. blocksize or end
        enc (m_enc, len[blocktype], blocklen);       ▷ encode size actual block
        foreach v ∈ {−blocklen + 1, ..., 0} do       ▷ no symbol if blocklen=0
            if blocktype = hot then
                i = index of m_{j+v} in H*;
                enc (m_enc, lk, i);                          ▷ encode hot symbol
            else
                enc (m_enc, 8, m_{j+v});                  ▷ encode normal symbol
        blocktype = 1 − blocktype;                         ▷ Switch block type
        blocklen = 0;
enc (m_enc, 8s − l_{k*} − 16, sec);            ▷ append piece of secret message
hv =Hash16(m_enc);                       ▷ create magic number as 16-bit hash
enc (m_enc, 16, hv);                                    ▷ append magic number
return m_enc;
```

Algorithm 2: Implementation for Runlength-inspired encoding. Computation and encoding of set H^* is similar to Alg. 1 and thus skipped.

our first example and payloads generated from Zipf distribution and near-equidistribution with $\varepsilon = 6$. We use different message lengths. The secret message is encrypted, i.e., comprised of random byte values. Table 5 depicts the results.

We see that the entropy of the transcoded message is generally higher than that of the original message, yet the dependence on length, distribution and compression ratio is complex. Thus, if the symbol distribution and the payload length is known in detail, so that the entropy of messages can be predicted well, it should be possible to detect the presence of such a covert channel. However, even then false positives and negatives seem possible. For example, the entropy of the message with near-equi-distribution and 150 byte length after compression and embedding of the secret message is still in the range of the entropy for the original message of the same distribution and length 250 bytes.

6 Conclusions

We have presented two algorithms for source encoding, i.e., lossless compression, of short payloads in network packets. We have demonstrated that, as soon as the symbol distribution deviates from equi-distribution such as in text-based payloads, a compression can be achieved even for 30 byte payloads. We have deployed this type of compression for the transcoding of network packet payloads in order to free some space in the payload. This freed space is used to embed a secret message or a part thereof. The steganographic bandwidth ranges from 5 to 30% of the payload length with higher bandwidth for longer messages and more uneven distributions. We have provided a combination of means that allow the covert receiver to reliably decide if a packet has been used for the payload transcoding and contains a part of the secret message, or if the payload has been left unmodified.

We have checked if this transcoding can be detected via measuring the entropy of network packet payloads and comparing to the entropy to be expected in "normal" traffic. If the normal traffic can be defined quite exactly via its symbol distribution for different message lengths, then a detection seems possible, at least for the considered case that the secret message is encrypted, i.e., looks like randomly chosen bytes. However, as the entropy varies notably between payloads of different distributions and lengths, without that knowledge a detection via entropy alone seems not reliable and must be accompanied by other countermeasures, such as ensuring that the covert receiver does not have the rights to re-construct the original payload after extraction of the secret message. Such countermeasures seem necessary as statistical detection alone seems difficult because the encoding of the secret message might try to mimic the look of the carrier, for example be adapted to the entropy of the original message by only encoding 6 secret message bits per payload byte and choosing an encoding that uses symbols that are used (or likely to be used) in the original payload. Still, even in this case, the countermeasures serve to reduce steganographic bandwidth, e.g., by a factor of 0.75 in the above example.

Future work will comprise an extension of the evaluation to more symbol distributions (e.g., geometric) and to more real-world payload examples. Furthermore, we will explore further countermeasures such as Cabuk's ε-similarity and compressibility score [1] adapted to storage channels, adversarial learning approaches and anti-countermeasures such as adapted secret message encodings For example, common bit patterns at the start of the encoding could be avoided by a light-weight encryption with a key derived from a packet parameter such as the sequence number. Finally, we will try to deploy our method in a prototype and compare it with state-of-the-art approaches for compression. Example scenarios could be a DNS-based covert channel or an MQTT transfer of short measurement packets where covert sender and receiver on the communication path can modify and re-construct message payloads.

References

1. Cabuk, S., Brodley, C.E., Shields, C.: IP covert channel detection. ACM Trans. Inf. Syst. Secur. **12**(4), 22:1–22:29 (2009). https://doi.org/10.1145/1513601.1513604
2. Huffman, D.A.: A method for the construction of minimum-redundancy codes. Proc. IRE **40**(9), 1098–1101 (1952). https://doi.org/10.1109/JRPROC.1952.273898
3. Lampson, B.W.: A note on the confinement problem. Commun. ACM **16**(10), 613–615 (1973). https://doi.org/10.1145/362375.362389
4. Lee, C., Chen, H.: Lossless text steganography in compression coding. In: Pan, J., Huang, H., Jain, L.C., Zhao, Y. (eds.) Recent Advances in Information Hiding and Applications, Intelligent Systems Reference Library, vol. 40, pp. 155–179. Springer, Heidelberg (2013). https://doi.org/10.1007/978-3-642-28580-6_8
5. Malik, A., Sikka, G., Verma, H.K.: A high capacity text steganography scheme based on LZW compression and color coding. Eng. Sci. Technol. Int. J. **20**(1), 72–79 (2017). https://doi.org/10.1016/j.jestch.2016.06.005
6. Mazurczyk, W., Szaga, P., Szczypiorski, K.: Using transcoding for hidden communication in IP telephony. Multimedia Tools Appl. **70**(3), 2139–2165 (2014). https://doi.org/10.1007/S11042-012-1224-8
7. Mazurczyk, W., Szary, P., Wendzel, S., Caviglione, L.: Towards reversible storage network covert channels. In: Proceedings of the 14th International Conference on Availability, Reliability and Security, ARES 2019, Canterbury, UK, 26–29 August 2019, pp. 69:1–69:8. ACM, New York (2019). https://doi.org/10.1145/3339252.3341493
8. Mazurczyk, W., Wendzel, S., Cabaj, K.: Towards deriving insights into data hiding methods using pattern-based approach. In: Doerr, S., Fischer, M., Schrittwieser, S., Herrmann, D. (eds.) Proceedings 13th International Conference on Availability, Reliability and Security (ARES), pp. 10:1–10:10. ACM (2018). https://doi.org/10.1145/3230833.3233261
9. Mazurczyk, W., Wendzel, S., Zander, S., Houmansadr, A., Szczypiorski, K.: Information Hiding in Communication Networks: Fundamentals, Mechanisms, and Applications. Wiley-IEEE Press (2016)
10. Piantadosi, S.T.: Zipf's word frequency law in natural language: a critical review and future directions. Psychon. Bull. Rev. **21**(5), 1112–1130 (2014). https://doi.org/10.3758/s13423-014-0585-6
11. Salomon, D.: Data Compression-The Complete Reference, 4th edn. Springer, Heidelberg (2007)
12. Satir, E., Isik, H.: A Huffman compression based text steganography method. Multimedia Tools Appl. **70**(3), 2085–2110 (2014). https://doi.org/10.1007/S11042-012-1223-9
13. Wendzel, S., et al.: A revised taxonomy of steganography embedding patterns. In: Reinhardt, D., Müller, T. (eds.) Proceedings 16th International Conference on Availability, Reliability and Security (ARES), pp. 67:1–67:12. ACM (2021). https://doi.org/10.1145/3465481.3470069

14. Wendzel, S., Zander, S., Fechner, B., Herdin, C.: Pattern-based survey and categorization of network covert channel techniques. ACM Comput. Surv. **47**(3), 50:1–50:26 (2015). https://doi.org/10.1145/2684195
15. Zipf, G.K.: The Psychobiology of Language. Houghton-Mifflin, New York (1935)
16. Ziv, J., Lempel, A.: A universal algorithm for sequential data compression. IEEE Trans. Inf. Theory **23**(3), 337–343 (1977). https://doi.org/10.1109/TIT.1977.1055714

Examining the Influence of Cultural Diversity in Cybersecurity Culture

Mohammed Aljohani(✉), Steven Furnell, Xavier Carpent, and Nicholas Gervassis

Cyber Security Research Group, University of Nottingham, Nottingham, UK
{mohammed.aljohani1,steven.furnell,xavier.carpent,
nicholas.gervassis}@nottingham.ac.uk

Abstract. Global organisations face complex cybersecurity challenges that require robust, culturally inclusive strategies. This study addresses the need to understand cultural diversity's role in strengthening cybersecurity practices, an area often overlooked in cybersecurity practice. We conducted a thematic literature review across 18 studies, focusing on publications from databases in cybersecurity and organisational behaviour. The study found themes of education and awareness, human and organisational behaviour, cybersecurity strategy, national culture in cybersecurity, and human-centric aspects. The findings suggest incorporating perspectives that enhance resilience, enrich cybersecurity policies, and support a collaborative and inclusive cybersecurity environment. The findings enable organisations to tailor and expand their strategies for each threat or attack from those identified themes.

Keywords: Cultural Diversity · Cybersecurity Culture · Thematic Analysis · Organisational Behaviour · Security Awareness · Human-Centric Factors

1 Introduction

Modern organisations commonly include a diverse mix of staff from a cultural perspective. For example, multinational businesses may have operations across a range of geographic locations, with the natural consequence that they will have an internationally diverse staff base. Similarly, any organisation of sufficient size has the potential to find that it has a culturally diverse staff community, even with a single location. From a security perspective, the potential for staff from different backgrounds brings potential considerations in terms of the attitudes and beliefs that they may hold (e.g. in relation to issues such as rules, authority, and privacy) and the implications that this may have for their default security behaviours. Similarly, with a diversity of staff, it would be reasonable to assume that certain related security efforts, such as awareness and training, could usefully be framed differently to suit different recipients.

The variety of perspectives and ways of communicating within a group can affect how people perceive security measures and training results, where organisations need to recognise and utilise these interactions to enhance their security practices [1, 2]. Thus, cybersecurity solutions should account for team and user diversity within an organisation,

I. Praça et al. (Eds.): EICC 2025, CCIS 2500, pp. 151–165, 2025.
https://doi.org/10.1007/978-3-031-94855-8_10

as these factors can significantly influence cybersecurity strategies and practices. Cultural diversity, which encompasses ethnic, language, and national background differences, influences how employees perceive and respond to cybersecurity measures [3, 4].

Developing cybersecurity culture involves recognising employee diversity and understanding that cultural diversity could influence cybersecurity practices [5]. Moreover, recognising and integrating existing diversity within the organisation can enhance cybersecurity by fostering innovative problem-solving [3, 4]. However, it also presents challenges, such as varying perceptions of security measures and differing responses to training. This review investigates the impact of cultural diversity on cybersecurity culture, contributing by (1) identifying diversity elements that influence practices, (2) examining diversity's role in policy and training, and (3) assessing its effects on awareness.

This review applies thematic analysis and explores multi-dimensional themes of cybersecurity culture in global organisations. Understanding and actioning diverse viewpoints to strengthen cybersecurity awareness, policy implementation, and training effectiveness across various cultural contexts. The following research questions guide this review:

(1) What specific cultural diversity elements may impact cybersecurity practices within organisations?
(2) How does cultural diversity influence the design, implementation, and effectiveness of cybersecurity policies and training programs in organisations?
(3) How does organisational cultural diversity enhance and challenge cybersecurity awareness, resilience, and response to cyber threats?

Overall, this study enhances the understanding of cultural diversity in cybersecurity culture and offers a structured thematic analysis to support organisations in developing more inclusive and effective security strategies.

The paper is organised as follows: Sect. 2 offers a comprehensive review of related work that discusses the impact of cultural diversity on cybersecurity culture. Section 3 describes the review methodology, where criteria for selecting studies, data collection, and analysis are presented. Analysis results are presented in Sects. 4 and 5, and research questions are discussed. The study concludes in Sect. 6 with practical implications summarised and suggestions for future research provided.

2 Related Work

Understanding the relationship between cultural diversity and cybersecurity culture is an emerging area of research that integrates insights from cybersecurity, organisational behaviour, and cultural studies [6, 7]. This section reviews the existing literature on these topics.

2.1 Cultural Diversity in Organisations

Cultural diversity refers to various cultural backgrounds, including race, gender, age, colour, religion, and disability. Race refers to broad social categories based on ancestry and shared physical traits, while colour pertains to variations in skin tone within racial groups, which can influence social perceptions and socioeconomic outcomes [8]. Culturally diverse workplaces feature a mix of workers with different ethnic backgrounds, ages, genders, and abilities. A widely regarded important factor is the influence of diversity on behaviour [3].This study explores the relationship between cybersecurity culture and individuals' ethnic and cultural backgrounds by examining how the organisational makeup (in terms of staff composition) impacts cybersecurity practices.

Recent diversity reports show that many organisations have a significant presence of ethnic and cultural minorities [9]. For example, Google's workforce is 51.7% White, 39.3% Asian, 4.9% Latinx (a term that includes all those who identify as Latinx, Latino, Latina, or Hispanic), and 3.1% Black [10]. Similarly, Meta's U.S. workforce consists of 37.6% White, 46.5% Asian, 6.7% Hispanic, and 4.9% Black employees, with smaller representations of other groups [9]. Twitter and Pinterest also display varied ethnic compositions, reflecting broader organisational diversity trends in staff composition.

Understanding the relationships between cybersecurity culture and individuals' ethnicity or cultural background is crucial, as these factors can affect how cybersecurity measures are implemented and perceived. Language, ethnic, and racial characteristics can reveal vulnerabilities that may expose certain groups to cyberattacks more [11]. For instance, non-native speakers may ignore phishing emails written in complex language, making them more susceptible to such attacks [12]. Similarly, cultural norms related to authority may lead individuals from high power distance cultures to comply with phishing attempts that appear to come from senior figures, highlighting the need for tailored interventions. However, in evaluating cybersecurity vulnerabilities, one should consider national culture and human factors to assess how culture influences cybersecurity practices [13]. The literature discusses the impact of cultural diversity on organisational practices and the necessity of increasing inclusion to improve overall performance [3, 4]. Studies have shown that including diverse viewpoints, such as ethnic, linguistic, and gender viewpoints, can enhance decision-making, innovation, and resilience within organisations [14, 15]. However, achieving these benefits needs dedicated attempts at employee engagement, tailored training, and leadership support for an inclusive culture [1]. These foundational principles give knowledge of the organisational aspects of writing strong cybersecurity strategies [2, 13, 16]. Thus, understanding cultural diversity cultivates a resilient cybersecurity culture that leverages leadership, accountability, and employee participation.

2.2 Types of Diversity and Their Impact on Cybersecurity Culture

By recognising the diverse aspects of organisational composition—such as workforce demographics and team dynamics—we can develop more effective and inclusive cybersecurity strategies by employing all employees' unique perspectives and backgrounds. This inclusive strategy leverages each diversity subgroup's distinct capabilities and experiences to foster a robust cybersecurity culture and address complex cybersecurity challenges. Building on the understanding of diversity subcategories. Several recent reviews have explored factors influencing cybersecurity culture within organisations, such as leadership and training programs [6, 13]. While some studies [14, 17] have touched on the role of cultural diversity, highlighting its potential to enhance collaboration, innovation, and adaptability, they often lack a specific focus on cybersecurity practices.

Therefore, a gap in the literature exists regarding the exact role of cultural diversity in shaping cybersecurity practices.

3 Methodology

This review follows the guidelines of Tranfield method [19], which is well-suited for studies requiring integrating diverse perspectives and evidence across interdisciplinary fields. Initially developed for management research, the method is particularly valuable for addressing complex, organisational-level issues like cultural diversity in cybersecurity, where evidence needs to be systematically gathered and analysed. The literature search was conducted at the end of 2023. The search string spans 2005 and 2023, ensuring a balance of foundational and contemporary perspectives. Given the exploratory nature of this study, the goal is not to conduct a comprehensive systematic review but rather to identify key factors shaping cybersecurity culture in the context of cultural diversity. The method includes three main phases: planning, conducting the review, and synthesising findings. The detailed reporting of results is postponed to the results section.

3.1 Phase 1: Planning the Review

Scoping Review. This review explores how cultural diversity affects cybersecurity culture. The primary research questions guide the review by focusing on (a) identifying cultural diversity elements that influence cybersecurity practices, (b) understanding the role of diversity in shaping cybersecurity policies and training, and (c) examining how diversity contributes to or challenges cybersecurity awareness.

Outline Review Protocol. Only peer-reviewed studies indexed in Scopus and IEEE Xplore were included to ensure methodological rigour. Studies from Google Scholar were cross-verified against these databases to confirm their peer-reviewed status before inclusion. The search strategy involved specific terms such as 'cybersecurity culture,' 'organizational cybersecurity,' 'cultural diversity in cybersecurity,' 'ethnicity and cybersecurity,' and 'human factors in cybersecurity,' combined using Boolean operators (e.g., AND, OR) to refine search results across multiple databases. This protocol includes:

- *Inclusion and Exclusion Criteria.* Studies were included if they (a) focused on cultural diversity within the context of cybersecurity, (b) were published in peer-reviewed journals, and (c) explored how cultural diversity influences organisational cybersecurity practices, training, or decision-making. These inclusion criteria were chosen to ensure the studies directly addressed the core objectives of the review, leveraging existing research that explicitly connects cultural diversity and cybersecurity. Older studies were included if they were foundational or highly cited in cybersecurity culture research, ensuring relevance. Additional articles were identified through citation tracking and expert recommendations to complement the database search results. Studies were excluded if they lacked an organisational focus, did not explore cultural diversity in a cybersecurity context, or were not peer-reviewed, as such studies were less likely to provide robust evidence relevant to this interdisciplinary focus. While this approach may exclude papers making new cultural diversity links outside the cybersecurity context, the aim was to focus on established intersections to build a targeted and cohesive review.

- *Source Selection.* We selected Google Scholar, IEEE Xplore, and Scopus for their extensive coverage of cybersecurity, organisational behaviour, and cultural diversity research, with Scopus prioritised over Web of Science for broader interdisciplinary indexing. The reviewed studies were published between 2005 and 2023, ensuring a balance of foundational and contemporary perspectives on cultural diversity in cybersecurity. Google Scholar and Scopus, both indexes, provided broad interdisciplinary coverage, with Scopus focusing on peer-reviewed sources. IEEE Xplore was chosen as a repository for its emphasis on cybersecurity within engineering and technology. Organisational behaviour literature was accessed primarily through Google Scholar and Scopus. Keywords were selected to capture the review's focus on cybersecurity culture and diversity. These included terms like "cybersecurity culture," "organisational cybersecurity," "cultural diversity in cybersecurity," "ethnicity and cybersecurity," and "human factors in cybersecurity." Boolean operators (e.g., AND, OR) were used to refine research and include variations of critical terms.
- *Screening, Quality Assessment, and Extraction.* We follow the following process to increase the quality of reviewed articles. The collected study titles and abstracts were screened for relevance to the research questions through two independent authors to minimise selection bias and ensure alignment with inclusion criteria. Discrepancies were addressed through discussions between authors. Thus, our review paper's quality assessment was conducted based on research design clarity and reliability aligned with research questions. During the screening and quality assessment, titles, publishers, and year of publication are recorded. This structured extraction enabled consistent data collection across studies. Studies from the last 5 years were prioritised to ensure an up-to-date review, with a few seminal works included to provide foundational perspectives. These works were identified based on their frequent citation, foundational contributions, and relevance to cultural diversity and cybersecurity, providing necessary context and theoretical grounding.

3.2 Phase 2: Conducting the Review

After screening titles and abstracts to exclude irrelevant studies, we collect each study's extracted information in a table aligned with the research questions; therefore, significant findings on diversity's impact on cybersecurity are collected. Furthermore, the studies were evaluated for quality before synthesis by assessing their methodological rigour—such as the clarity of research design, the robustness of data collection, and the validity of findings—and their direct relevance to addressing the research questions. In addition, we took a simple approach to validation by checking reliability, gathering some feedback, and getting expert input to ensure the identified themes made sense, as described in our methods [20]. Only studies with reliable data, clear inclusion criteria, and a systematic thematic analysis (Fig. 1) were synthesised to ensure methodological rigour and alignment with the research focus. The final papers underwent an inductive coding approach, where themes were iteratively identified and manually coded by two researchers. To enhance the study's reproducibility, the extracted data were categorised using a predefined codebook, ensuring consistency in theme identification and alignment with the research questions. The codebook structure facilitated systematic classification, enabling comparisons across studies and ensuring traceability in data synthesis.

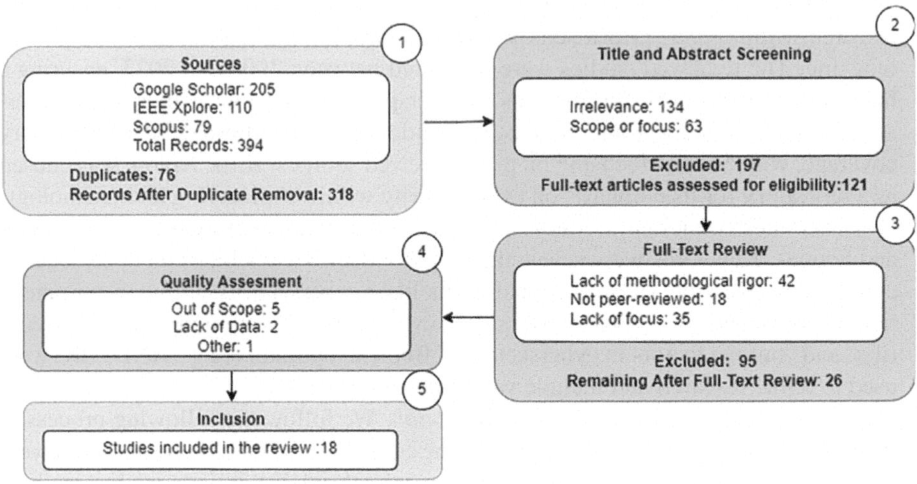

Fig. 1. Systematic review process illustrating records identified, screened, and included, with final selections based on quality criteria, including methodological rigour, relevance, and contribution to thematic analysis.

3.3 Phase 3: Synthesizing Findings

The synthesis prioritised studies with clear links to organisational security outcomes, focusing on diverse workforce impacts on cybersecurity awareness and policy adherence. While this study does not aim to provide an exhaustive systematic review, it identifies key factors shaping cybersecurity culture through cultural diversity. We acknowledge that limitations include potential publication bias due to the choice of databases, possible keyword bias, and the constraint of using primarily English-language sources. These limitations were considered when interpreting the findings. This approach allows for a focused and comprehensive presentation of the review's cybersecurity culture and diversity insights.

4 Results

Following the Tranfield method discussed in Sect. 3, Fig. 1 illustrates the systematic review process, detailing the identification, screening, and selection of studies at each stage. The initial set of 394 articles was progressively refined based on relevance, methodological rigour, and alignment with the research focus, culminating in the synthesis of 18 high-quality studies.

Table 1 identifies factors that impact cybersecurity culture within Organisations through a literature survey. These factors were carefully derived from a thematic analysis of key studies, emphasising diversity dimensions documented in research exploring their influence on organisational cybersecurity practices. While this list is not exhaustive, it represents the most commonly discussed categories in the literature that are crucial for fostering an inclusive and effective cybersecurity culture.

Table 1. Relationship of diversity with cultural diversity, where "core aspects" directly influence inclusion and policies, while "related aspects" enhance problem-solving.

Type of Diversity	Description
Ethnic and Racial Diversity [14]	It involves people of different ethnic and racial origins, crucial to understanding and respecting the various aspects
Linguistic Diversity [14]	Besides the diversity of languages spoken within the organisation, it improves the combination and inclusion of multicultural environments
Socioeconomic Diversity [17]	Recognises differences in the social and ecological contexts among employees, including perspectives and experiences in their approach to cybersecurity
Demographic Diversity [18]	It includes differences in age, gender, and ethnicity and directly affects organisational policies and practices in cybersecurity
Cognitive and Educational Diversity [15] (Related Aspect)	It brings a variety of thinking processes and educational backgrounds to contribute to innovative solutions for complex cybersecurity challenges

These studies provided the foundation for the thematic analysis, as depicted in Fig. 2, which identifies key themes (and their decompositions) as follows: Education and Awareness [6, 20–23], Human and organisational Behaviour [5, 7, 16, 24], Cybersecurity Strategy [3, 6, 17, 25], National Culture in Cybersecurity [15, 20, 26], and Human-Centric Aspects [14, 18, 20, 27]. These themes were derived through an inductive coding process, where recurring concepts and relationships were identified across the 19 studies included in the review. Each theme reflects varying degrees of support based on the number of studies addressing them and the depth of evidence provided. For instance, Education and Awareness emerged as a prominent theme, supported by multiple studies emphasising tailored training, while Human-Centric Aspects were highlighted for their role in improving adaptability and personal responsibility in cybersecurity practices.

4.1 Education and Awareness

Education and Training Awareness (SETA) builds the cybersecurity culture through awareness and proactive security measures. SETA training programs should be designed to address particular weak points and encourage a flexible response to threats from employees [20] where SETA activities should be systemic, embedded in the organisation to ensure that awareness is enhanced and people comply with the set security policies through frequent sensitisation and engagement [21]. The training should be such that it can meet the learning needs and wishes of most workers, regardless of their country, culture, or language [22]. However, effective SETA programs require content aligned with employees' linguistic and cultural backgrounds to address diverse perceptions [23]. Thus, cultural diversity should be incorporated into SETA strategies, for example, by providing multilingual resources or adapting training to regional contexts. This creates an inclusive environment to increase the likelihood of people supporting the cybersecurity mission and ensure consistency in applying policies and preparations to counter threats.

4.2 Human and Organisational Behaviour

The foundation of a good cybersecurity culture lies in human and organisational behaviour because it determines the attitude of employees, their willingness to follow the rules, and the security measures they enforce. Such a culture can be developed only through interventions that consider challenges encountered in a multicultural organisation to design strategies that fit different cultural orientations, including power distance and collectivism, in a way that promotes the enforcement of secure behaviour and trust [16]. These strategies must also incorporate cultural factors into the organisation's systems and processes because people's perception of risk and authority significantly influences the implementation of the cybersecurity policy [24]. In addition, the gaps of inadequate training, organisational misalignment, and insufficient awareness should be covered through inclusive initiatives. Training employees in culture and language can help overcome these gaps and increase employee engagement to understand security needs [7]. Organisations can develop strength by consistently evaluating and changing their cybersecurity culture through collaborative efforts and context-specific approaches to building resilience against evolving cyber threats while promoting accountability, trust, and alignment with their overall culture and goals.

4.3 Cybersecurity Strategy

Cybersecurity strategies are integral to managing risks, enhancing resilience, and fostering effective policy development in diverse organisational settings [13]. Effective strategies emphasise employee participation and strong management commitment, creating a security-conscious culture where all stakeholders contribute actively to safeguarding organisational assets [25]. Organisational cultural diversity enhances adaptability, encourages collaboration, and fosters innovation in strategic planning, which is vital for addressing diverse threats [3]. Additionally, contextualising cybersecurity frameworks to align with cultural and socio-economic realities ensures relevant policies resonate with the workforce, strengthening adherence and effectiveness [17]. A strong cybersecurity strategy must also prioritise continuous improvement through knowledge sharing, cross-sector collaboration, and inclusive approaches incorporating localised training and resources tailored to linguistic and cultural contexts. For instance, the EU Cybersecurity Strategy underscores the importance of enhancing cooperation between member states, fostering public-private partnerships, and investing in cybersecurity awareness and skills development to build long-term resilience [28]. These measures are crucial for bridging gaps in policy application and ensuring that strategies are robust and adaptable to the rapidly evolving threat landscape [6]. By embedding cultural awareness into cybersecurity strategies, organisations can transform diversity into an asset, fostering collective resilience and creating inclusive security cultures that promote innovation, trust, and shared accountability.

4.4 National Culture in Cybersecurity

National culture encompasses shared societal values and institutional practices within a country and shapes organisational cybersecurity practices by influencing leadership styles, resource allocation, and employee behaviour. Therefore, integrating national cultural values into cybersecurity strategies fosters organisational accountability and individual engagement [26]; flexible leadership frameworks ensure that security measures remain effective across diverse socio-economic and cultural contexts [15]. Moreover, government-led cybersecurity campaigns enhance public awareness of threats and foster compliance with standards, which are pivotal for societal engagement in cybersecurity initiatives [20]. Additionally, performance indicators (KPIs) could be employed to assess cybersecurity measures by identifying vulnerabilities [4]. Thereby, organisations could adjust policies to better align with evolving threats and workforce diversity [6]. In practice, this involves embedding cultural values into training programs and organisational policies, strengthening adherence to cybersecurity policies and fostering resilience and collaboration within diverse teams.

4.5 Human-Centric Aspects

Cultural Cybersecurity involves adopting heterogeneous industrial approaches incorporating sensitivities and individual liability toward building flexible security frames. Organisations foster creativity and innovation in cybersecurity by addressing varying technical expertise and cultural differences. This approach involves tailoring training programs to employees' cultural and linguistic backgrounds [7, 20]. The Leadership aspect embeds strategies by fostering an environment where employees feel empowered to take ownership of security practices and actively contribute to the organisation's cybersecurity mission [27]. Moreover, the human-centric perspective ensures that cultural nuances, such as communication styles and perceptions of authority, are accounted for in policy design and implementation, creating a cohesive and inclusive security culture [14].

5 Discussion

This section explores the development of cybersecurity culture and highlights its limitations.

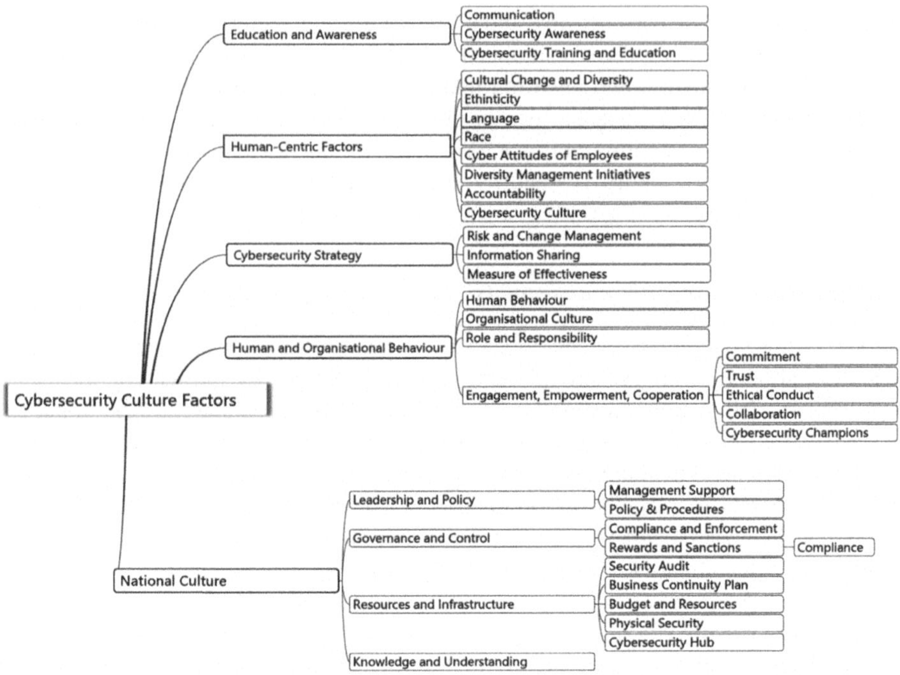

Fig. 2. Cybersecurity Culture Factors in Organisations.

5.1 Developing Cybersecurity Culture

This section addresses RQ1 by examining cultural diversity elements. These cultural themes are essential aspects that shape the development of a strong cybersecurity culture, which affects organisational factors, norms, and cybersecurity practices. The themes can influence different sectors and the size of an organisation differently. In an SME, the focus may be on protecting tangible assets due to limited resources, whereas larger organisations, despite having more significant resources, may still face constraints in terms of budget allocation, competing priorities, and scalability challenges. Hence, cybersecurity implementations must be reoriented toward the organisation's objectives and demands from the sectors and levels of operations, considering the workforce's cultural diversity to ensure that policies are applicable and practical. Leading cybersecurity professionals deliver continuous training specific to their workforce's varied educational needs and cultural settings to build a strong security culture within their organisations. Regrettably, though, support for cybersecurity among different companies is also a function of their size, structure, and cultural dynamicity. Hence, culturally aligned strategies across different scales of Organisations are needed to ensure that the plan is inclusive. In practice, less significant organisations would have a higher extent of adaptability of policies to make room for varied team dynamics, contrasting with more extensive organisations where bureaucratic structures necessitate an all-encompassing strategy that integrates cultural considerations in implementing policies. For example, healthcare and financial spaces, industries with specific requirements for data security and compliance, need to initiate training that caters to perceptions of risks and backgrounds of a workforce of various cultures. However, an effective cybersecurity culture for robust sectors, such as manufacturing, would entail building on the collaboration with cultural diversity and respect for which efforts to enhance security practices are based. Recognising that each institution has unique cultural characteristics and challenges is essential for identifying opportunities to improve cybersecurity practices, ultimately fostering a more inclusive and resilient cybersecurity culture [7].

Cultural diversity should be included in cybersecurity strategies at both organisational and national levels by shaping security policies, training programs, and awareness campaigns to account for diverse cultural perspectives, as it affects perceptions toward cybersecurity and leadership styles. However, we acknowledge that organisations often prioritise immediate security risks and resource constraints, which may limit the emphasis on cultural diversity within cybersecurity programs.

Moreover, diverse perspectives in accountability programs and advocacy teams enhance transparency and trust, fostering a culture of shared responsibility where individuals and teams collaborate effectively to achieve cybersecurity goals [22, 26]. Policies should encompass risk and communication styles to reach broader security goals, with leadership playing a key role in fostering an inclusive cybersecurity culture and ensuring policy adoption. Through that, organisations allocate resources effectively and build infrastructure that may best serve the specific needs of their diverse workforce [6, 21, 23]. A culture of cybersecurity resiliency would include industry-specific risks, leadership commitment, and culturally informed strategies that would help them effectively meet unique challenges and priorities [22]. Hence, such policies and practices should not just work towards eliminating threats but also create an environment in which diverse teams work actively to achieve the organisation's cybersecurity goals.

5.2 Integrating Cultural Perspectives in Cybersecurity Training and Awareness

This section relates to RQ2 by discussing how cultural diversity affects cybersecurity. Integrating different aspects of cultures in the cybersecurity training program may result in notable enhancements in security awareness and organisational preparedness. Cultural factors, for instance, trust and communication norms, can impinge on cybersecurity risks; for example, individuals are not confident of exposing any vulnerability, especially when it relates to the team or management [16]. Furthermore, leadership styles and balancing individualism with collectivism impact policy acceptance regarding cybersecurity and teamwork dynamics [29]. The infiltration of cultural factors in general risk assessment to heighten the organisation's commitment to security practice duly reinforces organisational resilience through engaging planned behaviour theory. Other factors include but are not limited to ethnicity, language, and cultural background influencing employee roles and ethical responsibilities [30]. Proactive diversity management, such as providing multilingual resources and culturally sensitive training, enhances employee engagement and strengthens the cybersecurity culture [20]. Additionally, factors such as ethnicity, language, and cultural background highlight the need for tailored cybersecurity training, as these factors shape employees' perceptions of their roles and ethical responsibilities [16]. However, Proactive diversity management, like offering multilingual resources and culturally sensitive training materials, turns challenges into strengths by enhancing employee engagement and building a strong cybersecurity culture [6].

5.3 Practical Strategies for Implementing Cultural Diversity in Cybersecurity

This section responds to RQ3 by identifying practical strategies that leverage cultural diversity. In enhancing cybersecurity, organisations should implement more inclusive strategies to accommodate different communication styles, risk perceptions, and attitudes among individuals with varied cultures. The earlier theme of human-centric aspects underscores the importance of tailoring training programs and communication to align with employees' diverse cultural and linguistic needs. Several approaches could potentially be taken, including, for example:

- The translation of training materials or interpreters can be utilised to prevent misunderstandings during single-language cybersecurity training sessions for explicit guarantees that all linguistically different employees understand security protocols [22]. This approach directly addresses the theme of education and awareness theme by ensuring inclusivity in training and enhancing employee engagement.
- The difference in risk perception due to culture also affects cybersecurity vigilance; workers in high-risk parts may be more sensitive about threats than their counterparts in low-risk regions. As such, regionally specific campaigns improved upon the existing differences by highlighting the entire team's level of risk awareness regarding these differences [7]. This aligns with the theme of national culture in cybersecurity, which emphasises adapting policies and initiatives to varying cultural and regional contexts.
- An anonymous reporting system would help to override those cultural practices that do not encourage questioning authority, assisting the employee to report security issues without fear of retaliation [13]. This recommendation connects to the theme of human and organisational behaviour by fostering an environment of trust and shared responsibility.

In aggregate, these steps enhance cybersecurity and help create an environment where heterogeneous opinions support organisational resilience [6, 20]. Organisations can foster a more inclusive, adaptive, and resilient cybersecurity culture by integrating these strategies within the broader themes of cultural diversity, human-centric aspects, and organisational behaviour.

5.4 Limitations

This review has limitations, including the potential biases brought about by the choice of databases and the over-representation of English-language publications. However, such limitations were constrained by the selection of cautionary, reputable databases such as Google Scholar, Scopus, and IEEE Xplore, among others, that guarantee the availability of high-quality peer-reviewed cybersecurity and cultural diversity research. The other reason is that most of the significant works in this field are published in English. Therefore, more impactful and relevant research has been captured. These limitations reflect the scope and methodology of this review, focusing on the selection of sources and language biases rather than broader claims about the overall development of cybersecurity culture. While there might be some information in non-English sources or smaller databases, their potential contribution is unlikely to significantly impact the findings, as this review focuses on the most authoritative and impactful research. Consequently, the inferences are sound and accurately reflect the basic patterns and challenges in infusing cultural diversity into cybersecurity practices.

6 Conclusion

This study has identified the importance of cultural diversity in enhancing organisational cybersecurity measures. Thus, incorporating cultural orientations will likely result in improved policy acceptance, increased employee involvement, and readiness for cyber risks. The findings from the analysed themes, including education, organisational behaviour, and human-centred aspects, illustrate how cultural differences in communication processes, risk acceptance, and authority imply a stronger culture of cybersecurity. Such a culture enables people to participate actively in the collective security effort, thus strengthening people-based and technology-based defences. Culturally inclusive strategies are promising areas for further exploration, offering practical guidance to organisations aiming to enhance their cybersecurity measures effectively.

References

1. Hassan, N.H., Maarop, N., Ismail, Z., Abidin, W.Z.: Information security culture in health informatics environment: a qualitative approach. In: 2017 International Conference on Research and Innovation in Information Systems (ICRIIS), pp. 1–6 (2017)
2. Van't Wout, C., Van't Wout, C.: Develop and maintain a cybersecurity organisational culture. In: 14th International Conference on Cyber Warfare and Security (ICCWS), pp. 457–466 (2019)
3. Cox, T.: Cultural Diversity in Organizations: Theory, Research and Practice. Berrett-Koehler Publishers (1994)
4. Ayega, E.N., Muathe, S.: Critical review of literature on cultural diversity in the workplace and organizational performance: a research agenda. J. Hum. Resour. Manag. **6**, 9–17 (2018)
5. Onumo, A., Ullah-Awan, I., Cullen, A.: Assessing the moderating effect of security technologies on employees compliance with cybersecurity control procedures. ACM Trans. Manag. Inf. Syst. (TMIS) **12**, 1–29 (2021)
6. Uchendu, B., Nurse, J.R.C., Bada, M., Furnell, S.: Developing a cyber security culture: Current practices and future needs. Comput. Secur. **109**, 102387 (2021). https://doi.org/10.1016/j.cose.2021.102387
7. Branley-bell, D., Coventry, L., Sillence, E.: Promoting cybersecurity culture change in healthcare. In: The 14th PErvasive Technologies Related to Assistive Environments Conference, pp. 544–549 (2021)
8. Burton, L.M., Bonilla-Silva, E., Ray, V., Buckelew, R., Hordge Freeman, E.: Critical race theories, colorism, and the decade's research on families of color. J. Marriage Family **72**, 440–459 (2010). https://doi.org/10.1111/j.1741-3737.2010.00712.x
9. Statista, Workforce Demographics of Social Media Companies (2022)
10. Google, Diversity Annual Report 2023 (2023)

11. Iuga, C., Nurse, J.R.C., Erola, A.: Baiting the hook: factors impacting susceptibility to phishing attacks. Hum.-Centric Comput. Inf. Sci. **6** (2016). https://doi.org/10.1186/s13673-016-0065-2
12. Hasegawa, A.A., Yamashita, N., Akiyama, M., Mori, T.: Why they ignore English emails: the challenges of non-native speakers in identifying phishing emails. In: Proceedings of the Seventeenth USENIX Conference on Usable Privacy and Security. USENIX Association, USA (2021)
13. Trim, P., Upton, D.: Cyber Security Culture: Counteracting Cyber Threats through Organizational Learning and Training. Routledge, New York (2016)
14. Koyuncu, C., Eda, Ö.: Religious, ethnic, linguistic and cultural diversity and female labor force participation. J. Econ. Bibliogr. **4**, 87–93 (2017)
15. Kabara, A.S., Khatib, S.F., Bazhair, A.H., Sulimany, H.G.H.: The effect of the board's educational and gender diversity on the firms' performance: evidence from non-financial firms in developing country. Sustainability **14**, 11058 (2022)
16. Gundu, T., Maronga, M.I., Boucher, D.: Industry 4.0 business perspective: fostering a cyber security culture in a culturally diverse workplace. In: Proceedings of 4th International Conference on the Internet, Cyber Security and Information Systems, pp. 85–94. Kalpa Publication in Computing (2019)
17. Kheiltash, O., Rust, V.D.: Inequalities in Iranian education: Representations of gender, socioeconomic status, ethnic diversity, and religious diversity in school textbooks and curricula. In: Inequality in Education: Comparative and International Perspectives, pp. 392–416. Springer (2009)
18. Khan, A., Yilmaz, M.K., Aksoy, M.: Does board demographic diversity affect the dividend payout policy in Turkey? EuroMed J. Bus. (2022)
19. Tranfield, D., Denyer, D., Smart, P.: Towards a methodology for developing evidence-informed management knowledge by means of systematic review. Br. J. Manag. **14**, 207–222 (2003)
20. Mountrouidou, X., et al.: Securing the human: a review of literature on broadening diversity in cybersecurity education. In: Proceedings of the Working Group Reports on Innovation and Technology in Computer Science Education, pp. 157–176 (2019)
21. Veiga, A., Eloff, J.H.P.: A framework and assessment instrument for information security culture. Comput. Secur. **29**, 196–207 (2010). https://doi.org/10.1016/j.cose.2009.09.002
22. Ciuperca, E.M., Vevera, V., Cirnu, C.: Social variables of cyber security educational programmes. In: The 15th International Scientific Conference ELearning and Software for Education Bucharest, pp. 190–194 (2019)
23. Thomson, K.-L.L., von Solms, R., Louw, L.: Cultivating an organizational information security culture. Comput. Fraud Secur. **2006**, 7–11 (2006). https://doi.org/10.1016/S1361-3723(06)70430-4
24. Henshel, D., Sample, C., Cains, M., Hoffman, B.: Integrating cultural factors into human factors framework and ontology for cyber attackers. In: Advances in Human Factors in Cybersecurity: Proceedings of the AHFE 2016 International Conference on Human Factors in Cybersecurity, pp. 123–137. Springer (2016)
25. Kraemer, S., Carayon, P.: Computer and information security culture: findings from two studies. In: Proceedings of the Human Factors and Ergonomics Society Annual Meeting, pp. 1483–1488. SAGE Publications Sage CA, Los Angeles, CA (2005)
26. Corradini, I.: Building a cybersecurity culture. In: Building a Cybersecurity Culture in Organizations, pp. 63–86. Springer International Publishing, Berlin/Heidelberg, Germany (2020). https://doi.org/10.1007/978-3-030-43999-6_4
27. AL-Nuaimi, M.N.: Human and contextual factors influencing cyber-security in organizations, and implications for higher education institutions: a systematic review. Global Knowl., Memory Commun. **73**, 1–23 (2024). https://doi.org/10.1108/GKMC-12-2021-0209

28. Commission, E.: The EU's Cybersecurity Strategy for the Digital Decade (2020). https://digital-strategy.ec.europa.eu/en/policies/cybersecurity-strategy
29. Hofstede, G.: Culture's consequences: Comparing values, behaviors, institutions and organizations across nations, 2nd ed. SAGE Publications, Inc. (2001)
30. Alanazi, M., Freeman, M., Tootell, H.: Exploring the factors that influence the cybersecurity behaviors of young adults. Comput. Hum. Behav. **136** (2022). https://doi.org/10.1016/j.chb.2022.107376

A Framework to Enable Cheating Detection in Multiplayer Online Games

Salman Shaikh[1(✉)], Hugo Bertin[2], Marc Dacier[1], and Yérom-David Bromberg[2]

[1] King Abdullah University of Science and Technology, Thuwal, Saudi Arabia
{salman.shaikh,marc.dacier}@kaust.edu.sa

[2] Univ Rennes, CNRS, INRIA, IRISA, Rennes, France
{hugo.bertin,david.bromberg}@irisa.fr

Abstract. The video games industry reached a revenue of US$455.27 billion in 2024 [6], with Multiplayer Online Games (MOGs) as the figurehead. However, this industry, which depends on user satisfaction, is threatened by cheating actions degrading the gaming experience. Cheaters can exploit the distributed nature of MOGs with **network flow disruption** attacks, manipulating the network to disrupt game synchronization. Detecting those attacks is challenging due to the difficulty of distinguishing between instabilities and latency inherent in the network and artificial manipulations. This paper proposes a framework that game developers can integrate into their games to generate data that helps understand whether it is possible to differentiate between honest and cheating behaviors. The framework enables developers to automate the execution of network flow disruption cheats in their MOG environment and incorporates a logging mechanism to record both cheating and honest behaviors. The framework is implemented within the game engine, making it generic to different MOGs. Early results of the generated data are presented, highlighting the effects of network flow disruption attacks on MOGs.

Keywords: Multiplayer · Online · Game · Security · Detection · Cheating

1 Introduction

Multiplayer Online Games (MOGs) immersive nature creates a sense of community and competition among players. In 2024, MOGs attracted 1.2 billion players, while Esports viewership rose to over 500 million viewers [16].

This popularity is threatened by cheaters seeking to gain unfair advantages against opponents. Cheating directly impacts game editors as it drives honest players away from games plagued by dishonest behaviors [9]. Furthermore, cheating undermines fairness in Esports tournaments, which poses a major threat, considering the importance of the rewards at stake [30].

MOGs are distributed systems that must cope with the delay inherent to networked communication while maintaining the user experience, which is crucial

I. Praça et al. (Eds.): EICC 2025, CCIS 2500, pp. 166–181, 2025.
https://doi.org/10.1007/978-3-031-94855-8_11

for a player. This introduces classes of cheats manipulating the flow of information exchanged on the network to impact the game synchronization. Technically, these cheats result in an increased communication latency between two nodes of a MOG system. Examples of such attacks are **Lag Switch** and **DDoS** [10,28]. Individually, these attacks can be prevented. However, their strength lies in their duality, making them unsolvable when considered together [2]: preventing one attack enables the other.

Since preventing such attacks is impossible, the last resort is to investigate the possibility of detection techniques. However, detecting network flow manipulation is challenging for a system which has no control over the network. The key question is whether genuine latency can be differentiated from cheating latency based on the pattern of packets received by the nodes of the system. To study this question, the access to labeled datasets, containing behavioral traces of a player, in a real MOG environment, is important. To the best of our knowledge, no dataset, labeling cheater and honest behavior, exist.

Therefore, in this paper, we propose a framework for developers to integrate and test cheats within their games, allowing them to safely evaluate cheating behaviors. This controlled inclusion of cheats, facilitates the generation of comprehensive labeled datasets containing both honest and cheating gameplay. The framework is configurable to different types of MOGs, making it applicable to a wide range of game development scenarios.

We provide early results of traces generated with the framework. They highlight the behavioral implications of network flow disruption attacks on the different game synchronization mechanisms. This is a stepping-stone for further work on network flow disruption cheats.

Section 2 presents the background regarding MOGs systems and the feasibility of network flow disruption attacks on real MOGs. Section 3 addresses the ethical considerations of this study. Section 4 presents the framework proposed to study those cheats. Section 5 shares the early results of the framework in generating behavioral traces. Section 6 concludes the paper and opens on future works.

2 Background

MOGs are distributed systems that can be implemented following a Client-Server, Peer-to-Peer [33] or Hybrid [13,31] architectures. This paper focuses on the Client-Server architecture due to its widespread adoption in the gaming industry [14] for its centralized control that simplifies management and updates, and its ability to deliver consistent and reliable experiences compared to other architectures.

In this model, the server maintains the authoritative game state based on updates received from clients and synchronizes the clients' states accordingly. However, players experience varying network latencies when communicating with the server. To preserve game responsiveness and enhance the user experience, MOGs employ latency compensation mechanisms [20], which can potentially introduce inconsistencies across players.

Game engines' interests are manifold, spanning graphics rendering, physics simulation, as well as components for networking and synchronization [24]. They enable developers to focus solely on creating engaging gameplay and storylines. Beyond gaming, game engines are also utilized in other fields such as Virtual Reality, Digital Twins, and are expected to shape tomorrow's virtual world with the Metaverse [34]. This paper focuses on the Unreal Engine (UE), a popular game engine used to power famous games such as Fortnite, PUBG, Rocket League, and Valorant. UE's source code is publicly available.

2.1 Real Examples of Network Level Cheats

Cheating is a major problem in the video game industry. Current research focuses on client-level cheats, which include malware exploiting client software vulnerabilities to enhance player performance, mainly with Bots [5,11,29,32] and Information Exposure cheats [12,32]. Software solutions exist to counter these threats with anti-cheats [19,21]. However, as in the anti-virus domain, a never-ending arms race persists between developers and cheaters.

On top of that, cheats can also be carried out at the network level. Network-level cheats have largely been overlooked, which emphasizes the reasons for this work. We classify them into the following categories:

Packet Tampering. These attacks, exploit the content of the packets and the related network protocols used by MOGs [18]. Attackers craft packets to deceive the server into accepting information, giving them an unfair advantage. For example, an attacker might trigger certain mechanisms at the server to disconnect other players [3]. Nonetheless, packet tampering cheats can be mitigated by encrypting the packet to protect its integrity and confidentiality [25]. Furthermore, mitigation at the receiving end of the packet is possible.

Network Flow Disruption. These attacks alter the flow of messages exchanged on the network link between different nodes, without modifying the messages' content. We identify three types of attacks, that can be exploited to disrupt a game's state synchronization. Each attack is illustrated by a video recording available in a publicly accessible Dropbox folder [17].

(i) *Fixed Delay (FD)*: Attackers artificially add a consistent delay to the network packets they are sending to the server [8,31,32]. By doing so, game actions performed locally by the attacker (e.g. shooting at an opponent) take more time than usual to arrive at the server, to be processed and replicated to other clients. Hence, opponents observe the attacker's game actions in a later version of the game which has potentially evolved. This can result in inconsistencies between the attacker's local state and the opponent's version of the game. For instance, an opponent seeing the attacker shooting at him can escape the shot by moving away. However, the shot may be considered on target by the server, because it was true according to the attacker's local version which was performed before the victim saw the attacker. This results in the victim being shot even if the victim moved away and was protected in its version of the game.

(ii) *Lag Switch (LS)*: Attackers buffer outgoing updates, for a specific amount of time before sending them to the server simultaneously [2,21,27]. This differs slightly from FD because in LS, packets arrive in a burst at the server. Upon receiving the packets, the server may apply different synchronization mechanisms to reconcile these updates and maintain consistency of the game state for all players [20]. In our experiments, we can observe that after an attack, the server considers objects individually to reconcile their states [17]. An attacker who buffers movement updates introduces inconsistencies in his avatar's location on his local version and at the server. When he releases the packets in a burst at the end of the attack, the reconciliation cancels those movements and replaces the attacker to a location dictated by the server. However, when the attacker buffers shooting actions and sends them in a burst, the shots are not canceled by the server. They are accepted by the server and replicated to the other clients. This happens even when the shot is not possible on the other clients' versions. This object-based synchronization makes LS effective as the opponent may not see the attacker because his movements were canceled, but still dies because the server accepted the attacker's shot.

(iii) *Distributed Denial-of-Service (DDoS)*: Attackers launch attacks that use multiple compromised devices to overwhelm a target [22] which is either the server or a specific player. Server DDoS can disrupt gameplay for all players. It can be used by cheaters to stop a game they are likely to lose [23]. Player-targeted DDoS delays game updates to and from the player [2]. The victim of this attack experiences conditions that are similar to suffering from a poor network infrastructure.

2.2 Network Flow Disruption Cheats are Unsolvable

Detecting network flow disruption attacks in MOGs is challenging, as the game server cannot differentiate between genuine network latency and delay artificially introduced by attackers.

The literature does not provide effective methods to detect FD or LS. Tandon et al. [27] propose to detect aberrant delays. The distinction between accidental and artificial delay is based on a delay-duration threshold. This presents limitations. First, setting a threshold value is challenging. Second, their approach risks misclassifying legitimate players experiencing abnormally high delay as cheaters. Chen et al. [4] propose to detect cheating by comparing Round Trip Time (RTT) with ping times. However, an attacker may also be able to manipulate the timing of ping messages, including, delaying them to match the delays applied to other packets.

Detecting DDoS attacks is also challenging as victims of DDoS experience regular connectivity issues like latency, packet loss, or disconnection.

Benhabbour et al. [2] highlight the duality between LS (and FD) and DDoS. With LS attacks, attackers add delay to their outgoing updates. On the other hand, DDoS attacks target other players, preventing them from sending or receiving updates on time. If a server accepts late incoming packets, it makes the game vulnerable to LS attacks. If a server rejects late incoming packets, it allows DDoS attacks to affect honest players, causing the server to invalidate their legitimate actions.

This duality hinders the development of mitigation strategies, as the prevention of one attack enables the existence of the other. Thus, the next step should be to develop effective detection techniques to identify network flow disruption cheats during the gameplay. However, as often in the security area, there is no real world representative dataset that would enable researchers to come out with applicable detection techniques and validate them experimentally.

3 Ethical Considerations

It is important to note that the network flow disruption attacks that we discuss in this paper are not new; rather, they are frequently used by cheaters. However, their analysis and mitigation is extremely difficult due to their complex nature and the absence of labeled datasets. The framework presented in this paper aims to address this challenge by providing a tool for game developers and researchers who are interested in studying cheating behaviors in their games.

The framework does not allow end users to execute attacks on existing game binaries. Therefore, they cannot leverage the framework to cheat in other games. The tests on Fortnite were not conducted with the framework, rather, they were implemented with external scripts, executed outside of the game process. Moreover, to not impact Fortnite's players, these experiments were carried out in private matches. However, it should be emphasized that both private and public matches share the same client-server architecture, where the game clients connect to a game server, hosted in the cloud. This ensures that the testing environment mirrors real-game conditions while maintaining ethical standards.

Lastly, to encourage further research and promote development of comprehensive datasets, the framework will be publicly available upon request, with appropriate documentation. Nonetheless, it is important to understand that since the framework is built on the Unreal Engine, it is subject to the licensing and copyright rules set by Epic Games [7].

4 Proposed Framework and Logging Architecture

The framework aims to experiment with network flow disruption attacks and generate detailed traces that can serve as a foundation for developing mitigation techniques.

4.1 Specifications and Design of the Framework

The primary goal of the framework is to enable developers to integrate cheating techniques into their games, facilitating the generation of a robust labeled dataset. These cheats will appear in the game as *superpowers* under the strict control of the developers. The game will allow players to launch network flow disruption attacks directly from within the game. These attacks must behave similarly to the observations realized in previous real life examples. The framework is interleaved with the underlying game engine, allowing researchers and developers to collect behavioral data from all nodes, whether from honest players or cheaters. Among attacks presented in Sect. 2.1 the framework implements fixed-delay, lag switch, and DDoS targeted against a player, as well as other client-side cheating techniques (e.g. aimbot and wallhack). Concerning DDoS, any player can attack any other player. Packet tampering attacks are not implemented in the framework, because mitigation techniques exist as discussed in Sect. 2.1.

The use of the framework can thus collect data that helps understand the system's behavior under attacks and paves the way for further dataset analysis to understand whether genuine latency can be differentiated from artificially added latency. The data includes logging the game events on the different nodes of the system and logging when a specific player is using a specific cheat. These logs are collected on each client and the server. At the client side, the framework pushes back the logs to the server, enabling synchronization in order to compare the collected logs over time.

The framework is designed to help researchers and developers automate both the provided attacks and the game scenarios. This includes automating the possible game events (e.g. player's movements, shooting action) as well as the attacks. This enables the reproducibility of the experimentation, allowing for testing of different scenarios and their variants to identify their respective implications. By automating these processes, we can generate a large dataset through multiple runs of each scenario, minimizing the impact of outliers and confirming the validity of the generated logs.

The framework is built to be generic with different MOGs. The interleaving of the framework in the game engine facilitates its deployment with any game implemented on top of the same game engine (Unreal Engine in our implementation). Developers do not have to consider further implementations to integrate the framework into their games. This feature can also be leveraged to determine if the same cheat presents the same effect on another type of game and to further understand if detection techniques can be approached in a generic way (a technique applicable to all types of games) or if they should target a specific type of game. Additionally, the framework is configurable and adaptable for any other types of cheats that researchers may want to implement for other purposes.

4.2 Implementation

Our implementation is made of three main modules, described briefly hereafter: i) the attacks, ii) the logging, and iii) the automation modules.

Attacks Module. To enable the attack within a MOG, the cheats are integrated into the framework. The framework facilitates the mappings between the game UI and the cheats through UE's console commands. We developed FD, LS, and DDoS with `NetEmulation`, a utility included within UE to simulate network conditions such as latency and packet losses. Other client-side cheats (e.g. aimbots and wall hacks) are also developed in the framework to highlight its adaptability with additional types of cheats.

Fixed Delay. Our framework automatically adds a specific delay to the communication from the attacker to the server. This lasts for a pre-defined duration.

Lag Switch. To achieve the effect of a LS, we follow an approach similar to FD. However, instead of applying the same delay to each outgoing packet, all packets are buffered and sent together once the delay has elapsed. This causes a burst of packets to be received by the other end.

DDoS. FD and LS are straightforward to execute through the framework as they are controlled by the cheater's machine. However, executing a DDoS attack is more complex. For ethical reasons, we do not want to overload players' networks with an excessive amount of packets, which could disrupt other services than the game, as it is the case in a real DDoS attack. Hence, we simulate the outcomes of a DDoS attack. It is directly realized by the victim's client who is told to drop packets (incoming and outgoing). The attacker starts the attack by sending the victim's game identifier to the server, which instructs the victim's client to simulate the DDoS. After a pre-defined timer ends, the server instructs the victim's client to behave normally again.

Additional Client-Side Attacks. The framework can be adapted to execute any client-side attacks. For instance, we developed two additional attacks: (i) a custom aimbot, (ii) a wall-hack. The aimbot automatically aims and shoots at opponents within a pre-defined field of view (FOV). The wall-hack alters the properties of a wall to see on the other side when a player shoots at it with a specific weapon, achieving the desired effect of a wall hack.

Logging Mechanism. The framework provides a logging mechanism to gather traces that can be analyzed to study the behavior at the different nodes of the system under an attack. These traces frequently serialize game objects, such as the player's location or health, to observe and compare their evolution on the different nodes.

The logging architecture records at pre-defined intervals, the game events taking place on each client and the server. The mechanism works as follows:

1. When the game begins, logging starts on the server and on each client at a pre-defined rate. The variables to log are configurable and can include player data such as *Location, Health, Inventory, etc.*. This process generates a labeled dataset of honest playing behavior.

2. When a player presses a key to activate a specific cheat, the logging rate increases to capture detailed data at a much higher rate on both the cheating client (or the victim's client in case of a DDoS attack) and the server. The reason to increase the logging rate is to capture detailed information, as players typically cheat for short durations to avoid detection. Their actions in this short amount of time are highly influential. Thus, the framework logs detailed information about the player during the cheating period to ensure the development of a large labeled cheating dataset.
3. Once the cheating activity ends, the logging rate reverts to the game's default logging rate.
4. The logs are continuously stored in separate JSON files: one on each client and one on the server. The clients periodically send their logs to the server where they are merged with the server-side logs. These combined logs are then organized by game time, which can be later used to understand the discrepancies between clients and server at different times.

Automation Module. To reproduce multiple runs without manual interference that may introduce discrepancies, we design a Domain Specific Language (DSL) to automate the game scenarios. The description of those scenarios in the DSL is based on JSON.

The DSL enables the description of game events containing fields such as times, positions, shots, and attacks. Each field is configured to ensure that the player's actions are executed at the desired time and in the intended manner, synchronized with other actions such as attacks or shooting. Once the run finishes, the logs are collected, and the same run can be repeated or modified by adjusting the game scenario.

5 Early Experimental Results

In this section, we present preliminary results obtained with the framework when launching the different types of network flow disruption cheats implemented as described in Sect. 2.1. To demonstrate the effectiveness of the framework interleaved with the game engine, we collected results using a homemade MOG that we designed. The MOG and the framework are developed with Unreal Engine.

The evaluation has been realized in two different network environments: (i) LAN and (ii) Cloud. In the LAN setting, we execute the network flow disruption cheats and observe their impacts. In the cloud environment, we analyzed the inherent latency and instabilities that occur even during normal gameplay. The results presented are examples of patterns that we observed consistently across multiple runs, using our automation module.

5.1 LAN – Threshold Conditions

The purpose of our first evaluation is to record traces in a controlled LAN environment. The testbed includes one server and two clients. We perform experiments on each network flow disruption attack. Our DSL enables us to write a specific game scenario for each attack (and their variants). The experiments in this section are expected to produce similar results to those observed in Fortnite test cases. These experiments are motivated by the following reasons: 1) If the results are similar, it demonstrates that Fortnite does not employ specific additional development techniques to handle client-server synchronization, but rather relies on the engine. 2) The remaining figures in the paper show the framework's logging mechanism by representing the player's state as observed on the client and the server, and highlighting discrepancies that occur between the two during an attack. All of these graphs are generated from the logs produced by this logging mechanism.

Maximum Duration of a FD Cheat. We started by testing if attackers could delay packets indefinitely with Unreal Engine's default settings. We define two game scenarios for this attack. Figure 1a illustrates the first scenario, which executes a fixed delay of 60 s. In this case, the framework logs the player's movement locally (the straight, blue, line), but at the server, the player remains stationary since it has not received packets containing the movement information. After the delay, packets are delivered to the server in the order they were executed locally, causing the server to update the player's position accordingly (dashed, red line). The second scenario applies a delay that exceeds 60 s (61 s), as shown in Fig. 1b. We can observe that the data is no longer logged after delaying the packets for more than 60 s. This is because the client got disconnected by the server. This highlights a timeout trigger disconnecting players if no packet is received from a client for more than 60 s.

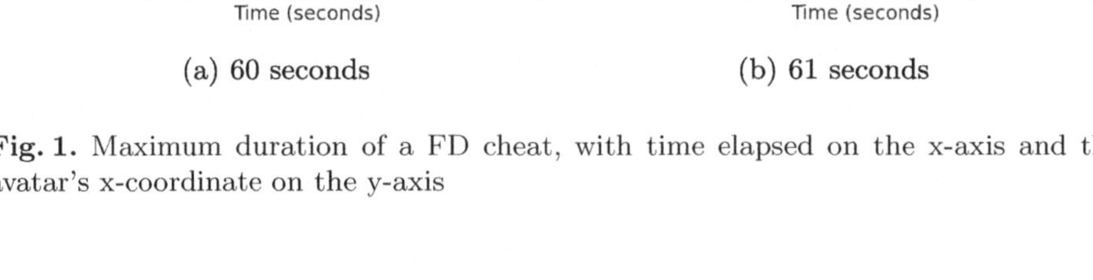

(a) 60 seconds (b) 61 seconds

Fig. 1. Maximum duration of a FD cheat, with time elapsed on the x-axis and the avatar's x-coordinate on the y-axis

Consequences of Lag-Switching. The second experiment presents the implications of LS attacks. We first highlight the mechanisms of extrapolation implemented for the movements in Unreal Engine.

Server extrapolation is a common technique used by MOGs to minimize the lag for players experiencing unstable network connectivity [20]. When the server does not receive information from the client (or receives it delayed), it extrapolates the player's position based on the last data it received. To demonstrate this behavior, we design multiple game scenarios. The first scenario demonstrates that if the player is not moving, the server assumes the player will remain stationary. The results of this scenario is presented in Fig. 2a, where the player starts to move (the straight, blue, line) only after having started to buffer its packets (starting a LS attack). The second scenario demonstrates an attacker who is already moving when starting the attack. In this case, the server assumes that the player continues moving in the same direction for some time. This scenario is illustrated in Fig. 2b, showing an attacker in motion before initiating the attack. It is worth noting that the extrapolation is realized following a pattern that resembles stairsteps, highlighting that the engine is waiting to receive information for a specific delay before deciding to extrapolate the player's position.

At the end of the LS attack (after 3 s), when the burst of delayed information arrives at the server, the server compares the position claimed by the attacker with its extrapolated position. If there is a discrepancy, the server forces the attacker to match the server's recorded position, as seen in Figs. 2a and 2b.

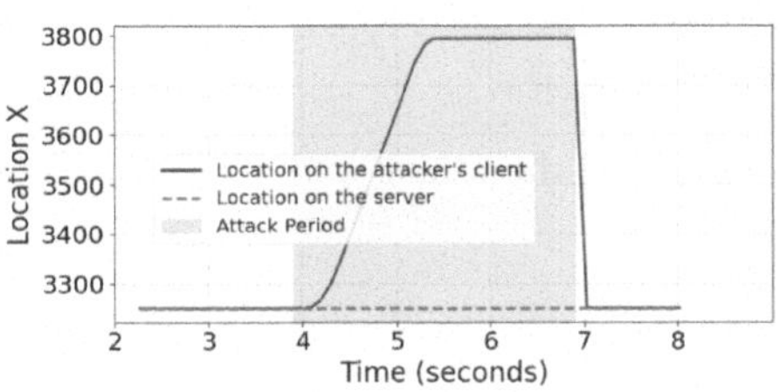

(a) Reconciliation of Attacker (Stationary)

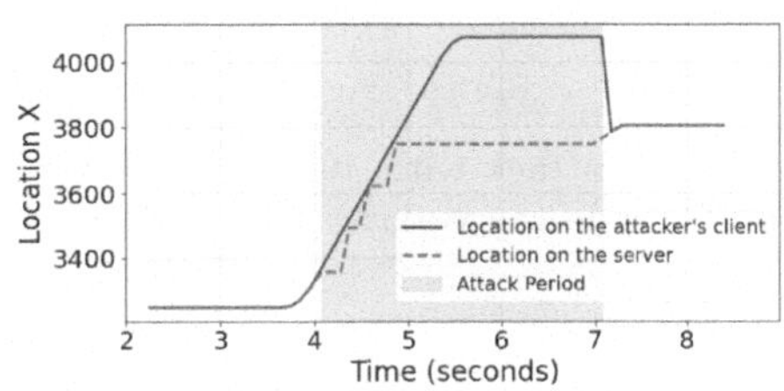

(b) Reconciliation of Attacker (Moving)

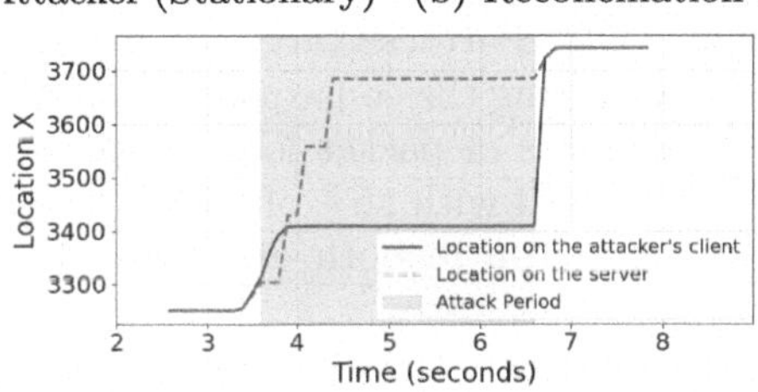

(c) Server forces attacker to match its state

Fig. 2. Variation of LS Cheats, with time elapsed on the x-axis and the avatar's x-coordinate on the y-axis

This extrapolation can be exploited by attackers to deceive the server into extrapolating a position they never reached. Figure 2c demonstrates another scenario where an attacker executes LS while moving and then quickly stopping. In this case, the server extrapolates the attacker's position to an unrealistic one, and once the attack ends, the server forces the attacker to "teleport" to the server's extrapolated location.

A scenario involving the LS attack for a much shorter amount of time (400 ms) is demonstrated in Fig. 3. We can see the server extrapolating the movements during the attack, as observed in previous illustrations. At the end of the attack, the player stops pressing movement commands on his client. However, the behavior we observe at this moment is particularly interesting, as the attacker's position oscillates for approximately 0.5 s before being re-synchronized with the server. Until now, we have not understood the cause of this behavior, which we have reproduced multiple times. This emphasizes the need for accumulating and studying logs, particularly across different variants of the same cheating techniques, using our framework.

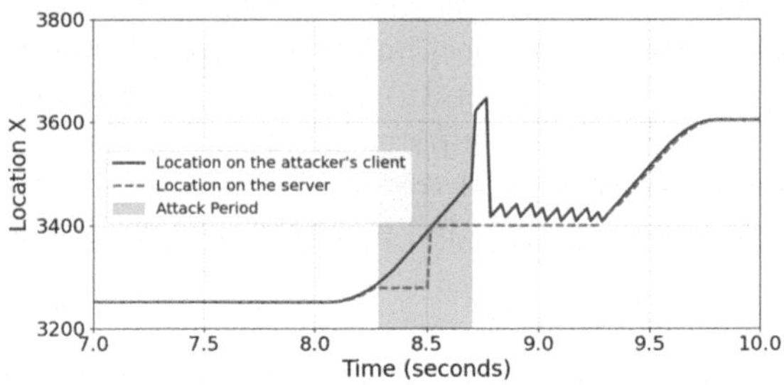

Fig. 3. LS attack with 400 ms, with time elapsed on the x-axis and the avatar's x-coordinate on the y-axis

Consequences of DDoS on a Targeted Player. We recorded the behavior of the MOGs system under DDoS attacks targeting a player. We present in this paper, two scenarios demonstrating the consequences of DDoS attacks, which are simulated with different amounts of packet losses applied. Figure 4a shows the results of DDoS attacks simulated with 85% of packet losses on the network link between the server and the victim in both directions. Figure 4b demonstrates DDoS attacks with 95% of packet losses.

Through these figures, we observe the evolution of the DDoS victim's location with time, on the victim's client (straight green line), as well as on the server (dashed red line). We can also notice the moment the shot executed by the attacker is registered by the server (vertical red 'x' line) and the victim's client (vertical green 'o' line). The first discrepancy is seen within the movements; when the victim moves during the attack, her movements are not seen on the server instantly. Sometimes some of her movement updates are considered by the server but the major effect observed is the victim to be replaced to her original position whenever packets from the server reach her. In particular, when the victim eventually gets the packet from the server informing her of the shot, one can also see that the server moves the client back where, in the server world's view, the victim was when the shot happened. This results from information embedded in the small amount (respectively 15% and 5%) of packets being transmitted. The figures demonstrate that higher outgoing packet losses cause significant deviations between the server's position of the victim and the victim's position on her client. On the other hand, increasing incoming packet losses significantly delays the time between when the server registers an attacker's shot and when the victim's client perceives it.

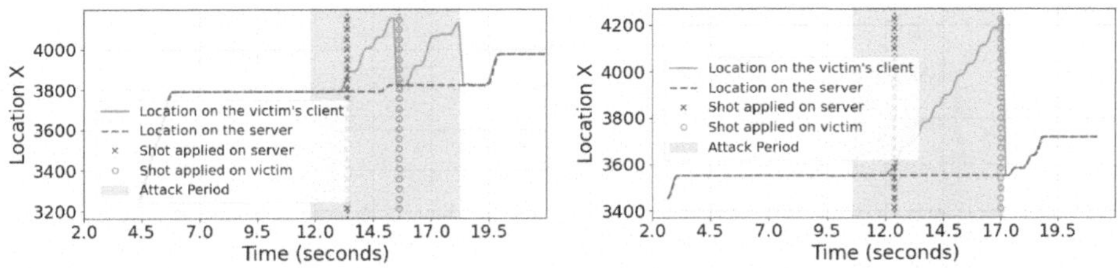

(a) 85% incoming and outgoing packets losses (b) 95% incoming and outgoing packets losses

Fig. 4. Consequences of DDoS attacks, with time elapsed on the x-axis and the avatar's x-coordinate on the y-axis

These results, confirmed through our homemade MOG, are consistent with the view that Fortnite does not employ any additional development practices and relies on the Unreal Engine for client-server synchronization. Furthermore, the figures demonstrate that our framework functions as intended and that our logging mechanism is effective.

5.2 Server on Cloud

In our previous experiments, the server and the client were in the same LAN, which minimizes network latency. However, this is not the case for real MOGs which rely on the Internet, introducing inherent latency and instability. These experiments aim to highlight whether the order of magnitude of this additional latency is large enough to make it challenging to differentiate between artificially added latency, and typical network latency. Our second experimental campaign compares the traces obtained in LAN setup to a second setup using the Internet to communicate. For this purpose, we rent a server in the cloud, located in Stockholm. The clients are located in the Middle East. The average RTT latency between a client and the server, measured with `ping` (15 runs), is 98 ms. Thus, the one-way latency is 49 ms on average.

The experiments done with the cloud setup aim to observe the inconsistencies introduced in the game state by the network. Firstly, we use the client to join a game hosted on the cloud server and we run a game scenario where the player is continuously moving for 30 s. We gather the logs on both ends to compare them and highlights the magnitude of the inconsistencies. The client and the server are both synchronized to a NTP server. Figure 5 shows an example of the traces obtained while running this experiment. The straight, blue, line represents the player's position (x-value) over time on the client. The dashed, red, line represents the position on the server. We can observe that there is a delay needed for the server to reflect the player's movement at any point of the curve. Additionally, the server approximates the moves.

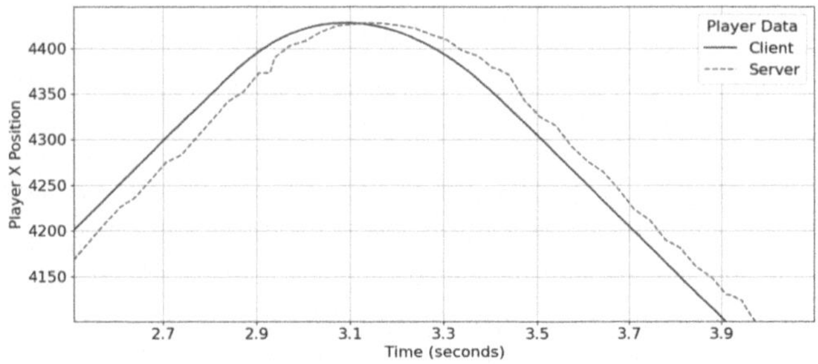

Fig. 5. Player Position (x-value) over time in the cloud setup

To characterize these discrepancies and the impact of the network, we run the game scenario again in the LAN and the Cloud setup. In the logs gathered, we calculate the delta time between two similar log entries, i.e. the delay for observing the move performed on the client to be reflected on the server. We repeat 15 runs of the experiment in each setup.

Figure 6 presents the results of all the delta times measured, separating the two setups. The mean delta time is lower in a LAN (7.69 ms) than when the server is hosted in the cloud (55.31 ms). This is consistent with the magnitude of network latency RTT measured. Furthermore, the latency is less stable in the cloud, showing a standard deviation of 8.38ms against 4.15 ms. We can also observe outliers reaching a delta time of 107 ms in the cloud. This highlights

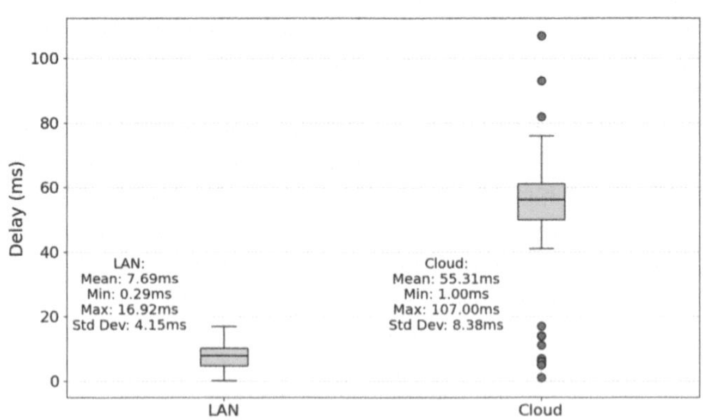

Fig. 6. LAN vs Cloud Latency Distribution

that genuine network conditions introduce inconsistencies between replicas of the game state.

These results show the expected additional network latency when the game state is communicated over the Internet. This additional latency is of the same order of magnitude as the time required to execute a successful network flow disruption cheat [26], further emphasizing the difficulty in mitigating such cheats.

6 Conclusion

Network flow disruption cheats manipulate the latency to disrupt game synchronization, and we highlight their applicability in MOGs. To streamline research for detection techniques, we propose a dedicated framework providing attacks implementations, an automation module and a logging system to generate labeled datasets of MOGs behavior under cheating attacks. The framework presented is implemented in the game engine with the goal to be enriched and used in a real game to develop large datasets. We validate the efficacy of the framework using the automation module ensuring the reliability of the datasets generated. It is generic and deployable with all the games based on the Unreal Engine. Our early results, leveraging our framework, emphasize the complexity of the implications network flow disruption cheats have on game synchronization. The framework is available to researchers and game developers who are interested in it.

As part of the future work, we are exploring the possibility of deploying the framework with Open Source games, such as Lyra Starter Game [1] or Eternal Crusade: Resurrection [15], which are Unreal Engine based multiplayer games presenting a complete gameplay.

References

1. Andy: Lyra starter game (2022). https://dev.epicgames.com/community/learning/paths/Z4/lyra-starter-game. Accessed 13 Jan 2025
2. Benhabbour, I., Bromberg, Y., Dacier, M., Dietrich, S., Rodrigues, R., Esteves-Verissimo, P.: Attacks on tomorrow's virtual world. In: 2023 53rd Annual IEEE/IFIP International Conference on Dependable Systems and Networks - Supplemental Volume (DSN-S), pp. 105–110. IEEE Computer Society (2023). https://doi.org/10.1109/DSN-S58398.2023.00033
3. Bertin, H., Benhabbour, I., Dacier, M., Bromberg, Y.D.: Disconnecting games with a single packet: an unreal untold story. https://www.youtube.com/watch?v=ZpMHGBULZgU. hack.lu 2024. Accessed 12 Mar 2025
4. Chen, B.D., Maheswaran, M.: A cheat controlled protocol for centralized online multiplayer games. In: Proceedings of 3rd ACM SIGCOMM Workshop on Network and System Support for Games, pp. 139–143 (2004). https://doi.org/10.1145/1016540.1016554
5. Choi, M., Ko, G., Cha, S.K.: BotScreen: trust everybody, but cut the aimbots yourself. In: 32nd USENIX Security Symposium (USENIX Security 23), pp. 481–498 (2023). https://www.usenix.org/conference/usenixsecurity23/presentation/choi. Accessed 28 Mar 2025

6. Clement, J.: Global video game industry revenue 2029 (2024). https://www.statista.com/forecasts/1344668/revenue-video-game-worldwide. Accessed 12Mar 2025
7. Games, E.: Unreal® Engine End User License Agreement. https://www.unrealengine.com/en-US/eula/unreal. Accessed 12 Mar 2025
8. GauthierDickey, C., Zappala, D., Lo, V., Marr, J.: Low latency and cheat-proof event ordering for peer-to-peer games. In: Proceedings of the 14th International Workshop on Network and Operating Systems Support for Digital Audio and Video, pp. 134–139 (2004). https://doi.org/10.1145/1005847.1005877
9. Granados, N.: Report: Cheating is becoming a big problem in online gaming (2018). https://www.forbes.com/sites/nelsongranados/2018/04/30/report-cheating-is-becoming-a-big-problem-in-online-gaming/. Accessed 12 Mar 2025
10. GTAFORUMS: Cheats: You know someone is using a lag switch.... https://gtaforums.com/topic/359863-cheats-you-know-someone-is-using-a-lag-switch/. Accessed 27 Mar 2025
11. Han, M.L., Park, J.K., Kim, H.K.: Online game bot detection in FPS game. In: Handa, H., Ishibuchi, H., Ong, Y.-S., Tan, K.-C. (eds.) Proceedings of the 18th Asia Pacific Symposium on Intelligent and Evolutionary Systems - Volume 2. PALO, vol. 2, pp. 479–491. Springer, Cham (2015). https://doi.org/10.1007/978-3-319-13356-0_38
12. Hao, J., Cai, W.: Measuring information exposure attacks on interest management. In: 2012 ACM/IEEE/SCS 26th Workshop on Principles of Advanced and Distributed Simulation, pp. 133–144. IEEE (2012). https://doi.org/10.1109/PADS.2012.24
13. Kabus, P., Terpstra, W.W., Cilia, M., Buchmann, A.P.: Addressing cheating in distributed MMOGs. In: 4th ACM SIGCOMM Workshop on Network and System Support for Games, NetGames '05, pp. 1–6. ACM (2005). https://doi.org/10.1145/1103599.1103607
14. Khan, A.M., Arsov, I., Preda, M., Chabridon, S., Beugnard, A.: Adaptable client-server architecture for mobile multiplayer games. In: Proceedings of the 3rd International ICST Conference on Simulation Tools and Techniques, pp. 1–7 (2010). https://doi.org/10.4108/ICST.SIMUTOOLS2010.8704
15. Knight, J.: Eternal crusade: Resurrection. https://eternal-crusade.com (2023). Accessed 13 Jan 2025
16. Kumar, N.: Esports viewership: 2025 (2024). https://www.demandsage.com/esports-statistics/. Accessed 12 Mar 2025
17. game-sec lab: Fd, ls and ddos attacks on fortnite (2024). https://www.dropbox.com/scl/fo/k1gjem9akbc1io7vw8ysk/AM_Uza91qNa9DTlHV_noCy0?rlkey=blaa5db9mcmjg816r0q9l0yhg&st=jdusdade&dl=0. dropbox folder. Accessed 12 Mar 2025
18. Laurens, P., Paige, R.F., Brooke, P.J., Chivers, H.: A novel approach to the detection of cheating in multiplayer online games. In: 12th IEEE International Conference on Engineering Complex Computer Systems (ICECCS 2007), pp. 97–106. IEEE (2007). https://doi.org/10.1109/ICECCS.2007.11
19. Lehtonen, S., et al.: Comparative study of anti-cheat methods in video games. University of Helsinki, Faculty of Science (2020). https://helda.helsinki.fi/items/b1141406-eb65-48a5-8922-d1b23d4cfe51. Accessed 28 Mar 2025
20. Liu, S., Xu, X., Claypool, M.: A survey and taxonomy of latency compensation techniques for network computer games. ACM Comput. Surv. (CSUR) **54**(11s), 1–34 (2022). https://doi.org/10.1145/3519023

21. Maario, A., Shukla, V.K., Ambikapathy, A., Sharma, P.: Redefining the risks of kernel-level anti-cheat in online gaming. In: 2021 8th International Conference on Signal Processing and Integrated Networks (SPIN), pp. 676–680. IEEE (2021). https://doi.org/10.1109/SPIN52536.2021.9566108
22. Mirkovic, J., Dietrich, S., Dittrich, D., Reiher, P.: Internet denial of service: attack and defense mechanisms (Radia Perlman Computer Networking and Security). Prentice Hall PTR (2004)
23. Moosa, S.: Overwatch 2 cheaters are crashing servers to avoid losing games (2024). https://www.ginx.tv/en/overwatch-2/cheaters-are-crashing-servers-to-avoid-losing-games. Accessed 12 Mar 2025
24. Phil: What is a game engine: An essential overview for beginners - draw & code (2024). https://drawandcode.com/learning-zone/what-is-a-game-engine/. Accessed 12 Mar 2025
25. Rescorla, E.: The Transport Layer Security (TLS) Protocol Version 1.3. RFC 8446 (2018). https://doi.org/10.17487/RFC8446
26. Shaikh, S.: Network Flow Disruption in Multiplayer Online Games. Master's thesis, King Abdullah University of Science and Technology (2024). http://hdl.handle.net/10754/701288. Accessed 28 Mar 2025
27. Tandon, V., Devore, J.: Detecting lag switch cheating in game (2017). uS Patent 9,636,589
28. TARKOV: Lag switch hack is ruining the game. https://forum.escapefromtarkov.com/topic/116665-lag-switch-hack-is-ruining-the-game/. Accessed 12 Mar 2025
29. Véron, M.: Scalable services for massively multiplayer online games. Ph.D. thesis, Université Pierre et Marie Curie - Paris VI (2015). https://theses.hal.science/tel-01230852/. Accessed 28 Mar 2025
30. Weatherbed, J.: Apex Legends devs confirm esports hacking incident (2024). https://www.theverge.com/2024/3/18/24104666/apex-legends-postpones-algs-competition-hack-concerns. Accessed 27 Mar 2025
31. Webb, S., Soh, S., Lau, W.: Racs: a referee anti-cheat scheme for p2p gaming. In: Proceedings of the 17th International Workshop on Network and Operating Systems Support for Digital Audio and Video, pp. 34–42. Association for Computing Machinery (ACM) (2007). https://espace.curtin.edu.au/handle/20.500.11937/32304. Accessed 27 Mar 2025
32. Webb, S.D., Soh, S.: Cheating in networked computer games: a review. In: 2nd International Conference on Digital Interactive Media in Entertainment and Arts, pp. 105–112 (2007). https://doi.org/10.1145/1306813.1306839
33. Yahyavi, A., Kemme, B.: Peer-to-peer architectures for massively multiplayer online games: a survey. ACM Comput. Surv. (CSUR) **46**(1), 1–51 (2013). https://doi.org/10.1145/2522968.2522977
34. Zenou, T.: A novel predicted the metaverse (and hyperinflation) 30 years ago (2022). https://www.washingtonpost.com/history/2022/06/30/snow-crash-neal-stephenson-metaverse/. Accessed 12 Mar 2025

Framework and Roadmap for Secure Design and Development of Applications in the Cloud and Mobile Ecosystem

Francisco T. Chimuco[1,2,3(✉)], João B. F. Sequeiros[1,2], Mário M. Freire[1,2], and Pedro R. M. Inácio[1,2]

[1] Universidade da Beira Interior, Covilhã, Portugal
franciso.chimuco@ubi.pt
[2] Instituto de Telecomunicações, Covilhã Delegation, Portugal
[3] Instituto Superior de Ciências de Educação da Huíla, Lubango, Huíla, Angola

Abstract. Mobile applications and devices became widely popular in the second decade of the 21st century, mostly driven by technologies like cloud computing, Internet of Things (IoT), blockchain, and big data, leading to and fostering very fast paced development cycles of cloud-based mobile apps, which offer enhanced convenience and accessibility for end users. However, significant security challenges remain, as cybersecurity is often overlooked in favor of a shorter time-to-market. This paper presents the *Security by Design for the Cloud and Mobile Ecosystem* framework, consisting of a set of tools and underlying logic aimed at helping mobile application developers build secure applications in the specific mobile and cloud ecosystem. It features a modular architecture and includes five main tools built in Python, and its outputs are enhanced through the use of ChatGPT. The paper also discusses real-world applications and presents usability evaluation results, showing the framework is effective and meets its goals.

Keywords: Cloud Computing · Mobile Computing · secD4CloudMobile · Security · Security Best Practice Guidelines · Security Framework · Security Requirement Elicitation · Security Attack Models · Security Tests Specification and Tools · Security Mechanisms · Software Implementation

1 Introduction

Computing paradigms have changed significantly in the last few decades. Hardware systems with little processing, communication, memory, and storage capacities evolved from occupying considerably large spaces into both small and larger devices (including personal devices such as smartphones, tablets, laptops and hybrid devices) that seamlessly operate in a networked environment to achieve complex cyberphysical objectives. This evolution has simultaneously fueled the development and benefited of the emergence and intersection

I. Praça et al. (Eds.): EICC 2025, CCIS 2500, pp. 182–200, 2025.
https://doi.org/10.1007/978-3-031-94855-8_12

of many advances and technologies along the years, namely, Robotics, Cloud Computing [24], Mobile Computing [12,18,21,37], IoT [4,42], Fog Computing [9,10], and Blockchain [16,19,25]. The Internet was instrumental in this process. Moreover, the emergence of new network interfaces such as Third Generation(3G)/Fourth Generation (4G)/Fifth Generation (5G) technology and Wi-Fi, with speeds above 1 Gb per second, has caused the widespread adoption of Cloud-based mobile applications, giving rise to new ecosystems such as Cloud and Mobile [39], Mobile IoT, Mobile IoT and Cloud [18], among others. Unfortunately, the fast pace at which technology has been evolving (namely in terms of mobile application development) has not been necessarily accompanied with solutions aimed at developing applications that are secure by design, where security engineering principles are taken into account in all phases of the Software Development Life Cycle (SDLC).

This article introduces an improved version of the SECD4CLOUDMOBILE [13] framework, which aims to integrate security engineering into software engineering. The framework addresses the rising security issues in mobile applications that rely on Cloud support. Our main argument is that integrating software and systems engineering with security engineering will make it easier to develop secure applications for the Cloud and mobile ecosystem. This will be facilitated by tools that assist programmers and systems architects, specially if they do not have in-depth cybersecurity expertise. The outputs of the developed tools of the framework depend on the options selected by a user, taking as input the descriptions of the system and the Cloud-based mobile application to be developed or already developed, via a questionnaire.

The main contributions of this article are as follows:

- A detailed and in-depth description of the framework, namely the design and implementation of its tools, named *Cloud Mobile Security Requirements Elicitation* (CMSRE), *Cloud Mobile Security Best Practices Guidelines* (CMSBPG), *Cloud Mobile Security Mechanism Elicitation* (CMSME), *Cloud Mobile Attack Modeling Elicitation* (CMAME), and *Cloud Mobile Security Tests Specification and Automation Tools* (CM2ST);
- Development of human expert knowledge representation in the form of an expert decision-making procedure within the fields of security aspects, mechanisms, and software testing in Cloud and mobile ecosystem;
- Thorough evaluation of the performance of the SECD4CLOUDMOBILE framework, using the System Usability Scale (SUS) questionnaire alongside Copilot, Gemini, and ChatGPT Large Language Models (LLMs).

The remainder of the paper is structured as follows: Sect. 2 presents the background and related work of the research; Sect. 3 presents the proposed framework and its implementation details; the performance evaluation of the framework is addressed in Sect. 4; Sect. 5 evaluates the usability of the tools of framework using the SUS questionnaire and compares their effectiveness to that of the LLMs Copilot and Gemini; and the research is concluded by presenting the conclusions and future work in Sect. 6.

2 Background and Related Work

This section provides an overview of the studies and projects that preceded this research work, as well as those that are more closely related to it. It begins by briefly discussing the SECURIoTESIGN – *Towards the assurance of the SECURity by dESIGN of the Internet of Things* – project, including the Security Advising Modules (SAM) platform.

2.1 SECURIoTESIGN

The SECURIoTESIGN project focused on ensuring that security is easily embedded in the design and development of IoT devices. It aimed to integrate security considerations from the design phase through to testing. The project acknowledged that architects and developers may lack specialized security knowledge and seeks to provide the necessary tools and frameworks to address this gap, ensuring that security is built into IoT devices at every stage of their development. The main output of the project is SAM, which acts as a security assistant, simulating a consultation with a human expert through questionnaires. The framework is organized into six main tools, as shown in Fig. 1.

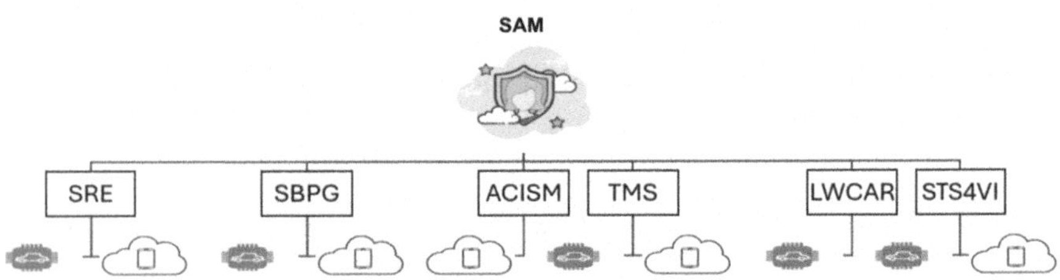

Fig. 1. Simplified representation of the SECURIoTESIGN SAM architecture (adapted from [7]).

Below is a brief description of the six main SAM tools (sometimes also referred to as *modules*, as the framework is built using modular design, to benefit expandability), as described in [7]:

- **Security Requirements Elicitation** – this module establishes security requirements for developing applications in Cloud, mobile, and IoT ecosystems based on user feedback from a questionnaire [32,33,39];
- **Security Best Practice Guidelines** – this module generates security guidelines for Cloud, mobile, and IoT applications based on user questionnaire responses [32,33,39];
- **Assessment of the Correct Integration of Security Mechanisms** – this module covers cryptographic algorithms in hardware, existing software, and planned software, indicating whether an application uses hardware or software and recommending algorithms based on processing power and security

needs; **Threat Modeling Solution** – this module selects a set of Common Weakness Enumeration (CWE) that the system may be susceptible to, given its characteristics.
- **Lightweight Cryptographic Algorithms Recommendation** – this module outlines cryptographic algorithms in hardware and software, identifying application types and recommending algorithms based on processing power and security requirements coming from the SRE module;
- **Security Tests Specification for Verification of Implementation** – this module provides security test recommendations for cloud, mobile, and IoT applications, ensuring security integration during coding and implementation based on user responses.

Two versions of applications (tools) were developed for each module – one for the IoT ecosystem, and a more extended version for the Cloud and Mobile ecosystem. The questionnaires developed for capturing the user inputs for each module are the result of an extensive analysis of the current paradigm of IoT [15,27–31], and of the Cloud and Mobile ecosystem [13,14], and the SAM platform has been continuously improved and analyzed [23]. The tools focused on Cloud and Mobile, contrary to the IoT ones, have only an overarching (though longer) questionnaire, whose answers serve as the input for all the tools.

2.2 Security Challenges in the Cloud and Mobile Ecosystem

Although efforts have been made to address security and privacy issues in the Cloud and Mobile ecosystem, these challenges remain prevalent. According to Sequeiros et al. [39], key challenges include:

- *Lack of Specific Modeling and Design Tools for the Cloud and Mobile ecosystem* – currently there are no modeling and design tools specifically for the Cloud and Mobile ecosystem. What does exist is a set of generic standards, such as ISO/IEC 27034, a set of generic application security guidelines, and cybersecurity reports and best practices;
- *Heterogeneity of Attack Vectors and Taxonomy* – the diverse technologies (e.g., network interface, Operating System (OS), device) in the Cloud and Mobile ecosystem lead to various attack vectors, posing major security challenges [40];
- *Stronger focus on Attack Modeling rather than Threat Modeling* – the emphasis on attack modeling over threat modeling hinders the ability to secure applications by design or by default during the SDLC [39].

2.3 Related Work

Many studies explore frameworks for integrating security mechanisms, but few address their use in cloud and mobile computing. This section discusses automated frameworks to ensure that Cloud-based mobile applications include security throughout development, from design to testing.

According to Al-Omary [2], before 2019, most of the proposed solutions and frameworks emphasized downloading secure data from the Cloud [26,44]. This study highlights security and privacy issues in the Cloud and mobile ecosystem and proposes a framework to address them. The framework focuses on runtime security mechanisms without guaranteeing inherent security for Cloud-based mobile applications and relies on statistical and manual methods.

Srivastava et al. [41] developed a security framework for mHealth applications that ensures authentication, authorization, and secure handling of medical data. It is easily integrated into existing and new apps to reduce security risks. However, it only applies to mHealth apps and does not cover all phases of software development. Although it allows for quick implementation, it does not guarantee consistent security considerations from design to testing.

Hussain et al. [20] present a conceptual framework for the security of mobile health applications on the Android platform. The mHealth Apps Security Framework (MASF) is designed to protect sensitive medical data by addressing threats such as data leakage, privilege escalation, and misuse of app permissions on Android.

The works of Samaila et al. [31], Costa [15] and Lopes [23] present, respectively, a framework with three tools (SRE, SBPG, LWCAR) that assist engineers in building secure IoT systems, prioritizing security and user privacy from the start, a TMS module from the SAM platform that identifies threats during IoT system development, ensuring security considerations are integrated from the design phase, and a SAM STS4VI solution that creates a tool to assess vulnerabilities, generate relevant tests, and identify security tools to evaluate system susceptibility to these threats.

Awaysheh et al. [5] explored "security by design" for Big Data in cloud computing. They created a methodology for automated security assessments that aligns with best practices. Tested with Apache Hadoop, the framework increased security awareness and reduced design time.

2.4 Comparison of Related Work

Table 1 compares the works mentioned in the previous sections based on nine aspects, namely: security by design, cloud, mobile, SRE Tool, SBPG Tool, ACISM Tool, TMS Tool, STS4VI, and automatic prescriptive model. Several symbols are used for this comparison on the table: ✓ is used to denote that the description of a subject is addressed on the respective work, + or ++ are used to emphasize special attention given to a specific aspect (symbol ++ is used to denote a higher degree of emphasis), X is used to denote the absence of the subject.

As it can be seen, although there are a number of studies that focus on guaranteeing security when building mobile applications, IoT systems, and the Cloud, as far as we know, there are no studies that focus specifically on building and proposing a prescriptive and complete semi-automatic framework with several tools that can be incorporated into all phases of the secure application development process for the Cloud and mobile ecosystem. This research work,

Table 1. Comparison of related work concerning security frameworks for the Cloud and Mobile ecosystem.

Research	Year	Security by Design	Cloud	Mobile	SRE Tool	SBPG Tool	ACISM Tool	TMS Tool	STS4VI Tool	Automatic Prescriptive Model
Srivastava et al. [41]	2009	++	X	✓	X	X	X	X	X	X
Hussein et al. [20]	2018	✓	X	✓	X	X	X	X	X	✓
Samaila et al. [31]	2020	++	✓	X	++	++	X	X	X	++
Costa [15]	2021	++	+	+	X	X	X	++	X	++
Lopes [23]	2021	++	+	+	X	X	X	X	++	++
Awaysheh et al. [5]	2022	++	X	✓	✓	✓	X	X	X	X
This article	**2024**	++	++	++	++	++	++	++	++	++

being an extension of the previous [13,38] works, is precisely motivated by the need to bridge this gap between software engineering and security engineering.

3 The Proposed Framework

The proposed framework is composed of a set of components (e.g., the *user interface*, the *security manager* or the *output interface*), and incorporates a series of different tools that implement or use several of those components. Figure 2 presents an overview of the different components of the framework, which set up the overall architecture of all of the implemented tools. The main components of the SECD4CLOUDMOBILE are discussed with more detail in subsequent subsections.

The SECD4CLOUDMOBILE improves and particularizes the SECURIoTESIGN framework for a new ecosystem, encapsulating a semi-automatic prescriptive model, made up of five main tools that can be incorporated into specific phases of the SDLC of mobile cloud based applications, whether they are specific to mobile devices or specific to IoT devices (that use mobile devices as an IoT gateway or as a thin client between them and the Cloud for data processing and storage). Below is a brief description of each of them:

- **Cloud Mobile Security Requirement Elicitation** (CMSRE), which generates necessary security requirements for mobile applications to safeguard user and system data, based on a description provided by the user through a questionnaire;
- **Cloud Mobile Security Best Practices Guidelines** (CMSBPG), which provides security recommendations for mobile app development to ensure the privacy of end-user and system data. Users fill out a questionnaire to receive tailored guidance;

- **Cloud Mobile Security Mechanisms Elicitation** (CMSME), which recommends security mechanisms for cloud-based mobile applications during coding and verification phases to meet security and non-functional requirements;
- **Cloud Mobile Attack Modeling Elicitation** (CMAME), which generates attack models for cloud-based mobile applications using user descriptions and questionnaires, producing documentation for requirements, verification, testing, and auditing;
- **Cloud Mobile Security Test Specification and Tools** (CM2ST), which generates recommendations for security tests and testing tools needed for cloud-based mobile applications. It uses a questionnaire to gather information about the application, whether it has already developed or still in development.

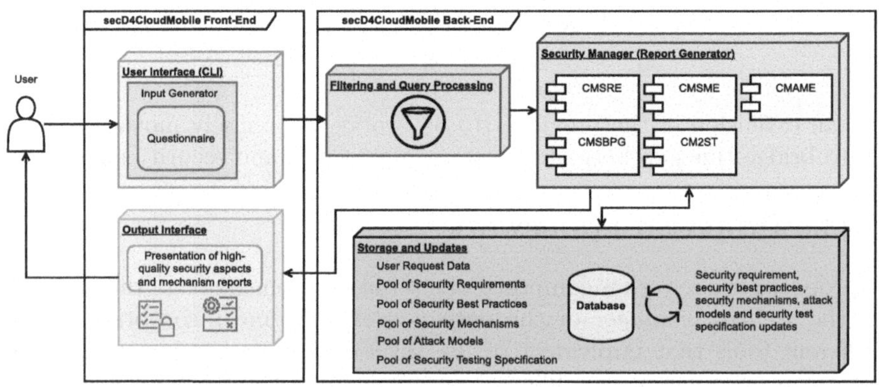

Fig. 2. High-level architecture of the secD4CloudMobile framework.

3.1 User Interface

The front end consists of a console *user interface*, from which the user interacts with the framework. Through it, the user has access to the main menu, which allows them to access the desired module. Figure 3 contains a representative screenshot of a part of the *user interface*.

As highlighted above, the answers to the questionnaire constitute the inputs to the framework. This component is responsible for the consistency of the information presented to the user and for the interaction.

```
Welcome to SecD4CLOUDMOBILE Framework!
What would you like to do (1-8)?

1 - Answer the Questions Required for Processing
2 - Security Requirement Elicitation Request
3 - Secure Development Best Practices Request
4 - Secure Development Mechanisms Request
5 - Attack Model Mapping Request
6 - Security Test Specification and Tool Request
7 - Full Report Request
8 - Exit
Input >
```

Fig. 3. Framework main menu presentation.

3.2 Filtering and Query Processing

The *Filtering and Query Processing* component manages user input by screening for errors and validating requests. It ensures that input data is accurate and conforms to specified formats. Any non-compliant entries are rejected, requiring users to re-enter their requests. Each request is stored in a database for further processing. The component includes 22 functions, each linked to a specific question in the questionnaire. After this, slices of the data are ready to be supplied, depending on the specific request from a user, to the generators of either security requirements, security mechanisms, security best practices, attack models and specific security tests and tools.

3.3 Security Manager

This component defines the blueprint of the tools developed for the framework and it contains three main similar components: 1) *User Answer Checker* – evaluates user input and extracts key information, such as `yes`, `no`, `personal`, `confidential`, `programming languages`, or `storage type` (local or remote); 2) *Selector* – Chooses the relevant recommendations and answers from a database based on the extracted information; and 3) *Aggregator* – Compiles the individual security requirements into a complete set and generates `PDF` reports for output. The implementation of this module differs for each of the tools.

To illustrate how the modules work, example questions and responses from the CMSRE module are provided in Table 2, along with the generated recommendations in Table 3, as well as its implementation logic, presented in Algorithm 1, with the function header (prototype) and accompanying pseudo-code that demonstrates the associations in this logical file. Algorithm 1 is just a series of if/then instructions. A significant part of the work towards achieving this logic involved reviewing existing literature and state-of-the-art practices to design this mapping. The complete implementations of all the modules are available in the GitHub repository of the project[1].

[1] https://github.com/SECURIoTESIGN/SECURIoTESIGN/tree/master.

Table 2. M-Payment application use case scenario example from the CMSRE module.

Q#	Questions	Answers
Q1	What will the system's mobile platform be?	Android App
Q2	What will the application's domain be?	m-Payment
Q3	Will the system use authentication?	Yes
Q4	What will be the authentication scheme to be implemented in the system?	Biometric-based & Factors-based
Q5	Will the system use a Database?	Yes
Q6	What will be the type of data storage?	SQL DBMS
Q7	Which Database (DBMS) will be?	SQLite
Q8	What type of information will the application handle?	Critical Data
Q9	Where will the system store the data?	Hybrid (Hybrid Storage)
Q10	Will there be a user registration?	Yes
Q11	Which way of user registration will be used?	The user will register themselves
Q12	Which programming languages will be used in the implementation of the system?	Java
Q13	Will the system allow user input forms?	Yes
Q14	Will the system allow upload files?	No
Q15	Will the system record event logs?	Yes
Q16	Will the system allow regular updates?	Yes
Q17	Will the system use third-party apps?	Yes
Q18	What is the environment in which the app will be used?	Hybrid Cloud (Mix connection)
Q19	Do you want to further specify hardware details concerning the system?	Yes
Q20	What will be the type of authentication implemented in hardware?	Basic Authentication (user/pass)
Q21	What wireless technologies will be on the hardware?	3G/4G/5G, Bluetooth, Wi-Fi, GPS, NFC
Q22	Can someone gain physical access to the machine where the system will operates?	Yes

Table 3. M-Payment use case recommendation from CMSRE.

Recommendations		
• Confidentiality	• Data Origin	• Integrity
• Reliability	• Availability	• Physical Security
• Authenticity	• Forgery Resistance	• Authorization
• Tamper Detection	• Non-repudiation	• Data Freshness
• Accountability	• Confinement	

3.4 Output Interface

The *Output Interface* component structures an output in the form of a complete report, aggregating all the individual outputs into a single file in PDF format. It also encapsulates services for storing intermediate data generated by the tools, namely MD, PNG, and TXT files.

4 Performance Evaluation

The results of usability tests are presented in this section. These tests were performed to evaluate the performance of the proposed SECD4CLOUDMOBILE framework. The SUS questionnaire, proposed by Brooker [11], was used for this evaluation because it is objective, fast, free, and easy to use. Additionally, SUS is the most commonly used standardized questionnaire for evaluating perceived usability [22], which is on the basis of the presented work.

We also compared the results of two of the framework tools with the results generated by the ChatGPT, Copilot and Gemini LLMs based on the descriptions or characteristics of the Cloud-based mobile application and the Cloud and mobile ecosystem, since they are nowadays also perceived as human assistants for several tasks. The use of LLMs involved collecting security requirements or mechanisms from a brief application description. We compared the outputs from the LLMs with those from two tools in the proposed framework. A closer match was perceived as better efficiency and effectiveness of the framework tools. LLMs are chosen for their recognized potential in Natural Language Processing (NLP), code generation [17], and program understanding, along with their recent applications in requirements elicitation, analysis, specification, and validation [3].

This procedure is a crucial step in validating the effectiveness of the proposed framework. The tests also confirm that the tools within the SECD4CLOUDMOBILE framework function as intended, according to the design goals. The process of creating usability tests involves developing several test tasks that reflect typical real-world scenarios for users to complete using the tool. Following the completion of these tasks, three performance metrics will be utilized, as detailed in the next subsection, to analyze and evaluate the test results.

4.1 Usability Testing Set-Up

To evaluate the functionality of the framework and ensure it is user-friendly, we created five test scenarios. Each scenario focuses on a different module: one for the security requirements generator, one for the best practices generator, one for the security mechanisms generator, one for the attack model generator, and one for the security test specification generator, along with tools for automating these processes. Table 4 describes these scenarios briefly.

Algorithm 1: Summary of the Main Stages of the Cloud Security Requirements Elicitation Tool Implementation.

Input: User.

Output: A set of security requirements.

```
if database = yes and (storage = personal or storage = confidential
or storage = critical) then
    requirements ← confidentiality
if database = yes and (storage = personal or storage = confidential
or storage = critical) then
    requirements ← integrity
if authentication = yes then
    requirements ← authentication
if database = yes and (storage = personal or storage = confidential
or storage = critical) then
    requirements ← confidentiality
if database = yes and (or storage = confidential or storage = critical) then
    requirements ← authorization
if database = yes and (storage = personal or storage = confidential
or storage = critical) then
    requirements ← non − repudiation
if database = yes and (storage = personal or storage = confidential
or storage = critical) then
    requirements ← accountability
if database = yes and (or storage = confidential or storage = critical) then
    requirements ← privacy
if database = yes and (storage = personal or storage = confidential
or storage = critical) then
    requirements ← confidentiality
if database = yes and physicalAccess = yes and (or storage = confidential
or storage = critical) then
    requirements ← physicalSecutiry
if database = yes and physicalAccess = yes and (or storage = confidential
or storage = critical) then
    requirements ← forgeryResistance
if database = yes and physicalAccess = yes and (or storage = confidential
or storage = critical) then
    requirements ← tamperDetection
if database = yes and (storage = personal or storage = confidential
orstorage = critical) then
    and (uploads = yes or regularUpdates = yes or thirdPartyApp = yes)
    requirements ← confinement
return requirements
```

Table 4. Use case scenarios for the framework tools evaluation.

#	Use Case Scenario	Description
1	End-to-End Mobile Messenger	This case concerns a smartphone messaging app where users register with their mobile numbers and can add contacts and invite friends
2	Vaccum Cleaner	This vacuum cleaner features a built-in 360°C surveillance camera, functioning as both a cleaning device and a mobile security tool
3	Smart Heart Rate	This case features a heart rate monitor with a wearable sensor and a mobile app for secure patient-clinical staff communication via Bluetooth
4	Smart Temperature and Humidity Monitoring	This app monitors temperature and humidity in fields to optimize irrigation, using a mobile app and a drone with sensors. Data is stored on a cloud server
5	Mobile Wallet	This scenario covers the Mobile Wallet system, which enables secure transactions between individuals or organizations via mobile networks

Usability Metrics. The usability tests involved selecting impartial participants to evaluate the SECD4CLOUDMOBILE framework. After a brief introduction to the a given tool, participants were allowed to independently choose the scenarios they wanted to test, giving them time to initially think of the characteristics and potential security issues of those cases. Each user would interact with the framework to assess the scenario (s)he has chosen. To assess the usability of the system they just tested, we later asked them to answer the questionnaire and used the three usability metrics of SUS: *effectiveness*, *efficiency*, and *satisfaction*.

4.2 SUS Results Overview

The usability tests involved inviting impartial participants to test the tools within the SECD4CLOUDMOBILE framework. After a brief introduction to the tool at the beginning, participants were allowed to conduct the tests independently and choose the scenarios they wanted to examine. This approach utilized non-randomized sampling based on voluntary responses. The sample included 15 individuals.

In terms of specialization, the sample is heterogeneous. The majority of respondents had expertize in computer science (80%), followed by information systems (33%) and electrical and computer engineering (7%). In terms of occupation, the sample is heterogeneous also. 40% of participants were teachers, 20% were researchers or developers, 13% were master's or doctoral students, and 7% were security specialists. The majority of respondents took between 6 and 10 min (60%) to complete the framework questionnaire, followed by those who took 1 to 5 min (40%).

To analyze the data from the SUS questionnaire, we utilized the System Usability Scale Analysis Toolkit developed and presented by Blattgerste et al. [8]. This tool is user-friendly and allows for the automation of data processing using the SUS questionnaire [8] (including, e.g., the percentile curve [35], and the conclusiveness graph [43]).

4.3 LLMs Approach Overview

Two of the use case scenarios and two modules were chosen at random for obtaining and comparing results of the tools with the responses generated by Copilot, Gemini, and ChatGPT. The scenarios and tools selected were the *End-to-End Mobile Messenger* and *Smart Heart Rate With a Mobile Interface* scenarios and the CMSRE and CMSME modules, respectively. The first scenario relates to the first module and the second to the second module. The following procedures were used in these experiments:

1. Run each of the tools for their corresponding scenarios, answering to their questions and saving the respective outputs;
2. Questions were posed to the LLMs, based on the characteristics of the applications to be developed, to generate security requirements and security mechanisms, respectively, as follows:
   (a) What security requirements should a Cloud-based instant messaging mobile application that handles sensitive user data meet to guarantee the security and privacy of the user data?
   (b) What security mechanisms should be implemented in a Cloud-based mobile health application (Smart Heart Rate With a Mobile Interface) that handles sensitive user data?
3. Finally, the results generated by the LLMs and the results generated by the CMSRE and CMSME tools were humanly compared.

5 Result Discussion

This section presents the discussion on the achieved results, both from the SUS methodology, and the LLM approach.

5.1 SUS Result Discussion

The average SUS scores were compared across different conditions, and the iterative scores were analyzed for formative testing, to facilitate comparisons with the contextualization scales provided. Figure 4 and Table 5 present these comparisons.

The adjective scale [6,36], the grade scale [6], the quartile scale [6], the net promoter scale [34] and the acceptability scale [1] were all considered and calculated from the achieved results, and are presented in Table 5. The percentile curve contextualization graph, showing the calculated SUS study scores of the uploaded dataset on a percentile curve, is presented in Fig. 5. The per-item contribution graph, which provides a breakdown of each of the 10 questions in the SUS, and the individual contribution of each of them to the overall score, is presented in Fig. 6. The conclusiveness graph [43] analyzed various sample sizes of completed SUS questionnaires, and indicated that a sample size of 8 provided a 78% conclusive rate, 10 yielded an 80% conclusive rate, and sample sizes of 12 or more produced 100% conclusive results.

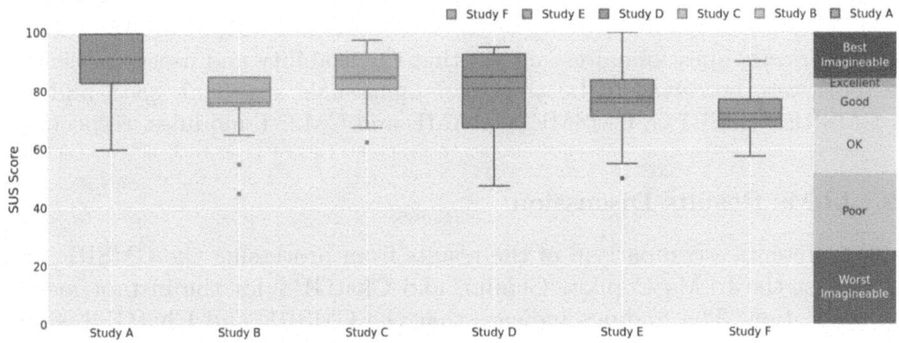

Fig. 4. SUS score comparison plot.

Table 5. Most important calculated metrics with all contextualization scale results.

Variable	SUS Score mean	SD	Min	Max	1.Quartile	Median	3.Quartile	Adjective Scale	Grade Scale	Quartile Scale	Acceptability Scale	NPS Scale
Study A	91.17	12.74	60.0	100.0	82.5	100.0	100.0	Best Imaginable	A	4th	Acceptable	Promoter
Study B	76.17	12.06	45.0	85.0	75.0	80.0	85.0	Good	B	3rd	Acceptable	Passive
Study C	84.50	10.70	62.5	97.5	80.0	85.0	92.5	Best Imaginable	A	4th	Acceptable	Promoter
Study D	80.83	15.23	47.5	95.0	62.5	85.0	92.5	Excellent	A	4th	Acceptable	Promoter
Study E	76.00	13.02	50.0	100.0	70.0	77.5	85.0	Good	B	3rd	Acceptable	Passive
Study F	72.17	8.34	57.5	90.0	67.5	70.0	77.5	Good	C	3rd	Marginal	Passive

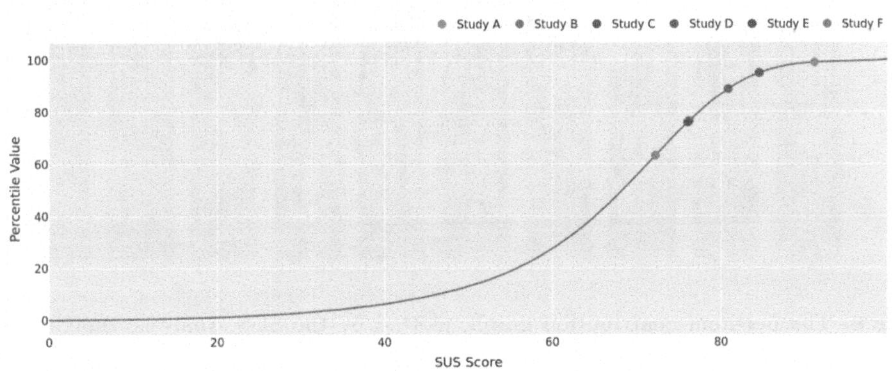

Fig. 5. The percentile curve graph, plotted by the SUS Analysis Toolkit.

In conclusion, the results presented in Fig. 4 and Table 5 indicate very favorable outcomes for all variables assessed. Studies A and C, which focus on the usability of the framework and the CMSPBG module, respectively, achieved the highest scores across all metrics, including the SUS *Score Mean*, *Adjective Scale*, *Grade Scale*, *Acceptability Scale*, and *NPS Scale*. They were followed by Study D, which evaluates the usability of the CMSME module, Study B for the CMSRE module, Study E for the CMAME module, and Study F for the CM2ST module. The results of the SUS (*Score* > 72; $8.33 < SD < 15.24$; *Scale Acceptability*, and *NPS Scale)* questionnaires suggest that the usability and user experience of the tools were perceived to be *good*, *best imaginable*, *excellent*, *good*, and *good* for CMSRE, CMSBPG, CMSME, CMAME and CM2ST modules, respectively.

5.2 LLMs Results Discussion

Table 6 presents a comparison of the results from processing the CMSRE module against the LLMs Copilot, Gemini, and ChatGPT for the instant messaging application. The findings indicate that the CMSRE and ChatGPT significantly outperformed both Copilot and Gemini. CMSRE and ChatGPT produced almost all the necessary security requirements for an instant messaging application, while Gemini and Copilot met only 50% of those requirements. Similarly, the same Table 6 presents the results of processing the CMSME module along with the LLMs Copilot, Gemini, and ChatGPT for the smart healthcare application. As shown, the CMSME and ChatGPT were the most effective, successfully identifying nearly all the necessary security mechanisms.

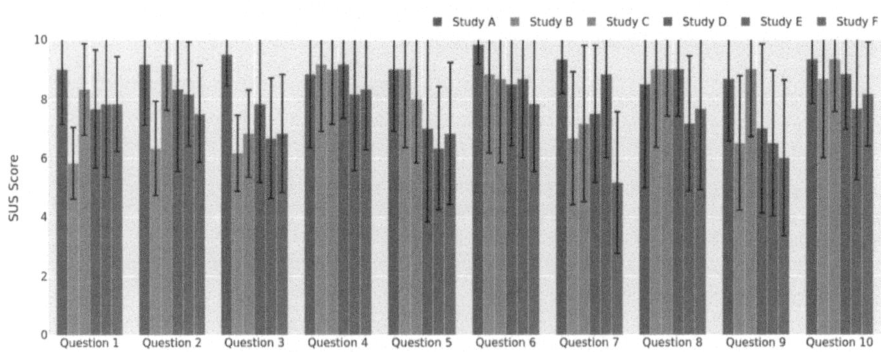

Fig. 6. The per-item contribution graph, plotted by the SUS Analysis Toolkit with data taken from this current study.

Table 6. Results of processing the instant messaging application use case scenario for the CMSRE module, Gemini, Copilot, and ChatGPT LLMs.

	Instant Messaging					Smart Health Care Use Case			
NFR	Copilot	Gemini	ChatGPT	CMSRE	**SM**	Copilot	Gemini	ChatGPT	CMSME
Authenticity	✓	✓	✓	✓	Cryptographic Primitives	✓	✓	✓	✓
Confidentiality	✓	✓	✓	✓	Cryptographic Protocols	✓	✓	✓	✓
Availability	✓	✖	✓	✓	Secure Backup	✓	✓	✓	✓
Integrity	✓	✓	✓	✓	Authentication	✓	✓	✓	✓
Authorization	✓	✓	✓	✓	Authorization	✓	✓	✓	✖
Non-repudiation	✖	✖	✖	✓	Access Control	✓	✓	✓	✓
Accountability	✓	✓	✓	✓	Inspection	✖	✓	✓	✓
Reliability	✖	✖	✓	✓	Security Audit	✓	✓	✖	✓
Physical Security	✖	✓	✓	✓	Logging	✖	✖	✓	✓
Forgery Resistance	✖	✖	✓	✓	Tamper Detection	✖	✖	✓	✓
Tamper Detection	✖	✓	✓	✓	Physical Location	✖	✖	✖	✓
Data Freshness	✖	✖	✓	✓	Confinement	✖	✖	✓	✓
Confinement	✖	✖	✓	✓					
Interoperability	✖	✖	✓	✓					
Data Origin Authentication	✓	✓	✓	✓					
Privacy	✓	✓	✓	✓					

Despite the limited effectiveness of the LLMs, the results generated by ChatGPT, CMSRE, and CMSME exhibit some similarities, indicating a level of agreement on the outcomes that should be presented to the user. This somewhat validates the effectiveness of the tools within the proposed framework.

6 Conclusions and Future Work

Despite security reports from major companies highlighting the rising security risks in the cloud and mobile ecosystems – particularly noting the application layer as a vulnerable point – security aspects and mechanisms continue to be overlooked during the design and development of applications, mostly in favor of faster time-to-market.

The version of the SECD4CLOUDMOBILE framework described herein serves as a virtual assistant for securely developing cloud-based mobile applications, integrating both software and security engineering and trying to act as effortlessly as possible. It helps identify security requirements, best practices, security mechanisms, and facilitates attack and system modeling, test specification, high-quality documentation generation, and auditing. This work outlines how expert knowledge can be represented for decision-making in security aspects and software testing within cloud and mobile environments. The performance of these tools was evaluated using the SUS questionnaire and assessments with ChatGPT, Copilot and Gemini LLMs. Usability tests have indicated that the target audience is interested in utilizing the framework tools with SUS score above 72 points.

For future development, and to ensure reusability and repeatability, plans are in place to integrate machine learning into the framework tools, including the use of Meta or OpenAI Application Programming Interfaces (APIs).

Acknowledgments. This work is funded by FCT/MECI through national funds and, when applicable, co-funded EU funds under UID/50008: Instituto de Telecomunicações. Part of this work was performed under the scope of Project SECURIoTESIGN, with funding from FCT/COMPETE/FEDER (reference number POCI-01-0145-FEDER-030657) and FCT research and doctoral grants BIM/n° 32/2018-B00582 and SFRH/BD/133838/2017, respectively.

Disclosure of Interests. The authors have no competing interests to declare that are relevant to the content of this article.

References

1. Aaron Bangor, P.T.K., Miller, J.T.: An empirical evaluation of the system usability scale. Int. J. Hum.-Comput. Interact. **24**(6), 574–594 (2008). https://doi.org/10.1080/10447310802205776
2. Al-Omary, A.: A secure framework for mobile cloud computing. In: 2019 International Conference on Innovation and Intelligence for Informatics, Computing, and Technologies (3ICT), pp. 1–6. IEEE, Sakhier (2019). https://doi.org/10.1109/3ICT.2019.8910294
3. Arora, C., Grundy, J., Abdelrazek, M.: Advancing Requirements Engineering Through Generative AI: Assessing the Role of LLMs, pp. 129–148. Springer, Cham (2024). https://doi.org/10.1007/978-3-031-55642-5_6
4. Ashton, K., et al.: That 'internet of things' thing. RFID J. **22**(7), 97–114 (2009)
5. Awaysheh, F.M., et al.: Security by design for big data frameworks over cloud computing. IEEE Trans. Eng. Manag. **69**(6), 3676–3693 (2022). https://doi.org/10.1109/TEM.2020.3045661
6. Bangor, A., et al.: Determining what individual sus scores mean: adding an adjective rating scale. J. Usabil. Studies **4**(3), 114–123 (2009)
7. Batista, P.M.M.: Applying and Testing Mult-Iclass and Multi-Output Algorithms in the Mapping of Security Requirements with Technologies and Best Practices. Master's thesis, Universidade da Beira Interior (Portugal) (2022)
8. Blattgerste, o.: A web-based analysis toolkit for the system usability scale. In: Proceedings of the 15th International Conference on PErvasive Technologies Related to Assistive Environments, PETRA '22, pp. 237–246. Association for Computing Machinery, New York (2022). https://doi.org/10.1145/3529190.3529216
9. Bonomi, F., Milito, R., Natarajan, P., Zhu, J.: Fog computing: a platform for internet of things and analytics. In: Bessis, N., Dobre, C. (eds.) Big Data and Internet of Things: A Roadmap for Smart Environments. SCI, vol. 546, pp. 169–186. Springer, Cham (2014). https://doi.org/10.1007/978-3-319-05029-4_7
10. Bonomi, F., et al.: Fog computing and its role in the internet of things. In: Proceedings of the First Edition of the MCC Workshop on Mobile Cloud Computing, MCC '12, pp. 13–16. Association for Computing Machinery, New York (2012).https://doi.org/10.1145/2342509.2342513
11. Brooke, J., et al.: Sus-a quick and dirty usability scale. Usabil. Eval. Ind. **189**(194), 4–7 (1996)
12. Buyya, R., Yeo, C.S., Venugopal, S., Broberg, J., Brandic, I.: Cloud computing and emerging it platforms: vision, hype, and reality for delivering computing as the 5th utility. Futur. Gener. Comput. Syst. **25**(6), 599–616 (2009). https://doi.org/10.1016/j.future.2008.12.001

13. Chimuco, F.T., Sequeiros, J.B., Simões, T.M., Freire, M.M., Inácio, P.R.: Expediting the design and development of secure cloud-based mobile apps. Int. J. Inf. Secur. **23**(4), 3043–3064 (2024). https://doi.org/10.1007/s10207-024-00880-6
14. Chimuco, F.T., et al.: Secure cloud-based mobile apps: attack taxonomy, requirements, mechanisms, tests and automation. Int. J. Inf. Secur. **22**(4), 833–867 (2023). https://doi.org/10.1007/s10207-023-00669-z
15. da Costa, J.C.A.N.: Threat Modeling Solution for Internet of Things in a WebBased Security Framework. Master's thesis, Universidade da Beira Interior (Portugal) (2021). http://hdl.handle.net/10400.6/11849
16. Di Pierro, M.: What is the blockchain? Comput. Sci. Eng. **19**(5), 92–95 (2017). https://doi.org/10.1109/MCSE.2017.3421554
17. Dong, Y., Jiang, X., Jin, Z., Li, G.: Self-collaboration code generation via chatgpt. ACM Trans. Softw. Eng. Methodol. **33**(7) (2024). https://doi.org/10.1145/3672459
18. Elazhary, H.: Internet of things (iot), mobile cloud, cloudlet, mobile iot, iot cloud, fog, mobile edge, and edge emerging computing paradigms: Disambiguation and research directions. J. Netw. Comput. Appl. **128**, 105–140 (2019). https://doi.org/10.1016/j.jnca.2018.10.021
19. Gupta, S.S.: Blockchain. IBM Onlone (http://www. IBM. COM) (2017). https://www.isical.ac.in/~debrup/slides/Bitcoin.pdf
20. Hussain, M., et al.: Conceptual framework for the security of mobile health applications on android platform. Telematics Inform. **35**(5), 1335–1354 (2018). https://doi.org/10.1016/j.tele.2018.03.005
21. Imielinski, T., Korth, H.F.: Mobile computing, vol. 353. Springer, Heidelberg (1996)
22. Lewis, J.R.: The system usability scale: past, present, and future. Int. J. Human-Comput. Interact. **34**(7), 577–590 (2018). https://doi.org/10.1080/10447318.2018.1455307
23. Lopes, C.G.: SemiAutomatic Generation of Tests for Assessing Correct Integration of Security Mechanisms in the Internet of Things. Master's thesis, Universidade da Beira Interior (Portugal) (2021). http://hdl.handle.net/10400.6/11844
24. Mell, P., Grance, T., et al.: The nist definition of cloud computing. National institute of science and technology, special publication (2011)
25. Nofer, M., et al.: Blockchain. Bus. Inf. Syst. Eng. **59**(3), 183–187 (2017). https://doi.org/10.1007/s12599-017-0467-3
26. Ren, W., et al.: Lightweight and compromise resilient storage outsourcing with distributed secure accessibility in mobile cloud computing. Tsinghua Sci. Technol. **16**(5), 520–528 (2011). https://doi.org/10.1016/S1007-0214(11)70070-0
27. Samaila, M.G., et al.: A quick perspective on the current state of iot security: a survey. In: Networks of the Future, pp. 431–464. Chapman and Hall/CRC (2017)
28. Samaila, M.G., et al.: Challenges of securing internet of things devices: a survey. Secur. Priv. **1**(2), e20 (2018). https://doi.org/10.1002/spy2.20
29. Samaila, M.G., et al.: Security threats and possible countermeasures in iot applications covering different industry domains. In: Proceedings of the 13th International Conference on Availability, Reliability and Security. ARES '18. Association for Computing Machinery, New York (2018). https://doi.org/10.1145/3230833.3232800
30. Samaila, M.G., et al.: "iot-harpseca: A framework for facilitating the design and development of secure iot devices". In: Proceedings of the 14th International Conference on Availability, Reliability and Security. ARES '19. Association for Computing Machinery, New York (2019). https://doi.org/10.1145/3339252.3340514

31. Samaila, M.G., et al.: Iot-harpseca: a framework and roadmap for secure design and development of devices and applications in the iot space. IEEE Access **8**, 16462–16494 (2020). https://doi.org/10.1109/ACCESS.2020.2965925
32. Samaila, M.G., et al.: A preliminary evaluation of the sre and sbpg components of the iot-harpseca framework. In: 2020 Global Internet of Things Summit (GIoTS), pp. 1–7. IEEE, Dublin (2020). https://doi.org/10.1109/GIOTS49054.2020.9119590
33. Samaila, M.G., et al.: Performance evaluation of the sre and sbpg components of the iot hardware platform security advisor framework. Comput. Netw. **199**, 108496 (2021). https://doi.org/10.1016/j.comnet.2021.108496
34. Sauro, J.: Predicting net promoter scores from system usability scale scores. MeasuringU **16**, 2022 (2012)
35. Sauro, J., Lewis, J.R.: Quantifying the User Experience, Second Edition: Practical Statistics for User Research, 2nd edn. Morgan Kaufmann Publishers Inc., San Francisco (2016)
36. Sauro, J., Lewis, J.: 5 ways to interpret a sus score (2018). https://measuringu.com/interpret-sus-score/
37. Senyo, P.K., Effah, J., Addae, E.: Preliminary insight into cloud computing adoption in a developing country. J. Enterp. Inf. Manag. (2016)
38. Sequeiros, J., et al.: Secure design and development of applications in the cloud and mobile ecosystem. In: Inforum - Simpósio de Informática (2023)
39. Sequeiros, J.B.F., et al.: Attack and system modeling applied to iot, cloud, and mobile ecosystems: embedding security by design. ACM Comput. Surv. **53**(2) (2020). https://doi.org/10.1145/3376123
40. Sherman, M.: Attack surfaces for mobile devices. In: Proceedings of the 2nd International Workshop on Software Development Lifecycle for Mobile. DeMobile 2014, pp. 5–8. Association for Computing Machinery, New York (2014). https://doi.org/10.1145/2661694.2661696
41. Srivastava, M., Thamilarasu, G.: Msf: A comprehensive security framework for mhealth applications. In: 2019 7th International Conference on Future Internet of Things and Cloud Workshops (FiCloudW), pp. 70–75. IEEE, Istanbul (2019). https://doi.org/10.1109/FiCloudW.2019.00026
42. Tan, L., Wang, N.: Future internet: the internet of things. In: 2010 3rd International Conference on Advanced Computer Theory and Engineering(ICACTE), vol. 5, pp. V5-376–V5-380 (2010). https://doi.org/10.1109/ICACTE.2010.5579543
43. Tullis, T.S., Stetson, J.N.: A comparison of questionnaires for assessing website usability. In: Usability Professional Association Conference, vol. 1, pp. 1–12 (2004)
44. Zkik, K., et al.: A new secure framework in mcc using homomorphic signature: application in banking data. In: Transactions on Engineering Technologies, pp. 413–427. Springer, Singapore (2016)

Enhancing JavaScript Malware Detection Through Weighted Behavioral DFAs

Pedro Pereira(✉), José Gonçalves, João Vitorino, Eva Maia, and Isabel Praça

GECAD, ISEP, Polytechnic of Porto, rua Dr. António Bernardino de Almeida, 4249-015 Porto, Portugal
{peesp,jpsgs,jpmvo,egm,icp}@isep.ipp.pt

Abstract. This work addresses JavaScript malware detection to enhance client-side web application security with a behavior-based system. The ability to detect malicious JavaScript execution sequences is a critical problem in modern web security as attack techniques become more sophisticated. This study introduces a new system for detecting JavaScript malware using a Deterministic Finite Automaton (DFA) along with a weighted-behavior system, which we call behavior DFA. This system captures malicious patterns and provides a dynamic mechanism to classify new sequences that exhibit partial similarity to known attacks, differentiating them between benign, partially malicious, and fully malicious behaviors. Experimental evaluation on a dataset of 1,058 sequences captured in a real-world environment demonstrates the capability of the system to detect and classify threats effectively, with the behavior DFA successfully identifying exact matches and partial similarities to known malicious behaviors. The results highlight the adaptability of the system in detecting emerging threats while maintaining transparency in decision making.

Keywords: JavaScript malware detection · Deterministic finite automata (DFA) · Behavior-based classification · Partial match detection

1 Introduction

Nowadays, ensuring client-side security is a critical challenge [2], as attackers increasingly exploit vulnerabilities to manipulate, inject, or alter script's execution [22]. Web scripts execute a range of actions, from simple tasks like locating and changing document object model (DOM) elements to harmful tasks such as injecting dynamic code or stealing sensitive user data [10,23]. With the increasing complexity of today's web applications and the development of sophisticated evasion techniques, conventional detection methods are not enough [25]. Signature-based detection and anomaly-based detection with machine learning (ML) classifiers are the conventional approaches which are usually not feasible

I. Praça et al. (Eds.): EICC 2025, CCIS 2500, pp. 201–214, 2025.
https://doi.org/10.1007/978-3-031-94855-8_13

when large datasets are not available, requiring better interpretable and adaptable classification techniques [6].

In order to address these challenges, this paper presents an automaton-based classification framework for analyzing script behaviors. Script behaviors refer to the observable actions performed by JavaScript code during its execution, which collectively reveal the script's functionality and potential for malicious activity. The classification of script behavior is a fundamental step in improving web security, as it allows a deeper understanding of how attacks work [19]. Understanding attack behavior supports the detection of new versions or variances of malware scripts, facilitating the development of generalized defenses [12]. Analyzing malicious scripts through behavior modeling makes it possible to find common patterns that all variants of each attack type share [8]. Then, the script's maliciousness can be determined by comparing its execution trace to established behavior models of known attacks.

The developed system uses scripts to model execution behaviors through sequences that demonstrate particular activities, such as DOM element generation, storage operations or dynamic code injection. By analyzing these sequences, it is possible to assess the degree to which the observed behaviors are aligned with a set of known malicious patterns. This is performed with a Deterministic Finite Automaton (DFA) system that tracks state transitions by monitoring observed script behaviors is constructed. This system provides a deterministic way to display action sequences so that it can identify whether scripts are benign, malicious, or partially malicious.

A key innovation of the system is the use of behavior weights, which assign significance to each script action based on its importance in an attack context. This weighting system ensures a finer-grained risk assessment by recognizing that certain individual behaviors inherently carry more risk than others. For example, actions like accessing cookies or interacting with the DOM may typically be benign, while behaviors such as injecting dynamic code or sending data are probably more suspicious due to their potential use in malicious activities. This weighting system has been developed in close collaboration with JavaScript experts, ensuring that the assigned weights accurately reflect the real-world threat dynamics. By assigning higher weights to high-risk behaviors, the system prioritizes their significance in the classification process, enabling a more accurate evaluation of scripts with potential threats.

Therefore, this work proposes a novel classification system that combines automata theory and behavior weighting. Unlike traditional classifiers, the proposed method offers structured and interpretable decision making, increasing transparency in script behavior analysis. Experimental evaluation shows that this method is effective in distinguishing benign from malicious scripts, opening a promising path to improve automated threat detection systems. Through this work, our goal is to complement existing solutions by offering a behavior-based classification system that enhances web application security against evolving threats.

This paper is divided into several sections that provide details that could help researchers replicate the baseline and compare it with their findings. Section 2 presents related works. Section 3 details the preparation steps and the methodology used to implement our system. Section 4 provides an in-depth analysis of the results obtained in a proprietary dataset. Finally, Sect. 5 reviews the primary conclusions reached and suggests potential directions for future research.

2 Related Work

The detection of JavaScript malware has relied on static, dynamic, and hybrid analysis techniques, each with inherent limitations. Static analysis approaches such as ZOZZLE [4] and PROPHILER [1] seek to predict maliciousness from syntactic and lexical features extracted from JavaScript code. For example, ZOZZLE detects obfuscated malware by using statistical models over Abstract Syntax Trees, while PROPHILER uses lightweight static features to classify scripts. However, these methods often fail at highly obfuscated or dynamically generated scripts which do not present themselves as obvious static patterns [1,3,4].

Dynamic analysis tools, such as JSAND [3], execute JavaScript in controlled environments to extract behavioral features like re-directions, de-obfuscation, and exploit attempts. This method detects unseen threats through runtime behavioral analyzes, as seen in the work of Gorji et al. [5] who clustered predictive function calls to create behavioral signatures. However, this analysis technique is not only computationally expensive, but also vulnerable to evasive malware that can detect and escape sandbox testing environments. Tools like FV8 [14] have been built to detect evasive actions in malicious scripts. FV8 changes the V8 JavaScript runtime in order to force code execution on API's that inject dynamic code, increasing the code coverage and visibility of malicious behaviors.

Hybrid analysis methods, such as CUJO [17] employ static and dynamic features, but still face scalability issues and difficulties in modeling sophisticated behavioral sequences. He et al. [7] developed a combined static and dynamic analysis system that uses Random Forest classifier to analyze feature-rich datasets achieving high detection accuracy. The context-sensitive characteristics of JavaScript malware remain a challenge for hybrid analysis approaches despite recent improvements.

The integration of ML techniques within JavaScript malware detection has significantly improved the detection process, showing superior results compared to traditional methods [21]. A systematic review of the literature on studies from 2009 to 2019 reveals that ML-based detection models achieve greater precision in detecting malicious JavaScript code [21], underscoring the growing reliance on ML for malware detection. For example, Jodavi et al. [9] utilized ensemble of support vector machine classifiers with a binary particle swarm optimization algorithm to identify vulnerable JavaScript code, reporting f-measure over 86%. However, despite these promising results, ML-based approaches have several inherent limitations. Successful operation of these models depends on extensive datasets and faces challenges when addressing new attacking methods [21]. The black-box operational nature of numerous ML algorithms leads to complicated understanding of detection decision [16] by analysts who need to interpret the rationale supporting classification results, thus diminishing practical usability in dynamic security environments.

In contrast, automata-based models offer a structured and transparent alternative by explicitly modeling the sequence of actions performed by scripts during execution. Automata-based models have been recently explored as ways to represent the malware's behavior. Previous studies have used Finite State Automaton (FSA) to represent the behavior of processes based on system calls sequences [11,13]. For example, Sekar et al. [20] used heuristic-based methods to describe normal behavior using FSAs, but the approach is limited to heuristics that are predefined. Xue et al. [26] introduced a DFA to abstract and summarize common behaviors of malicious JavaScript associated with specific attack types, effectively identifying variants of known malware through behavioral patterns rather than static signatures. This makes them particularly useful for detecting obfuscated or polymorphic malware. The results of these studies indicate a shift towards modeling the dynamic execution patterns of malware in order to improve detection capabilities. Lemberger et al. [24] explored the use of register automata and pushdown systems to describe malware specifications, in order to capture more complex and context sensitive interactions in execution traces. These models extend beyond DFA's by incorporating program stack and register values, though they still face limitations in addressing the dynamic and adaptive nature of JavaScript malware.

Therefore, to overcome the limitations of current detection methods in handling context-based behaviors and the dynamic nature of JavaScript malware, this work introduces a novel approach focused on enhancing script behavior analysis. Although AI techniques have shown some advantages in certain cases, their practical application in this project is not the most appropriate. In addition to requiring a large amount of data, these methods act as 'black boxes, making their decision-making processes opaque and difficult to interpret. In environments where transparent and timely analysis is key, lack of clarity in AI results can impede rapid response and expert assessment. This work enhances script behavior analysis by investigating sequential and contextual behavior patterns during execution, aiming to identify characteristics that distinguish benign from malicious JavaScript activities, thereby providing enhanced adaptability, robustness to emerging risks, and clarity in how decisions are made.

3 Methodology

To effectively detect malicious JavaScript sequences, we begin by analyzing the behaviors that scripts exhibit during execution. Our research started by analyzing a proprietary dataset which represents JavaScript execution details using structured attributes including behavior identifiers alongside timestamps and execution contexts and DOM interactions. The records include information about the invoked methods, their associated behaviors (e.g. *"Add DOM*

Element(s)", *"Access Cookies"*) and metadata about the element and script involved, labeled as malicious or benign. From this comprehensive dataset, we extract sequences of behaviors, which are organized into a relational database for further analysis and DFA construction. It should be noted that this research relies on our unique dataset because no existing public dataset provides equivalent dynamic execution details, making our proprietary dataset essential for this study.

Each sequence in the dataset is identified by a unique ID and is represented as an ordered list of behaviors. These behaviors capture key execution events, such as DOM manipulations (*"Find DOM Element(s)"*, *"Update DOM Element"*, *"Inject Code Dynamically"*), storage access (*"Access Browser Storage"*, *"Access Cookies"*, *"Change Cookies"*), and network interactions (*"Send Data"*). Each script behavior receives a weight value that determines how important it is to detect benign and malicious behaviors. The weight system assigns *"Find DOM Element(s)"* and similar behaviors lower values because their occurrence is harmless, typically doesn't exhibiting malicious behavior. On the other hand, the system assigns elevated weights to dangerous behaviors such as *"Inject Code Dynamically"* or *"Send Data"* because they act as important elements for carrying out malicious activity. This assignment reduces the influence of low-risk behaviors while amplifying high-risk behaviors to create sophisticated risk assessments through action weights in the evaluation process. The label of each sequence, indicating whether it is malicious or benign, is derived directly from the dataset.

To illustrate how the behavior weighting system works, let's consider two examples, one representing a benign script behavior and another demonstrating a malicious behavior. For a benign example, let's consider a script that performs routine DOM manipulations to dynamically update a webpage. The sequence of actions begins with locating a DOM element, such as a button or a div, represented by the behavior *"Find DOM Element(s)"* that has a weight of 2. The script then modifies element attributes through *"Update DOM Element"* which has a weight value of 3. To conclude, the script introduces fresh DOM content through element additions like banners or notifications which receive a weight value of 1. In our dataset this sequence will be represented as [*"Find DOM Element(s)"*, *"Update DOM Element"*, *"Add DOM Element(s)"*]. This sequence has the associated behavior weights of 2, 3 and 1 respectively. These standard activities have received low weight values because they maintain typical behavior patterns without creating significant risk.

For a malicious example inspired by a Magecart-style attack, we usually have behaviors associated with credit card skimming and data theft [18]. At the start of this sequence, the script normally adds harmful code to the webpage, represented by the *"Inject Code Dynamically"* behavior with a weight of 4. The injected code performs searches for sensitive form fields through *"Find DOM Element(s)"*, associated with a weight value of 2. Once these elements are identified, the script captures user data through the behavior *"Access Input"*, assigned a weight of 4, and subsequently sends the stolen data to an external server using *"Send Data"* with a weight of 5. The sequence [*"Inject Code Dynamically"*, *"Find DOM Element(s)"*, *"Access Input"*, *"Send Data"*], associated with the weights of 4, 2, 4 and 5 respectively, demonstrates a systematic progression from start to finish of malicious activities. The sequences risk level increases through weight assignments that focus on actions such as code injection and data exfiltration, which distinguish it from routine system behaviors.

To model the behavior of malicious JavaScript scripts, a DFA, called behavior DFA is built. Formally, a DFA is defined as a 5-tuple $(Q, \Sigma, \delta, q_0, F)$ where Q is the set of states, Σ is the input alphabet, δ is the transition function from $Q \times \Sigma$ to Q, q_0 is the initial state and F is the set of final (accepting) states. In the context of malicious JavaScript modeling, the state space Q represents the distinct states that a script can transition through during execution. These states encode significant intermediate conditions of the execution sequence, such as DOM manipulation, data exfiltration, or script injection. The input alphabet Σ consists of the possible execution behaviors, such as *"Find DOM Element(s)"*, *"Inject Code Dynamically"*, and *"Send Data"*. Transitions between states, governed by δ, occur based on the sequence of observed script behaviors.

The construction of the behavior DFA starts with an analysis of existing malicious JavaScript behavioral sequences. Identified sequences serve as basic elements in creating state transition definitions that help to recognize states with malicious behavior characteristics. Each execution behavior corresponds to a transition between states, where the path through the behavior DFA captures the sequence of actions. Any sequence that reaches a final state F, where malicious execution occurs, is flagged as a potential threat. Fig. 1 present a visual representation of the proposed behavior DFA system constructed using two JavaScript behavior sequences:

[*"Add Event Handler"*, *"Set Callback"*, *"Find DOM Element(s)"*, *"Find DOM Element(s)"*, *"Find DOM Element(s)"*, *"Add Event Handler"*, *"Add Event Handler"*, *"Set Callback"*, *"Find DOM Element(s)"*, *"Find DOM Element(s)"*, *"Find DOM Element(s)"*] and

[*"Set Callback"*, *"Find DOM Element(s)"*, *"Find DOM Element(s)"*, *"Find DOM Element(s)"*, *"Set Callback"*, *"Find DOM Element(s)"*, *"Find DOM Element(s)"*, *"Find DOM Element(s)"*, *"Find DOM Element(s)"*].

These encoded sequences will result in [7, 5, 1, 1, 1, 1, 7, 7, 5, 1, 1, 1] and [5, 1, 1, 1, 5, 1, 1, 1, 1], respectively. Each transition in the behavior DFA is labeled with a behavior represented by an identifier, 7 for example corresponds to *"Add Event Handler"*. The state space Q consists of labeled states q_0, q_1, ..., q_{10}, representing progression through the execution sequence. The behavior DFA starts in the initial state q_0, and transitions between states occur based on the detected behavior having two final states q_6 and q_{10}.

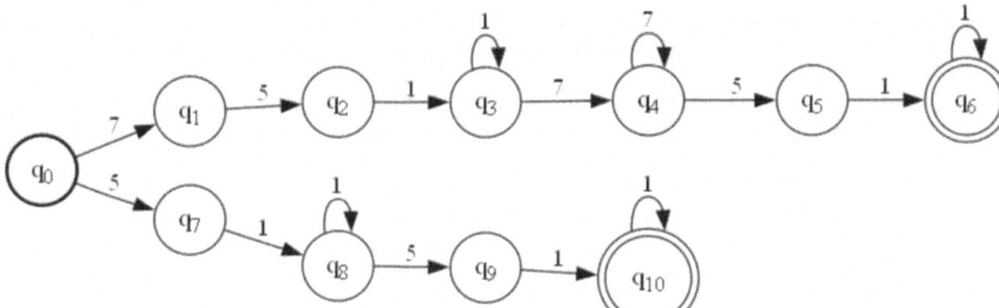

Fig. 1. Behavior DFA example.

To classify a new set of behaviors, the behavior DFA evaluates the sequence of actions performed during the JavaScript execution. The behavior DFA receives sequences of states as transitions that begin from its initial state q_0. During sequence processing, the behavior DFA advances through states according to input behaviors which follow the behavior DFA defined transitions. The behavior DFA detects potential maliciousness whenever a sequence reaches a final state and results in a partially malign behavior when it ends in a non-final state.

As it is essential to identify not only exact matches to known malicious patterns but also taking into account sequences that may share similar behaviors, we further refine the classification process using a distance to the nearest final state. We compute the distance to the closest final state using breadth-first search (BFS) when a sequence does not perfectly match any known malicious pattern. This approach calculates the shortest path from the sequence's last state to a confirmed malicious state, with each transition accumulating a cost based on behavior weights. Lower distances indicate that a sequence is more similar to a known malicious pattern, while higher distances suggest that the sequence is less likely to be malicious. For instance let's consider a sequence [7, 5]. Using the behavior DFA from Fig. 1 we can see that this sequence ends in a state that does not directly reach a confirmed malicious state. Using BFS, we calculate the cost of the shortest path from the sequence's last state, in this case q_2, to all malicious state. In this case, the closest final state is q_6.

Then the classification algorithm will give us the probability of a given sequence being malicious. To do that, we calculate the match percentage, which is the ratio of how well the sequence matches known malicious patterns. The formula for match percentage is:

$$\text{Match Percentage} = \left(\frac{\text{Total Matched Behavior Weight}}{\text{Total Weight to Nearest Final State}} \right) \times 100$$

In this formula, the "Total Matched Behavior Weight" represents the sum of transition weights that lead to the current state. The "Total Weight to Nearest Final State" represents the sum of transition weights which directly lead to the nearest known malicious state, starting on the initial state. For example let's consider the sequence [7, 5], knowing that behaviors 7 and 5 have a weight of 3.

Comparing the sequence with the behavior DFA in Fig. 1 we can see that this sequence matches the behavior DFA, ending in state q_2. The sum of behaviors weight matched, 7 and 5, will be the "Total Matched Behavior Weight", in this case 3 + 3, that is, 6. From this state the nearest final state is q_6, thus the "Total Weight to Nearest Final State" in this case will be the sum of all behaviors weight from state q_0 to q_6 [7, 5, 1, 7, 5, 1]. Assuming that behavior 1 has a weight of 2, this results in the following sum, 3 + 3 + 2 + 3 + 3 + 2 = 16. Applying the formula, we have 6 / 16 * 100, which will give us the results of 37.5%.

We classify a sequence as malicious if it matches a known attack pattern perfectly, partially malicious if the sequence matches some behaviors associated with malicious activity, or benign if there is no strong similarity. For partially malicious sequences, the classification includes a match percentage, reflecting the degree of similarity to known malicious patterns. In this case, the sequence is classified as partially malicious with a match percentage of 37.5%. This result indicates that the script exhibits some suspicious behaviors but has not yet fully demonstrated malicious intent. It is necessary to perform additional behavioral analysis and monitoring to assess any potential threat from the script.

The proposed approach employs known attack patterns to construct itself, but its ability to classify behaviors partially enables it to detect unknown malicious actions that are similar to existing attacks, since diverse attacks use similar behavioral sequences. This is consistent with behavior-based detection frameworks as demonstrated by Ravi et al. [15], where partial behavioral overlaps, common in adversarial campaigns due to shared tactics such as privilege escalation and code injection allowed one to generalize novel threats. The behavior DFA benefits from this shared behavior pattern between known attacks to detect malicious actions that are similar from recognized patterns, thereby strengthening its detection efficiency. Moreover, the proposed approach provides straightforward mechanisms for adding newly discovered malicious sequences. Through its adaptable design, the model maintains operational effectiveness by easily updating its threat detection capabilities without necessitating a complete redesign.

4 Results and Evaluation

The evaluation criteria for this work focus on classifying sequences based on their match percentage, which reflects how closely a sequence aligns with the defined malicious patterns within the behavior DFA, using the formula shown in Sect. 3. A sequence is classified as malign if it exhibits a perfect match with a known malicious pattern, as partially malign if it shows partial alignment, or as benign if no significant alignment is observed. The behavior DFA was initially built using known malicious sequences and subsequently tested on a proprietary dataset. The evaluation methodology combines real-world data assessment with a solid system to detect benign from malicious activities.

The system evaluated 1058 sequences, classifying 10 as malign, 288 as partially malign and 760 as benign. The results are consistent with expectations, since the behavior DFA was built from 10 malicious sequences. This design guarantees that the behavior DFA can detect known malicious patterns with high precision and can discriminate sequences that have partial similarities. Further categorization of the partially malign sequences found match percentages of 18.75% (165 sequences), 30% (73 sequences), 37.5% (25 sequences) and 50% (25 sequences) in the proprietary dataset. These variations show how closely aligned sequences are to different malicious patterns, demonstrating that the system can capture partial matches, demonstrating the importance of analyzing sequences that share sufficient similarities with existing patterns. Attack sequences frequently display shared behaviors including like event handler addition and DOM element interaction, but its important to note that these behaviors can also exist within benign sequences.

To demonstrate the functionality of the classification approach, we present a case study for the sequence ID 1058 containing the next sequence of behaviors: [7, 11, 3, 7, 11, 3]. Each number in the sequence represents an identifier for a behavior, where, for example, 7 represents *"Add Event Handler"*, 11 represents *"Send Data"*, and 3 represents *"Update DOM Element"*. This sequence indicates that a script has implemented actions to maintain continuous communication and data extraction methods. The initial stage involves the addition of event handlers that could serve either for user monitoring or interaction response purposes. The script follows these actions with data transmission and DOM element modifications, which indicates potential information exfiltration or preparation for extra malicious activity like dynamic injections or deceptive content display. Each of these script actions keeps repeating, which strengthens the hypothesis that the script attempts to sustain ongoing communication and repeated user interface modifications for interaction manipulation. However, this sequence is classified as benign on our dataset, the repetition of adding event handlers and subsequent interactions with the DOM suggests that the script can also be used to perform legitimate actions, such as dynamically updating page content or enabling user interactions. The inclusion of *"Send Data"* may correspond to benign functionalities, such as analytics tracking or communicating with a server for non-malicious purposes. So, it can be interesting to test this sequence on our behavior DFA.

When processed by the behavior DFA, the match percentage was calculated as 18.75%, classifying the sequence as partially malign. The sequence 1058, matched the behavior DFA only on the first behavior (7). The nearest final state found was the state q_{25}, with the sequence [7, 5, 1, 7, 5, 1,] representing the path from the state q_0 to this state, q_{25}. Knowing that behavior 7 and behaviors 5 had weights of 3 and behavior 1 has a weight of 2 we can calculate the final match percentage. The total matched behavior weight is 3 representing the matched behavior 7 and the total weight to the nearest final state is the sum of the weights for the sequence leading to state q_{25}, $3 + 3 + 2 + 3 + 3 + 2 = 16$. The final match percentage is then calculated as:

$$\frac{\text{Total Matched Behavior Weight}}{\text{Total Weight to Nearest Final State}} \times 100 = \frac{3}{16} \times 100 = 18.75\%.$$

The partial match outcomes demonstrate the behavior DFA's capability to analyze sequences, offering nuanced insights into their alignment with known malicious patterns. The behavior DFA produces partial matches that measure the varying degrees of similarity to known attack sequences. The partially malign sequences were further analyzed based on their match percentages, revealing different degrees of alignment with malicious behaviors and the importance of analyzing them thoroughly:

- 18.75% (165 sequences): 165 sequences of the dataset were classified as partially malign with a match percentage of 18.75%. They matched the behavior DFA on behavior 7 (*"Add Event Handler"*, weight: 3) with a total weight to the nearest final state of 16. The presence of this behavior points to script a functionality that mostly serve as standard user interface interactive mechanisms. However in malign scenarios its appearance may also represent a preparatory step in a broader malicious strategy if followed by more critical actions. The sequences show relatively low match percentage with established malicious patterns, but suggest a monitoring need for this emerging suspicious activity.
- 30% (73 sequences): In the entire dataset 73 sequences were classified as partially malign with a match percentage of 30%. These sequences matched the behavior DFA on behavior 5 (*"Set Callback"*, weight: 3) with a total weight to the nearest final state of 10. This indicates a script executing activities which can range from harmless preparation steps to benign tasks that connect event with callbacks for user interaction management or functionality activation on web pages. However, setting a callback can also be exploited in malicious contexts, where it serves as a step in executing further harmful actions, such as tracking user input or hijacking event-driven functionality. The absence of others high-risk behaviors in this sequence leans toward a benign classification, though the potential for future malicious activity justify a detailed analysis.
- 37.5% (25 sequences): A total of 25 sequences in the dataset were identified as partially malign, exhibiting a 37.5% similarity threshold to known malicious patterns. In these sequences the behaviors matched were 7 and 5 (*"Add Event Handler"* and *"Set Callback"*, weights: 3 each) with a total weight to the nearest final state of 16. The addition of an event handler suggests the script is interacting more closely with the user or the DOM. This could be entirely benign, such as enabling custom events to improve user experience or monitor page functionality. On the other hand, in a malign scenario, adding event handlers may be used to monitor sensitive user inputs or trigger harmful actions dynamically in response to user interactions. The sequence's match percentage reflects its potential to be either benign or malicious, depending on its broader execution context.
- 50% (25 sequences): Of the total dataset, 25 sequences were classified as partially malign with a match percentage of 50%. From these sequences 19 matched behaviors 5 and 1 (*"Set Callback"* and *"Find DOM Element"*, weights: 3 and 2, respectively) with a total weight to the nearest final state of 10. These behaviors are often benign, as scripts frequently search for DOM elements and set callbacks for event-driven programming. For example, they

might be part of a legitimate script dynamically loading content or enhancing interactivity. However, the relatively high match percentage with malicious patterns signals caution, as these same actions could also be used to locate elements for data exfiltration or as part of a setup to inject harmful code. The classification as partially malign highlights its dual potential. Other 6 sequences matched behaviors 7, 5 and 1 (*"Add Event Handler"*, *"Set Callback"* and *"Find DOM Element"*, weights: 3, 3 and 2, respectively) with a total weight to the nearest final state of 16. For benign activity, these actions might represent routine DOM manipulation and event management to enable dynamic functionality. For instance, adding event handlers and setting callbacks could be part of a framework managing user inputs. The same sequence of actions can lead to severe attacks in malign situations because it indicates vulnerabilities exploitation and targeted payload delivery against specific DOM elements. The 50% match rate indicates that script behavior analysis is essential to distinguish between attacks with benign intent and those with malicious objectives.

Overall, the system successfully shows its ability to detect different malicious pattern alignments, measuring effectively diverse threat levels from partially matching behaviors. Using the partial match functionality as its key strength, the system is able to identify exact and similar patterns from known attacks, remaining effective against evolving attack approaches. Application match percentage calculations from the system provide essential security information regarding partly suspicious sequences, supporting early threat detection without unnecessary false alarms. The findings align with the behavior DFA's design, based on 10 malicious sequences, ensuring the identification of attack components while retaining sensitivity to partial overlaps. This enables operators to flag suspicious activities even when attacks exhibit only partial similarity to known threats to known threats, as reviewed and validated by domain experts.

5 Conclusion

This study presents a behavior-based approach to detect malicious JavaScript through a behavior DFA combined with behavior weighting. By modeling script execution as sequences of weighted actions, the system detects both exact and partial matches based on known malicious patterns, identifying evolving threats. The weighting system ensures that high-risk behaviors, such as code injection or data exfiltration, have greater impact on the classification process. Even though typical benign actions are also taken into account, they have less impact on the final classification.

The evaluation process demonstrated the system's ability to generalize beyond its known malicious pattern, classifying sequences as benign, partially malicious, or fully malicious based on a match percentage that represents their similarity to known malicious states. This capability highlights the adaptability of the system, as it can recognize new threats that share behavioral similarities with known attacks. Compared to traditional AI techniques, this approach offers greater flexibility, resilience against new threats, and transparency in the decision-making process. The modular design structure of the behavior DFA enables the improvement of its set of known attacks with new attack patterns, preserving its ongoing utility against new malware approaches, without requiring complete retraining.

The next steps to improve over this work are focused on the implementation of subpattern detection in the behavior DFAs to enhance it's capabilities. By identifying malicious subquences embedded within larger, seemingly benign workflows, the system's granularity and sensitivity will be improved. This will involve redesigning the automaton along with matching logic reset functionality at each input step, letting the system identify modular attacks wherever they occur in the sequence. Future development will also include the implementation of external elements which use ML models or API's to examine scripts' visited URLs and resources. The integrated system will enhance threat detection capabilities through its ability to identify malicious redirects or compromised external resources. These improvements will further contribute to the security and robustness of web applications against the evolving JavaScript-based threats.

Acknowledgements.. This work was done and funded in the scope of the BEHAVIOR project (NORTE2030-FEDER-00576300 no. 14391). This work was also supported by UIDB/00760/2020.

References

1. Canali, D., Cova, M., Vigna, G., Kruegel, C.: Prophiler: a fast filter for the large-scale detection of malicious web pages. In: Proceedings of the 20th International Conference on World Wide Web. WWW '11, pp. 197–206. Association for Computing Machinery, New York (2011). https://doi.org/10.1145/1963405.1963436
2. Chen, P., Desmet, L., Huygens, C., Joosen, W.: Longitudinal study of the use of client-side security mechanisms on the european web. In: Proceedings of the 25th International Conference Companion on World Wide Web. WWW '16 Companion, pp. 457–462. International World Wide Web Conferences Steering Committee, Republic and Canton of Geneva (2016). https://doi.org/10.1145/2872518.2888605
3. Cova, M., Krügel, C., Vigna, G.: Detection and analysis of drive-by-download attacks and malicious javascript code. In: Proceedings of the 19th International Conference on World Wide Web, pp. 281–290 (2010). https://doi.org/10.1145/1772690.1772720
4. Curtsinger, C., Livshits, B., Zorn, B., Seifert, C.: Zozzle: fast and precise in-browser javascript malware detection. In: Proceedings of the 20th USENIX Conference on Security. SEC'11, p. 3. USENIX Association (2011)
5. Gorji, A., Abadi, M.: Detecting obfuscated javascript malware using sequences of internal function calls. In: Proceedings of the 2014 ACM Southeast Conference. ACMSE '14. Association for Computing Machinery, New York, NY (2014). https://doi.org/10.1145/2638404.2737181
6. Guo, Y.: A review of machine learning-based zero-day attack detection: challenges and future directions. Comput. Commun. **198** (2022). https://doi.org/10.1016/j.comcom.2022.11.001

7. He, X., Xu, L., Cha, C.: Malicious javascript code detection based on hybrid analysis. In: 2018 25th Asia-Pacific Software Engineering Conference (APSEC), pp. 365–374 (2018). https://doi.org/10.1109/APSEC.2018.00051
8. Jing, C., Wu, Y., Cui, C.: Ensemble dynamic behavior detection method for adversarial malware. Futur. Gener. Comput. Syst. **130**, 193–206 (2022). https://doi.org/10.1016/j.future.2021.12.013
9. Jodavi, M., Abadi, M., Parhizkar, E.: Jsobfusdetector: a binary pso-based one-class classifier ensemble to detect obfuscated javascript code. 2015 The International Symposium on Artificial Intelligence and Signal Processing (AISP), pp. 322–327 (2015). https://api.semanticscholar.org/CorpusID:14848014
10. Johns, M.: Code injection vulnerabilities in web applications - exemplified at cross-site scripting. it - Inf. Technol. **53**, 256 (2011). https://doi.org/10.1524/itit.2011.0651
11. Kosoresow, A., Hofmeyer, S.: Intrusion detection via system call traces. IEEE Softw. **14**(5), 35–42 (1997). https://doi.org/10.1109/52.605929
12. Ma, W., Duan, P., Liu, S., Gu, G., Liu, J.C.: Shadow attacks: automatically evading system-call-behavior based malware detection. J. Comput. Virol. **8**(1), 1–13 (2012). https://doi.org/10.1007/s11416-011-0157-5
13. Martignoni, L., Stinson, E., Fredrikson, M., Jha, S., Mitchell, J.C.: A layered architecture for detecting malicious behaviors. In: Lippmann, R., Kirda, E., Trachtenberg, A. (eds.) RAID 2008. LNCS, vol. 5230, pp. 78–97. Springer, Heidelberg (2008). https://doi.org/10.1007/978-3-540-87403-4_5
14. Pantelaios, N., Kapravelos, A.: Fv8: a forced execution javascript engine for detecting evasive techniques (2024). https://arxiv.org/abs/2405.13175
15. Ravi, S., Balakrishnan, N., Venkatesh, B.: Behavior-based malware analysis using profile hidden markov models. In: 2013 International Conference on Security and Cryptography (SECRYPT), pp. 1–12 (2013). https://api.semanticscholar.org/CorpusID:18622802
16. Ribeiro, M.T., Singh, S., Guestrin, C.: "why should i trust you?": explaining the predictions of any classifier (2016). https://arxiv.org/abs/1602.04938
17. Rieck, K., Krueger, T., Dewald, A.: Cujo: efficient detection and prevention of drive-by-download attacks. In: Proceedings of the 26th Annual Computer Security Applications Conference, pp. 31–39 (2010). https://doi.org/10.1145/1920261.1920267
18. Rus, C., Sarmah, D.K., El-Hajj, M.: Defeating magecart attacks in a naiss way. In: SECRYPT, pp. 691–697 (2023)
19. Sebastio, S., et al.: Optimizing symbolic execution for malware behavior classification. Comput. Secur. **93**, 101775 (2020). https://doi.org/10.1016/j.cose.2020.101775
20. Sekar, R., Bendre, M., Dhurjati, D., Bollineni, P.: A fast automaton-based method for detecting anomalous program behaviors. In: Proceedings 2001 IEEE Symposium on Security and Privacy. S&P 2001, pp. 144 – 155 (2001). https://doi.org/10.1109/SECPRI.2001.924295
21. Sohan, M.F., Basalamah, A.: A systematic literature review and quality analysis of javascript malware detection. IEEE Access **8** (2020). https://doi.org/10.1109/ACCESS.2020.3031690
22. Stock, B., Johns, M., Steffens, M., Backes, M.: How the web tangled itself: uncovering the history of client-side web (in)security. In: Proceedings of the 26th USENIX Conference on Security Symposium. SEC'17, pp. 971–987. USENIX Association (2017)

23. Toreini, E., Mehrnezhad, M., Shahandashti, S.F., Hao, F.: Domtegrity: ensuring web page integrity against malicious browser extensions (2019). https://arxiv.org/abs/1905.12951
24. Touili, T.: Register automata for malware specification. In: Proceedings of the 17th International Conference on Availability, Reliability and Security. ARES '22. Association for Computing Machinery, New York (2022). https://doi.org/10.1145/3538969.3544442
25. Wang, X., Zhai, J., Yang, H.: Detecting command injection attacks in web applications based on novel deep learning methods. Sci. Rep. **14**(1), 25487 (2024). https://doi.org/10.1038/s41598-024-74350-3
26. Xue, Y., Wang, J., Liu, Y., Xiao, H., Sun, J., Chandramohan, M.: Detection and classification of malicious javascript via attack behavior modelling. In: Proceedings of the 2015 International Symposium on Software Testing and Analysis. ISSTA 2015, pp. 48–59. Association for Computing Machinery, New York (2015). https://doi.org/10.1145/2771783.2771814

Advances in Interdisciplinary Cybersecurity: Insights from Funded Research Projects - CyFRP 2025 Special Session

Maritime Cyber Resilience: Bridging Cybersecurity and Regulatory Frameworks

Sandra Lielbārde[1], Agnė Brilingaitė[2], Linas Bukauskas[2], Evita Roponena[3], Elizabete Citskovska[3], and Rūta Pirta[3](✉)

[1] Latvian Maritime Academy, Riga Technical University, Riga, Latvia
sandra.lielbarde@rtu.lv

[2] Institute of Computer Science, Vilnius University, Vilnius, Lithuania
{agne.brilingaite,linas.bukauskas}@mif.vu.lt

[3] Institute of Information Technology, Riga Technical University, Riga, Latvia
{evita.roponena,elizabete.citskovska,ruta.pirta}@rtu.lv

Abstract. The maritime industry is a leader in the transportation sector. Shipping companies play key roles in a global supply chain. Moreover, the maritime sector is part of the critical infrastructure. Due to the adaptation of diverse digital technologies, new threats and risks led to the introduction of a cybersecurity dimension into maritime legal frameworks and management processes. Typical cybersecurity frameworks provide general cybersecurity workforce development approaches and do not cover specialized recommendations for particular sectors. However, maritime cybersecurity needs more attention and efforts to ensure national and regional cyber resilience, especially in the tense geo-political situation. Safe and secure shipping is operationally resilient to cyber risks. The most effective method of dealing with an incident is to prevent it from ever happening. Therefore, this paper aims to lay the foundation for developing the national maritime cybersecurity ecosystem, considering good practices from multiple disciplines.

Keywords: Maritime cybersecurity · Maritime legal framework · Regional resilience · Maritime cybersecurity skill development

1 Introduction

The maritime cybersecurity landscape is evolving due to rapid digitalisation. The market size of global maritime digitalisation is growing each year and is expected to reach 266.5 billion dollars by 2028 [39]. The application of IT technologies in the maritime sector improves efficiency, transparency, and reduces errors. However, it also leads to cybersecurity-related challenges. The Marlink's Security Operations Centre [3] reported cybersecurity-related incidents statistics of 2024 for 21800 monitored vessels. The report concluded that malicious activity increased compared to 2023. The most common threats included phishing fraud, C&C/Blacklist malicious traffic, SPAM abusive content, and intrusion attempts.

I. Praça et al. (Eds.): EICC 2025, CCIS 2500, pp. 217–228, 2025.
https://doi.org/10.1007/978-3-031-94855-8_14

The Baltic Maritime has also faced cybersecurity-related threats recently. The fourth cable-cutting incident in recent months was reported in January of this year [16]. A Bulgarian-owned vessel, *The Vezhen*, caused an incident that was reported as unintentional damage. However, recent attacks on undersea cables between Latvia and Sweden and Estonia and Finland align with the patterns of sabotage seen in the region that cause power disruptions in these regions, further investigation by the authorities and the "Baltic Sentry" mission launched by NATO. The disruption of mobile communication and ship-tracking data in the Baltic Sea was experienced in December last year [14] and caused by Russia. The authorities suspect these disruptions were used to hide the movement of Russian-owned vessels and disrupt the operations of NATO allies.

Beyond the Baltic region, more maritime-specific cybersecurity incidents were observed. In 2022, a GPS jamming incident delayed a ferry for 2 h from leaving a Danish port. The GPS malfunction was caused by two lorries onboard the ferry with GPS jamming equipment to block tracking devices [43]. A mass AIS spoofing event occurred in early June of 2024. Because of the incident, vessels appeared to be sailing at absurd speeds and locations - nearly 50 vessels appeared to be at the Simferopol International Airport in Crimea. In contrast, 30 appeared to be at the Gelendzhik Airport in Russia. The incident was likely connected to the ongoing Russian-Ukrainian war [17].

The goal of this paper is to highlight cybersecurity requirements in the maritime sector, considering international, European Union, and national legal acts. The research emphasizes strengthening cyber resilience in the maritime sector by integrating good practices from multiple disciplines, with a particular focus on workforce development and regulatory gaps. The paper presents the progress of the state-funded project at Riga Technical University, Latvia, aimed at strengthening maritime cybersecurity resilience and training.

The rest of the paper is structured as follows. Section 2 reviews cybersecurity frameworks and scientific related work regarding workforce development and existing gaps. Section 3 highlights the cybersecurity dimension in the maritime regulatory frameworks. Section 4 discusses the specifics and existing limitations of maritime cybersecurity skill development. Section 5 shapes the future work as the following activities of the ongoing project.

2 Background

General cybersecurity roles and competences are presented in several industry frameworks, such as the National Initiative for Cybersecurity Education (NICE) Workforce Framework (NIST NICE framework) [42] and the European Union Agency for Cybersecurity (ENISA) skill framework (ENISA framework) [15]. Meanwhile, it is observed that in smaller nations, such as Baltic states, the key cybersecurity competences may be concentrated in a single or few roles. Besides, several cybersecurity competences are missing. There are few research studies, concerning cybersecurity work profiles and competences in smaller nations, such as [2,44], however, existing research primarily focuses on industry-agnostic cybersecurity roles. Therefore, it might pose challenges for educators in designing maritime-specific cybersecurity educational programs.

Research studies highlighted the challenges and deficiencies in maritime cybersecurity education, relating them to the industry's unique operational environment, regulatory complexity, lack of training opportunities, and the continuously evolving cyber threat landscape [4,5,18,52]. The recent study on maritime cybersecurity education [4] highlights lack of customized training programmes impacting training quality, lack of impact visibility and time constraints. Other challenges include low participation rates, unclear training delivery methods, and the evolving nature of cyber threats, which require frequent updates to training programs. The study underscores the need for customized, continuous, and practical training approaches to improve cybersecurity resilience in the maritime. Similarly, another study [18] finds that there are no standardized requirements for including cybersecurity practices in maritime education and training curricula. Additionally, the slow adaptation of regulatory frameworks, the lack of practical training, and limited awareness among seafarers contribute to the vulnerability of maritime operations. Also, Stoker et al. [52] share the opinion that maritime education and training institutions face challenges in rapidly addressing the pressing cybersecurity education needs. Others [5] emphasise that maritime cybersecurity training must effectively address cyber-attack risks on operational technology associated with vessel or port operations.

Given these challenges and a context of small national states, there is a need to investigate and design tailored maritime cybersecurity education, considering maritime regulatory requirements and industry recommendations.

3 Highlighting Cybersecurity Dimension in Maritime Regulatory Frameworks

This section analyses the existing maritime regulatory framework to identify to what extent it addresses maritime cybersecurity. This analysis is based on a doctrinal analysis of the relevant international and national legal instruments and documents. Figure 1 provides a timeline of legal instruments. Numbers are given to instruments based on the issue date. The numbers represent bibliographic items: 1, 2, 4, 5, 9, and 18 represent EU security requirements that do not consider the cybersecurity dimension [6–8,45,46,49], numbers 3, 6, 7, 8, 10, 13, 14, 24, 26, and 31–32, 34 represent the International Maritime Organization (IMO) regulatory framework documents [24–28,30–36], numbers 11, 12, 16, 17, 19, 21, 29, 30, and 35 represent EU Security requirements including cybersecurity aspects [9–13,47,48,50,51], and numbers 15, 20, 22, 23, 27, and 28 represent national and industrial recommendations [1,20,22,37,40,53], respectively. Numbers 25 and 33 represent the revision and amendments of other documents available in the timeline (connected by the arcs in the figure). Numbers 10 and 24 represent the amendment and revision of IMO documents issued earlier than 2002.

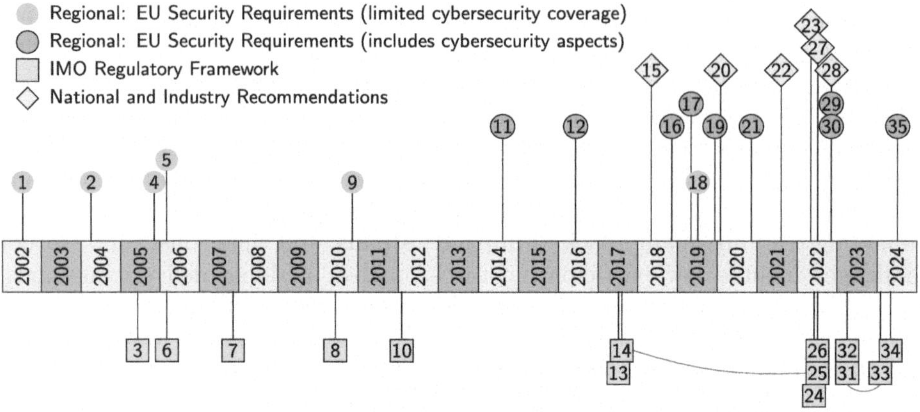

Fig. 1. Timeline of legal instruments

3.1 IMO Regulatory Framework

IMO is the global standard-setting authority for the safety, security, and environmental performance of international shipping. Its primary role is to create an effective, universally adopted, and universally implemented regulatory framework for the shipping industry [23]. Cybersecurity considerations are included in the International Ship and Port Facility Security Code (ISPS Code) and the International Safety Management (ISM) Code. ISPS Code does not contain explicit reference to cybersecurity. However, the fundamental purpose of the ISPS Code is to address all security threats to the maritime industry, including cybersecurity [21,29].

The purpose of the ISM Code is to provide an international standard for the safe management and operation of ships and pollution prevention. The cybersecurity was placed as a safety practice under the ISM Code after IMO Maritime Safety Committee (MSC) in June 2017 adopted Resolution Maritime Cyber Risk Management in Safety Management Systems [36, MSC.428(98)] and affirmed that an approved safety management system should take into account cyber risk management and encouraged administrations to ensure that cyber risks are appropriately addressed in existing safety management systems no later than the first annual verification of the company's Document of Compliance after 1 January 2021. In July 2017, the Guidelines on Maritime Cyber Risk Management were approved by the IMO Facilitation of International Maritime Traffic Committee and the MSC. Additional guidance and standards, included in paragraph 4.2 of these Guidelines, were updated in 2022 [31, MSC-FAL.1/Circ.3/Rev.2]. These guidelines are not mandatory. The Convention on Facilitation of International Maritime Traffic (FAL) was completely revised by amendments of 13 May 2022 [35]. FAL lists the documents that can be requested by the Member State's authorities regarding ship arrival, staying, and departure from the port. These documents cover information on the ship, cargo, passengers, and other objects, including dangerous goods, postal and health requirements. The amendments made the Maritime Single Window (MSW) for data exchange by electronic means mandatory in ports around the world from 1 January 2024. As a result of cyberattacks, electronic communication used in the management and operation of the MSW can be damaged, thus affecting many maritime stakeholders. Accordingly, the MSW should incorporate information security measures to ensure the cybersecurity of the MSW system. The MSW system includes processes for electronic certificates and data exchange between ships and shore, processes involving ship-port interface, and other facilitation aspects of the maritime transport network.

Additionally, there are several Guidelines addressing MSW cybersecurity, e.g., guidelines on setting up MSW [33], secure information exchange in MSW [30], harmonized communication with ports [32], and port community systems [34]. Cybersecurity is addressed only in few sentences in these Guidelines.

The important role in managing cybersecurity risks is the human factor and the personnel's professional skills. The International Convention on Standards of Training, Certification and Watchkeeping for Seafarers (STCW) [24] prescribes minimum standards relating to training, certification, and watchkeeping for seafarers. STCW Code prescribes mandatory minimum requirements for the issue of certificates of proficiency for ship security officers regarding security-related training and instruction for all seafarers. These mandatory minimum requirements, currently in force and applicable for the mentioned seafarers and other seafarers, do not make explicit reference to cybersecurity. Additionally, there are several guidelines that specify security-related knowledge and proficiency for personnel on ships and ports, e.g., for security-related training of the shipboard and port facility personnel [27,28] and training and certification of port facility security officers and company security officers [25,26]. The mentioned guidelines do not refer to cybersecurity. Accordingly, the IMO regulatory framework comprises mandatory and non-mandatory legal instruments - IMO Conventions, Codes, Resolutions, and Guidelines.

3.2 Regional—EU Requirements and Maritime Cybersecurity

Within the EU, there are several legal instruments adopted that have a significant impact on cybersecurity in the maritime industry, such as Regulation (EU) 910/2014 on electronic identification and trust services for electronic transactions [47], General Data Protection Regulation [48], Cybersecurity Act [50], Directive on establishing EU electronic communications code [9], NIS2 Directive [10], Directive (EU) 2022/2557 on the resilience of critical entities [11], and Artificial Intelligence Act [51].

Regulation (EU) 910/2014 aims to ensure an adequate level of security of electronic identification means, electronic transactions, electronic documents, etc., which are used by natural and legal persons also in maritime sector. Standards relating to the protection of natural persons with regard to the processing of their personal data established by General Data Protection Regulation are applicable also to processing of personal data in maritime industry. Cybersecurity Act prescribes a principles for the establishment of European cybersecurity certification schemes innorder to ensure an adequate level of cybersecurity for Information and communications technology (ICT) products, ICT services and ICT processes which support everyday activities also in such key sector as maritime. Inland, sea and coastal passenger and freight water transport companies, managing bodies of ports and their port facilities, and entities operating works and equipment contained within ports are considered transport entities with high criticality according to the NIS 2 Directive and accordingly are objects of standards aimed at enhancing their cybersecurity according to the Directive (EU) 2022/2557.

The EU maritime legal instruments relating to the operation of the MSW and implementation of the ISPS and FAL requirements within the EU (for example, the Directive 2002/59/EC on a Community Vessel Traffic Monitoring and Information System [6], Directive 2005/65/EC on enhancing port security [7], Directive 2010/65/EU on reporting formalities for ships arriving in and/or departing from ports of the Member States [8], Regulation 725/2004 on enhancing ship and port facility security [45], Regulation 336/2006 on the implementation of the International Safety Management Code within the Community [46], and Regulation 2019/1239 on establishing a European Maritime Single Window environment [49])do not address cyber security.

However, to address cybersecurity in maritime, ENISA has issued maritime-oriented guidelines on good cybersecurity practices [12] and cyber risk management in ports [13].

3.3 National and Industry Recommendations

International law, particularly IMO legal instruments, and EU law are interlinked with national law, implementing international and EU requirements nationally.

The more general and non-specific international standards are, the more different approaches different countries will have in their national implementation. The maritime industry and digitalization, in general, have an international dimension. Therefore, challenges related to the digitalization of maritime business operations need to be addressed through the uniform implementation of internationally applicable standards.

At the national level, some states have also prepared guidelines on cybersecurity. In the UK, there is a good practice guide on cybersecurity code for port systems [53]. Moreover, another code of practice is intended to complement existing standards and guidance within maritime security [53]. In the USA, NIST provides a voluntary risk-management framework to improve critical infrastructure cybersecurity [40].

Industrial maritime stakeholders issued many recommendations regarding cybersecurity in the maritime sector. They are linked to the existing mandatory IMO legal instruments. However, the difference is that industry recommendations contain more detailed strategies and methods for protecting maritime operations from cyber threats.

There are some of the most known industry recommendations, such as Guidelines on Cyber Security Onboard Ships [1], produced and supported by BIMCO, the Digital Containership Association, International Association of Dry Cargo Shipowners, and other stakeholders, IAPH Guidelines for Port and Facilities [22], IACS Recommendation on cyber resilience [20], and ISO standard on information security management systems [37].

4 Discussion

The regulatory framework regarding maritime cybersecurity, consisting of international maritime law, EU regulation regarding cybersecurity or maritime, national law implementing international and EU standards related to maritime cybersecurity, and national and industry recommendations, can be considered very complex. Because of overlapping requirements, different approaches and methods, different terminology, and different legal statuses of different legal instruments, it could create difficulties for maritime stakeholders to establish an effective maritime cybersecurity system and, at the same time, to ensure compliance with the relevant mandatory standards.

Because of the maritime industry's unique nature, maritime cybersecurity needs to be addressed by international and national legal instruments regulating specifically maritime. First of all, by applicable IMO legal instruments. After analysis of the IMO regulatory framework, it could be agreed with other authors that IMO legal instruments do not cover maritime cybersecurity sufficiently and that there is a lack of comprehensiveness and uniformity [38, 41].

The fast evolution of digitalization in the maritime industry requires the fast development of a modern regulatory environment regarding cybersecurity standards. Considering the significant time consumption for the development of new legislation, especially amendments to the international conventions, it is a challenge for the maritime industry to ensure that regulation is constantly updated to address emerging cyber threats effectively. The complexity of the existing regulatory framework is another challenge for legislators to ensure that individual legal instruments are always updated and do not contradict each other. One of the solutions could be creating a standalone Cyber Code, based on a framework created by previous IMO Codes such as the Polar Code, as suggested by some authors [19].

The complex regulation requires attention and effort in workforce development to ensure the safety of systems and humans from the perspective of cybersecurity. Maritime cybersecurity is cross-domain, and any implementation of legal regulation, making proposals for improvements, and applying frameworks in everyday activities require professional personnel.

Figure 2 presents the project scope with regulatory document relationships of cybersecurity and maritime domains for workforce development. Elements presented with dashed lines show the focus of the project. Cybersecurity and maritime frameworks imply guidelines on specialist training and professional development. We aim to develop a research-based skill development model that

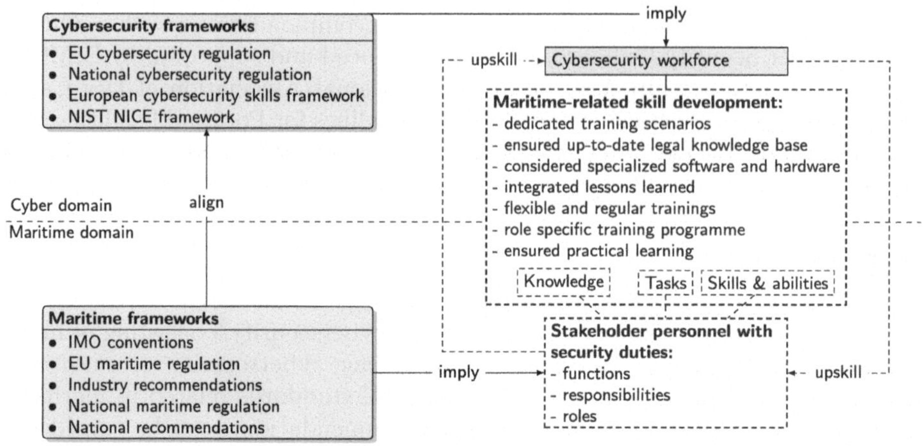

Fig. 2. Regulatory framework and maritime cybersecurity workforce development

considers maritime specifics, e.g., strict regulatory mechanisms and specialized digital systems, and follows general cybersecurity guidelines. The model should ensure bi-directional professional growth, i.e., upskilling towards cybersecurity or maritime specifics. Such an ecosystem would include diverse, transferable, and up-to-date scenarios in training programs. This paper presents the conceptual model, which will be validated in future research stages, taking into account stakeholder feedback and its implementation in educational programs.

5 Future Work

This paper presents the initial phase of a research project focused on strengthening maritime cybersecurity resilience and training. The project aims to propose a maritime ecosystem's cyber resilience knowledge model to enhance human performance in cybersecurity crisis situations, considering risk levels of organization, human, software, hardware/cloud, and communication elements most specific to the maritime field. The project will produce a cyber resilience knowledge model, define the typical cyber kill chain for ports, and align with scenarios and experiences of several countries, contributing to the enhancement of maritime port infrastructures. Moreover, the project will complement cybersecurity playbooks for essential infrastructures used in ports by adding research-based elements to assess the risk, mitigate cybersecurity breaches, and relate to critical port events.

The paper outlines the foundation for the next research activities concerning the legal requirements of maritime cybersecurity. The future research includes the development of an integrated maritime cybersecurity management and educational model, human-centric educational artifacts, and artifact-based toolsets. The knowledge and performance model will address various conceptual elements, such as contents, competences, risks, measures, technological requirements, and dependencies. The model will be implemented, using a specialized, research-based training toolset, dedicated to maritime cybersecurity, including open educational resources, a serious game, and a handbook for trainers.

Acknowledgments. The work has been funded by Riga Technical University project "Enhancing Maritime Cybersecurity Resilience and Training," project No. RTU-PA-2024/1-0061, within the framework of the European Union Recovery and Resilience Mechanism funded project No. 5.2.1.1.i.0/2/24/I/CFLA/003 "Implementation of consolidation and management changes at Riga Technical University, Liepaja University, Rezekne Academy of Technologies and Latvian Maritime Academy and Liepaja Maritime College for Excellence in Higher Education, Science and Innovation."

Disclosure of Interests. The authors have no competing interests to declare that are relevant to the content of this article.

References

1. BIMCO and other industry organisations: The Guidelines on Cyber Security Onboard Ships, 4th edn. https://www.ocimf.org/publications/information-papers/the-guidelines-on-cyber-security-onboard-ship. Accessed 28 Jan 2025
2. Bukauskas, L., Brilingaitė, A., Juozapavičius, A., Lepaitė, D., Ikamas, K., Andrijauskaitė, R.: Remapping cybersecurity competences in a small nation state. Heliyon **9**(1), e12808 (2023). https://doi.org/10.1016/j.heliyon.2023.e12808
3. Centre, M.: Global maritime cyber threat report. Technical report, Marlink (2024)
4. Chupkemi, D.C., Mersinas, K.: Challenges in maritime cybersecurity training and compliance. J. Marine Sci. Eng. **12**(10) (2024). https://doi.org/10.3390/jmse12101844
5. D'Amelio, E., Abuzneid, S.: All hands on tech: effectively addressing maritime operational technology risks through cybersecurity awareness training. In: 2024 IEEE Long Island Systems, Applications and Technology Conference (LISAT), pp. 1–6 (2024). https://doi.org/10.1109/LISAT63094.2024.10808032
6. Directive 2002/59/EC of the European Parliament and of the Council of 27 June 2002 establishing a Community vessel traffic monitoring and information system and repealing Council Directive 93/75/EEC. https://eur-lex.europa.eu/eli/dir/2002/59
7. Directive 2005/65 of the European Parliament and of the Council of 26 October 2005 on enhancing port security. https://eur-lex.europa.eu/eli/dir/2005/65
8. Directive 2010/65/EU of the European Parliament and of the Council of 20 October 2010 on reporting formalities for ships arriving in and/or departing from ports of the Member States and repealing Directive 2002/6/EC. https://eur-lex.europa.eu/eli/dir/2010/65
9. Directive (EU) 2018/1972 of the European Parliament and of the Council of 11 December 2018 establishing the European Electronic Communications Code. https://eur-lex.europa.eu/eli/dir/2018/1972
10. Directive (EU) 2022/2555 of the European Parliament and of the Council of 14 December 2022 on measures for a high common level of cybersecurity across the Union, amending Regulation (EU) No 910/2014 and Directive (EU) 2018/1972, and repealing Directive (EU) 2016/1148 (NIS 2 Directive). https://eur-lex.europa.eu/eli/dir/2022/2555

11. Directive (EU) 2022/2557 of the European Parliament and of the Council of 14 December 2022 on the resilience of critical entities and repealing Council Directive 2008/114/EC. https://eur-lex.europa.eu/eli/dir/2022/2557
12. ENISA: Port security: Good practices for cybersecurity in the maritime sector (2019). https://doi.org/10.2824/328515
13. ENISA: Guidelines - cyber risk management for ports (2020). https://doi.org/10.2824/671060
14. Erling, B., Strzelecki, M.: Russia disrupting Baltic security in test for NATO, Polish admiral says. Reuters (2024). https://www.reuters.com/world/europe/russia-disrupting-baltic-security-test-nato-polish-admiral-says-2024-12-16. Accessed 24 Feb 2025
15. European Union Agency for Cybersecurity, ENISA: European cybersecurity skills framework (2022). https://www.enisa.europa.eu/topics/cybersecurity-education/european-cybersecurity-skills-framework/ecsf-profiles-v-0-5-draft-release.pdf
16. Gongadze, M.: NATO, EU on high alert as undersea cable attacks escalate in Baltic. Voice of America (2025). https://www.voanews.com/a/nato-eu-on-high-alert-as-undersea-cable-attacks-escalate-in-baltic/7956600.html. Accessed 24 Feb 2025
17. Gowans, G.: Police suspect jammer devices on HGVs caused ferry's GPS system failure. Trans.INFO (2022). https://trans.info/en/jammer-ferry-282005. Accessed 25 Feb 2025
18. Heering, D., Maennel, O.M., Venables, A.N.: Shortcomings in cybersecurity education for seafarers. In: Soares, C.G., Santos, T.A. (eds.) Developments in Maritime Technology and Engineering, Marine Technology and Ocean Engineering Series, vol. 6. CRC Press (2021). https://doi.org/10.1201/9781003216582-06
19. Hopcraft, R., Martin, K.M.: Effective maritime cybersecurity regulation - the case for a cyber code. J. Indian Ocean Reg. **14**(3), 354–366 (2018). https://doi.org/10.1080/19480881.2018.1519056
20. IACS: Recommendation on cyber resilience (rec. 166) (2022). https://iacs.org.uk/resolutions/recommendations/161-180/rec-166-new-corr2-cln
21. IAPH: Port community cyber security (2020). https://sustainableworldports.org/wp-content/uploads/IAPH-Port-Community-Cyber-Security-Report-Q2-2020.pdf
22. IAPH: Cybersecurity guidelines for ports and port facilities (2021). https://events.iala.int/content/uploads/2021/09/MSC-104-7-1-IAPH-Cybersecurity-Guidelines-for-Ports-and-Port-Facilities-IAPH.pdf
23. IMO. https://www.imo.org/. Accessed 26 Feb 2025
24. IMO: International Convention on Standards of Training, Certification and Watchkeeping for Seafarers (STCW). https://www.imo.org/en/ourwork/humanelement/pages/stcw-conv-link.aspx. Accessed 26 Feb 2025
25. IMO: Guidelines on training and certification for company security officers (2005). https://wwwcdn.imo.org/localresources/en/OurWork/Security/Documents/MSC.Circ.1154.pdf
26. IMO: Guidelines on training and certification for port facility security officers (2006). https://wwwcdn.imo.org/localresources/en/OurWork/Security/Documents/MSC.Circ.1188.pdf
27. IMO: Guidelines on security-related training and familiarization for shipboard personnel (2007). https://wwwcdn.imo.org/localresources/en/OurWork/Security/Documents/MSC.Circ.1235.pdf
28. IMO: Guidelines on security-related training and familiarization for port facility personnel (2010). https://wwwcdn.imo.org/localresources/en/OurWork/Security/Documents/MSC.1%20Circ.1341.pdf

29. IMO: Guide to Maritime Security and the ISPS Code, 2021 edition edn. (2021). https://doi.org/10.62454/KB116E
30. IMO: Guidelines on authentication, integrity and confidentiality of information exchanges via maritime single windows and related services (2022). https://wwwcdn.imo.org/localresources/en/OurWork/Facilitation/FAL%20related%20nonmandatory%20documents/FAL.5-CIRC.46.pdf. fAL.5/Circ.46. Accessed 25 Feb 2025
31. IMO: Guidelines on maritime cyber risk management (2022). https://wwwcdn.imo.org/localresources/en/OurWork/Security/Documents/MSC-FAL.1-Circ.3-Rev.2%20-%20Guidelines%20On%20Maritime%20Cyber%20Risk%20Management%20(Secretariat)%20(1).pdf. mSC-FAL.1/Circ.3/Rev.2. Accessed 25 Feb 2025
32. IMO: Guidelines for harmonized communication and electronic exchange of operational data for port calls (2023). https://wwwcdn.imo.org/localresources/en/OurWork/Facilitation/FAL%20related%20nonmandatory%20documents/FAL.5-CIRC.52.pdf. fAL.5/Circ.52. Accessed 25 Feb 2025
33. IMO: Guidelines for setting up a maritime single window (2023). https://wwwcdn.imo.org/localresources/en/OurWork/Facilitation/FAL%20related%20nonmandatory%20documents/FAL.5-Circ.42-Rev.3.pdf. Accessed 25 Feb 2025
34. IMO: Guidelines on port community systems (2024). https://wwwcdn.imo.org/localresources/en/OurWork/Facilitation/FAL%20related%20nonmandatory%20documents/FAL.5-Circ.54.pdf. fAL.5/Circ.54. Accessed 25 Feb 2025
35. IMO: Resolution "amendments to the annex to the convention on facilitation of international maritime traffic, 1965", res.fal.14(46). In: Convention on Facilitation of International Maritime Traffic, 1965. IMO (2024). https://doi.org/10.62454/KE350E
36. IMO MSC: Maritime cyber risk management in safety management systems (2017). https://wwwcdn.imo.org/localresources/en/OurWork/Security/Documents/Resolution%20MSC.428(98).pdf. mSC.428(98). Accessed 25 Feb 2025
37. ISO/IEC 27001:2022 Information security, cybersecurity and privacy protection - Information security management systems - Requirements (2022). https://www.iso.org/standard/27001
38. Karim, M.S.: Maritime cybersecurity and the IMO legal instruments: sluggish response to an escalating threat? Mar. Policy **143**, 105138 (2022). https://doi.org/10.1016/j.marpol.2022.105138
39. KBV Research: Global Maritime Digitization Market Size, Share & Industry Trends Analysis Report By Application, By End User (Ports & Terminals, Shipping Companies, and Maritime Freight Forwarders), By Technology, By Regional Outlook and Forecast, 2022–2028. Technical Report. KBV-13961, Knowledge Based Value (KBV) Research (2023). https://www.kbvresearch.com/maritime-digitization-market/
40. NIST: Framework for improving critical infrastructure cybersecurity. NIST (2018). https://doi.org/10.6028/NIST.CSWP.04162018
41. Ogundare, B., Akinwande, G.: International maritime organisation framework on cyber risk management - a case for a comprehensive legal framework. Maritime Saf. Secur. Law J. **9**, 14–36 (2021)
42. Petersen, R., Santos, D., Smith, M.C., Wetzel, K.A., Witte, G.: Workforce Framework for Cybersecurity (NICE Framework). NIST Special Publication 800–181 (2020). https://doi.org/10.6028/NIST.SP.800-181r1

43. Pili, G., Armenzoni, A., Kessler, G.C.: Mass AIS Spoofing Event "Moves" Dozens of Ships to Crimean Airport. The Maritime Executive (2024). https://maritime-executive.com/editorials/mass-ais-spoofing-event-moves-dozens-of-ships-to-crimean-airport. Accessed 25 Mar 2025
44. Pirta, R., Veigurs, M.: Latvian cyber workforce of tomorrow: closing the gap. Econ. Bus. **39**, 26–37 (2025). https://doi.org/10.7250/eb-2025-0003
45. Regulation (EC) No 725/2004 of the European Parliament and of the Council of 31 March 2004 on enhancing ship and port facility security). https://eur-lex.europa.eu/eli/reg/2004/725
46. Regulation (EC) 336/200624 of the European Parliament and of the Council of 15 February 2006 on the implementation of the International Safety Management Code within the Community and repealing Council Regulation (EC) No 3051/95. https://eur-lex.europa.eu/eli/reg/2006/336
47. Regulation (EU) 910/2014 of the European Parliament and of the Council of 23 July 2014 on electronic identification and trust services for electronic transactions in the internal market and repealing Directive 1999/93/EC. https://eur-lex.europa.eu/eli/reg/2014/910
48. Regulation (EU) 2016/679 of the European Parliament and of the Council of 27 April 2016 on the protection of natural persons with regard to the processing of personal data and on the free movement of such data, and repealing Directive 95/46/EC (General Data Protection Regulation). https://eur-lex.europa.eu/eli/reg/2016/679
49. Regulation (EU) 2019/1239 of the European Parliament and of the Council of 20 June 2019 establishing a European Maritime Single Window environment and repealing Directive 2010/65/EU. https://eur-lex.europa.eu/eli/reg/2019/1239
50. Regulation (EU) 2019/881 of the European Parliament and of the Council of 17 April 2019 on ENISA (the European Union Agency for Cybersecurity) and on information and communications technology cybersecurity certification and repealing Regulation (EU) No 526/2013 (Cybersecurity Act). https://eur-lex.europa.eu/eli/reg/2019/881
51. Regulation (EU) 2024/1689 of the European Parliament and of the Council of 13 June 2024 laying down harmonised rules on artificial intelligence and amending Regulations (EC) No 300/2008, (EU) No 167/2013, (EU) No 168/2013, (EU) 2018/858, (EU) 2018/1139 and (EU) 2019/2144 and Directives 2014/90/EU, (EU) 2016/797 and (EU) 2020/1828 (Artificial Intelligence Act). https://eur-lex.europa.eu/eli/reg/2024/1689
52. Stoker, G., Greer, J., Clark, U., Chiego, C.: Considering maritime cybersecurity at a non-maritime education and training institution. Cybersecur. Pedagogy Pract. J. **2**(1), 66–76 (2023)
53. The Institution of Engineering and Technology: Good practice guide: cyber security for ports and port systems (2020). https://www.gov.uk/government/publications/ports-and-port-systems-cyber-security-code-of-practice

Evaluating KASLR Break on RISC-V Using Gem5: Microarchitectural Side-Channel Analysis of Page-Table Walks

Mahreen Khan(✉), Maria Mushtaq, Renaud Pacalet, and Ludovic Apvrille

Telecom Paris, Institut Polytechnique de Paris, Palaiseau, France
{mahreen.khan,maria.mushtaq,renaud.pacalet,
ludovic.apvrille}@telecom-paris.fr

Abstract. This paper leverages the gem5 simulator to analyze a microarchitectural KASLR break on RISC-V systems. Previous research [2] demonstrated the feasibility of KASLR breaks on RISC-V hardware platforms (C906 and U74). Our paper aims to provide insights that are not easily attainable through traditional hardware experiments. By employing gem5, we gain access to fine-grained metrics such as cycle counts, cache behavior, branch prediction statistics, and TLB accesses, among others. These detailed insights give a deeper analysis of the KASLR bypass and help understand the attack mechanics better.

Keywords: Microarchitectural Security · Side-Channel Attacks · gem5 Simulator · Embedded Systems · Cache Timing Analysis

1 Introduction

Kernel Address Space Layout Randomization (KASLR) is designed to hinder kernel exploitation by making memory layout prediction difficult. However, current RISC-V implementations employ only **coarse-grained** randomization rather than the **fine-grained** per-function randomization which is under development [3]. Instead of randomizing the locations of individual kernel functions, RISC-V KASLR merely shifts the starting position of the entire kernel image. This means that once the starting address is determined, the locations of all kernel functions can be inferred [1].

Microarchitectural side channels can expose these address mappings. The timing differences during a page table walk can indicate whether an address is mapped. For mapped addresses, the walk accesses all levels and performs permission checks. For unmapped addresses, the walk aborts early, leading to lower latency.

In our paper, we leverage the gem5 simulator [5] to emulate a RISC-V system and analyze the timing behavior of page table walks. Previous work [2] demonstrated the feasibility of KASLR breaks on hardware platforms such as

I. Praça et al. (Eds.): EICC 2025, CCIS 2500, pp. 229–235, 2025.
https://doi.org/10.1007/978-3-031-94855-8_15

the C906 and U74. However, hardware experiments do not capture detailed internal state or provide comprehensive performance metrics. The gem5 simulator supplies cycle counts, cache behavior, branch prediction data, and TLB accesses. These metrics offer a deeper view of the attack mechanics and support a thorough analysis of the KASLR bypass on RISC-V systems.

The main contributions of this paper are:

- **Simulation-based validation of Page-table walk vs no-walk impact:** We demonstrate a working proof-of-concept for breaking RISC-V KASLR in gem5, showing a 44% timing difference in `rdcycle` counts between successful and aborted page table walks (Fig. 2).
- **Microarchitectural side-channel analysis using gem5:** We provide detailed measurements of:
  - TLB behavior, showing 1.5–2× more misses during page walks (Figs. 3, 4)
  - Branch prediction impacts, with 30–40% increase in mispredictions during attacks (Table 2)
  - Simulation overhead characteristics (Figs. 5, 6)

The remainder of this paper is structured as follows: Sect. 2 details the attack methodology, explaining how page table walk timing differences can break KASLR on RISC-V systems. Section 3 describes our gem5 simulation environment, including configuration parameters and experimental setup. Section 4 presents our results, analyzing timing measurements, TLB behavior, and branch prediction metrics. Section 5 discusses potential future work, including gem5-based security enhancements and defense mechanisms. Finally, Sect. 6 concludes with implications of our findings and the value of simulation for microarchitectural security research.

2 Attack

This section describes the exploitation of Kernel Address Space Layout Randomization (KASLR) proof-of-concept (PoC) tested on gem5 RISC-V system. We focus on the microarchitectural side channels that can be leveraged to infer whether a kernel address is physically backed.

The attack assumes an **unprivileged** adversary who has execution capabilities in user space but no direct kernel access. Existing x86 defenses like KAISER, LAZARUS, and FLARE rely on x86-specific features and are inapplicable [2].

2.1 Exploiting Page Table Walk Timing for KASLR Breaks

A **page table** is a hierarchical data structure used by modern CPUs to translate virtual addresses into physical addresses in systems implementing virtual memory. The page table consists of multiple levels, with each level refining the translation until the final physical page frame is identified.

A **page table walk** refers to the process of navigating through the *multi-level page table hierarchy* to resolve a virtual address into a physical address.

This occurs whenever a **Translation lookaside buffer (TLB) miss** happens, requiring the memory management unit (MMU) to fetch the translation manually from the page tables stored in memory.

The duration of a page table walk depends on whether all required page table entries (PTEs) exist and are valid, leading to different timing characteristics that can be exploited for **KASLR breaks**.

- **Full Page Table Walk (Walk Case):** When all necessary PTEs are present and valid, the MMU performs a complete page table walk, traversing all levels of the hierarchy until it resolves the address. At this stage, a permission check occurs, ensuring that the accessing process has the necessary privileges to access the memory. This process incurs a measurable delay due to multiple memory accesses.
- **Early Termination (No-Walk Case):** If a required PTE is missing (e.g., the entry is not present, marked as invalid, or leads to an access violation), the page walk aborts early. This results in a page fault, which is detected significantly faster than a full page walk.

A page walk incurs a longer delay, which will happen for the mapped addresses, while no page-table walk is faster, which indicates unmapped addresses. This time measurement can be done by using an **unprivileged** `rdcycle` RISC-V instruction. These timing differences enable attackers to **derandomize** the kernel by detecting whether a given address is backed by a physical page. If an attacker can pinpoint the base address of the kernel image, they can effectively bypass RISC-V's KASLR implementation, exposing kernel structures to further attacks [1].

3 Gem5 Simulator

The gem5 simulator is a modular, open-source framework for computer architecture research. It enables detailed modeling and simulation of hardware components, offering insights into microarchitectural interactions. This section outlines the gem5's configurations and the methodology for attack analysis.

3.1 Configurations

gem5 supports multiple instruction set architectures (ISAs), including x86, ARM, RISC-V, SPARC, and MIPS [6], making it adaptable to diverse research scenarios. Simulations range from simple single-core systems to heterogeneous multi-core architectures. Users can select between two modes: full-system (FS) mode, which models the complete operating system (OS), and syscall emulation (SE) mode, which abstracts OS interactions for faster simulations [7]. CPU models include the `AtomicSimpleCPU` for fast functional emulation, the `TimingSimpleCPU` for basic timing analysis, and the out-of-order `O3CPU` [4], which models speculative execution and pipeline dynamics. Cache hierarchies, interconnects, and memory subsystems are defined via Python scripts, enabling customization of L1/L2 caches, coherence protocols, and DRAM controllers [5].

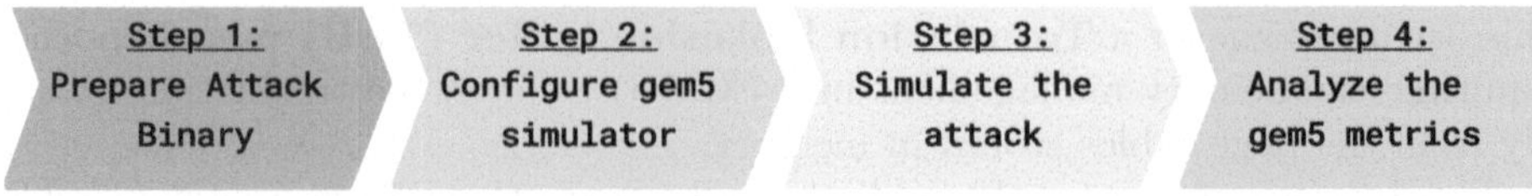

Fig. 1. The gem5 simulator attack analysis methodology.

3.2 Methodology

The gem5 workflow for analyzing page-table walk impact for breaking KASLR on RISC-V is illustrated in Fig. 1. First, the attack binary is prepared and integrated into a modified RISC-V disk image. Next, the gem5 simulator is configured via Python scripts to define hardware parameters such as the out-of-order `O3CPU` model and cache hierarchies, all running in full-system (FS) mode to capture OS interactions. The detailed configuration is given in Table 1.

The attack simulation is then executed, leveraging gem5's detailed timing models to track microarchitectural behaviors. Finally, metrics including branch predictor accuracy, instructions executed and TLBs behavior are analyzed to correlate the attack's execution with observable hardware-level side effects, enabling insights into vulnerabilities.

Table 1. Gem5 configuration for attack analysis.

Parameter	Value
Simulation Mode	Full-System (FS)
ISA	RISC-V 64-bit
CPU Model	`O3CPU` (out-of-order)
L1 Cache	32 KB I/D
L2 Cache	256 KB
Memory	4 GB DDR4
Kernel	`riscv-bootloader-vmlinux-5.10`

The gem5 simulator has certain limitations. Its execution speed is slower than real hardware, particularly for FS-mode simulations with detailed timing models. Resource demands (memory and CPU) scale with system complexity, limiting accessibility for large-scale experiments. Despite these limitations, gem5 configurable workflow balances granular data collection with flexibility, making it an excellent tool for microarchitectural research.

4 Results

Building on prior work demonstrating KASLR breaks on RISC-V hardware [2], we leverage gem5's granular metrics such as cycle counts, and TLB behavior to dissect attack mechanics in simulated environments. By exploiting the unprivileged `rdcycle` instruction, we measure page-table walk timing discrepancies and correlate them with microarchitectural side effects. Our analysis confirms that cycle counts, TLB misses, and simulation overheads differ significantly between mapped (Walk) and unmapped (No-Walk) scenarios, validating gem5's utility for studying KASLR vulnerabilities.

Figure 2 highlights the 44% increase in `rdcycle` count during page-table walks exposes timing variations exploitable for KASLR inference.

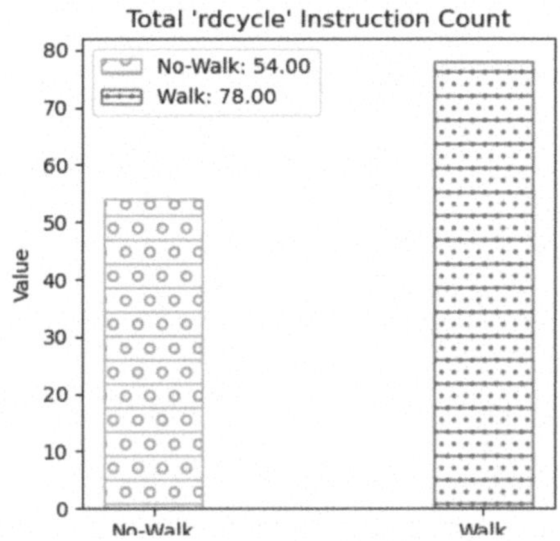

Fig. 2. `rdcycle` Counts during Page-walk and No-walk scenario.

Figures 3 and 4 illustrate increased data and instruction TLB misses in walk scenarios. This reflects address translation overheads that amplify timing leaks.

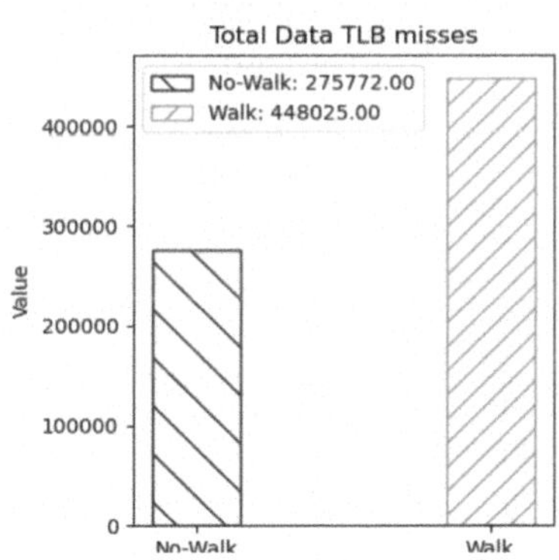

Fig. 3. Data TLB misses during Page-walk and No-walk scenario.

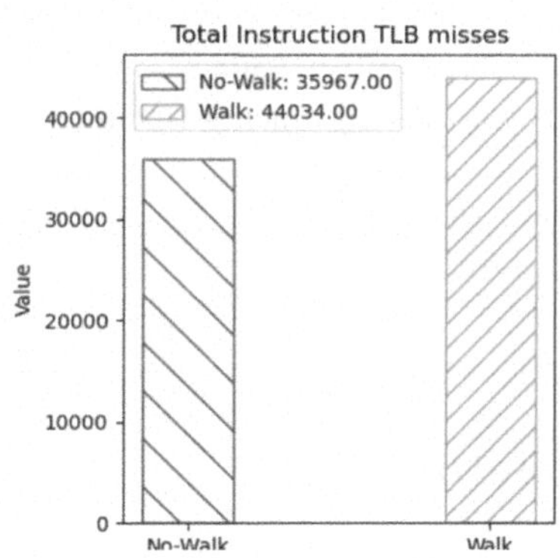

Fig. 4. Instr. TLB misses during Page-walk and No-walk scenario.

Figure 5 and 6 quantifies gem5's execution costs. Walk scenarios require longer simulation time and more instructions, reflecting the computational burden of page-table walks. These metrics validate gem5's ability to model real-world timing dynamics critical for side-channel analysis.

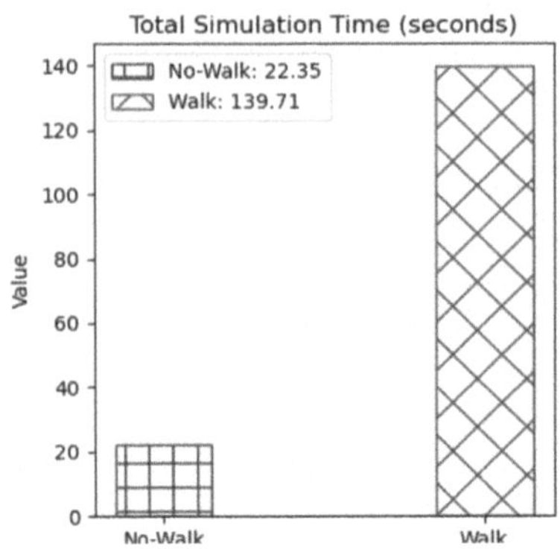

Fig. 5. Total Simulation Time.

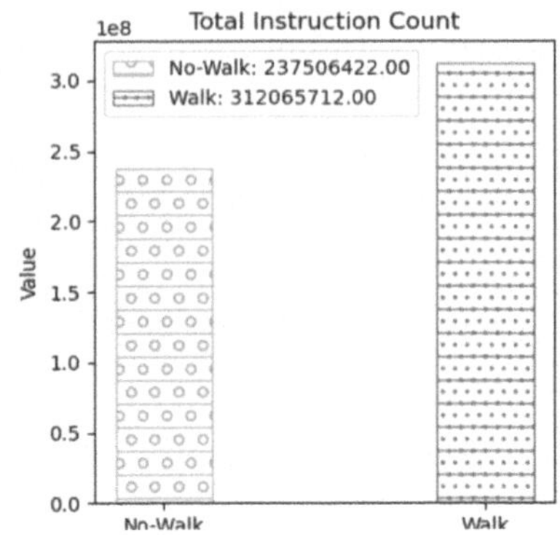

Fig. 6. Total Instruction Count.

Table 2 details branch predictor behavior. Walk scenarios experience an increase in branch prediction metrics. These metrics highlight Gem5's capacity to model complex microarchitectural interactions beyond timing.

Table 2. Comparison of Branch Predictor Metrics for Walk vs No-walk.

#	Metric Name	No Page-table Walk	Page-table Walk
1	Branch Predictor lookups (Count)	67621535	87952408
2	Squashed branches (Count)	395515	777397
3	Corrected branches (Count)	3466976	4795993
4	Committed branches (Count)	44251757	55011904
5	Mispredicted branches (Count)	2297497	3245251

5 Future Work

Future research can focus on advancing gem5 as a dedicated security platform. This involves creating enhanced visualization and tracing tools within gem5, enabling security researchers to observe and analyze microarchitectural events with greater granularity. Further enhancements could be to develop a vulnerability assessment framework leveraging gem5-simulated hardware performance counters, thereby providing an environment for refining new defenses. Potential countermeasures include finer-grained KASLR (e.g., per-function randomization), and improved cache and TLB randomization strategies. With gem5's reconfigurability and the repeatability of experiments, these countermeasures can be thoroughly tested, ultimately leading to more effective protective measures.

6 Conclusion

KASLR is a widely used defense mechanism that randomizes the kernel's base address to prevent reliable exploitation, but current RISC-V implementations lack dedicated countermeasures and rely on coarse-grained randomization. As shown by our gem5 metrics and prior work by Gerlach [2], attackers can exploit timing variations during page walks to bypass KASLR on RISC-V CPUs using via the `rdcycle` instruction. Unlike hardware platforms, which do not provide detailed microarchitectural insights, the gem5 simulator enables comprehensive analysis, making it a valuable tool for evaluating attacks and testing countermeasures in a controlled environment. In addition to `rdcycle` count, gem5 offers detailed metrics such as instruction counts, cache behavior, and TLB access patterns, facilitating a deeper understanding of attack mechanics.

References

1. Canella, C., Schwarz, M., Haubenwallner, M., Schwarzl, M., Gruss, D.: KASLR: break it, fix it, repeat. In: Proceedings of the 15th ACM ACCS (2020)
2. Gerlach, L., Weber, D., Zhang, R., Schwarz, M.: A security RISC: microarchitectural attacks on hardware RISC-V CPUs. In: 2023 IEEE S&P. IEEE (2023)
3. Larabel, M.: FGKASLR patches revised a 10th time for improving Linux kernel security (2022). https://www.phoronix.com/news/FGKASLR-Linux-v10
4. Li, J., Tufte, K.: Out-of-order processing: a new architecture for high-performance stream systems. Proc. VLDB Endow. **1**(1) (2008)
5. Lowe-Power, J.: Gem5 doc (2024). https://www.gem5.org/documentation/
6. Lowe-Power, J., Ahmad, A.M., Akram: The gem5 simulator. arXiv (2020)
7. Ta, T., Cheng, L., Batten, C.: Simulating multi-core RISC-V systems in gem5. In: Workshop on Computer Architecture Research with RISC-V (2018)

Complex Network Analysis for Cybersecurity -CNACYS 2025 Special Session

Spectral Graph Analysis of Bipartite Graphs for Advanced Attack Detection

Majed Jaber[1,2](✉), Nicolas Boutry[1], and Pierre Parrend[1,2]

[1] Laboratoire de Recherche de l'EPITA, 14-16 Rue Voltaire, Le Kremlin-Bicêtre, France

{majed.jaber,pierre.parrend}@epita.fr, nicolas.boutry@lrde.epita.fr

[2] ICube, UMR 7357, Université de Strasbourg, CNRS, IDEM, Strasbourg, France

Abstract. Spectral graph theory offers powerful tools for understanding graph properties through spectral signatures. This work leverages the inherent link between graph topology and spectral characteristics to enhance anomaly detection in network traffic, particularly in medical IoT networks. We introduce *SPECTRA*, a spectral graph analysis technique designed to detect anomalies in dynamic and complex network structures. This method incorporates five spectral metrics, including the newly proposed BiFlowness metric derived from *Singular Value Decomposition (SVD)*, which captures the flow dynamics within bipartite graph topologies. By combining these spectral metrics, *SPECTRA* provides a comprehensive model for detecting and analyzing advanced cyberattack patterns, such as multistep intrusions, in critical systems. Focusing on hybrid topologies that integrate star and bipartite structures, this technique applies spectral analysis to evolving networks, enabling the detection of attacks (port scanning, fingerprinting) over time. Performed experiments validate the effectiveness of *SPECTRA* across IoT datasets, demonstrating its superiority in identifying attack behaviors. The proposed approach aligns with the critical demands of medical IoT environments by providing a good threat detection procedure to enhance security in sensitive networks.

Keywords: Spectrum · Spectral graph · Bipartite topology · Attack detection · Laplacian matrix · Singular Value Decomposition

1 Introduction

In recent years, medical networks have emerged as critical targets for cyberattacks due to the sensitivity of the data they handle. These networks, often comprised of interconnected devices, servers, and systems, face sophisticated and multi-step attacks that exploit vulnerabilities in their structure. A multistep attack is a cyberattack that involves a series of interconnected and sequential stages, each designed to achieve specific objectives, such as gaining access, followed by privilege elevation, and finally data theft or disruption. Defending

I. Praça et al. (Eds.): EICC 2025, CCIS 2500, pp. 239–254, 2025.
https://doi.org/10.1007/978-3-031-94855-8_16

against these advanced attacks requires understanding their behavior to capture the evolving connectivity patterns within these networks. Through spectral graph analysis, it is possible to uncover the spectral properties of the networks over time, where spectral properties are important in capturing dynamic and evolving relationships between the devices in the medical networks. To address the complexity of these hybrid topologies. The contribution of this work is the introducing of a new metric called *BiFlowness* derived from singular values obtained through *Singular Value Decomposition (SVD)*. This metric is specifically designed to measure the flow within bipartite topologies, capturing critical aspects of network connectivity and behavior. The highlights of this paper are as follows: (1) we introduce a *BiFlowness* metric based on singular values; (2) *BiFlowness* metric is designed to measure the flow within the bipartite subgraph of the network; (3) we extend the graph conception beyond star topologies to a combination of bipartite and star graphs; (4) we adapt the previous methodology *SPECTRA* [14] to handle these hybrid topologies; and (5) we provide extensive benchmarks demonstrating the effectiveness of combining the five metrics for analyzing and understanding hybrid network topologies.

2 State-of-the-Art

Spectral Graph Analysis for Attack Detection: Several spectral analysis approaches have been applied in cybersecurity. For example, power spectral density (PSD) was used in [8] to detect periodic changes in Transmission Control Protocol (TCP) transport, aiding in defending against DoS attacks. Another approach introduces an adversarial attack on Graph Convolutional Networks (GCNs) by perturbing graph structures to maximize spectral distance. This disrupts spectral filters, compromising GCN performance in both test-time and training-time scenarios. Methods such as MC-GPCA and MC-GDL in [6] reduce graph complexity and perform dictionary learning, utilizing spectral decomposition through Singular Value Decomposition (*SVD*). Techniques like discrete Fourier transform (DFT) combined with hypothesis testing [7] and dynamic graph modeling [2] further expand anomaly detection and pattern discovery in static and evolving networks.

2.1 Machine Learning for Anomaly Detection

Anomaly detection methods are categorized into statistical and machine learning (ML)-based approaches. Statistical methods like ReTiNA [19] offer real-time and scalable anomaly detection but face accuracy limitations. ML-based platforms, such as CAMLPAD [13], use diverse data sources (e.g., logs and network traffic) and algorithms like Logistic Regression, Decision Trees, and k-Means to improve detection. However, issues such as limited labeled data [3], skewed datasets, label noise, and rapidly evolving cyber threats remain significant. Graph-based ML methods, such as random walks [12] and StrGNN [5], address these challenges by leveraging graph structures for anomaly detection. Deep learning techniques, including Graph Convolutional Networks (GCNs) [15], and GNN model for analyzing malicious web requests [17], further enhance detection by capturing complex relationships in graph data.

3 Network Graph Analysis

This section examines network graph analysis through structural and spectral approaches, focusing on matrix representations and eigenvalue-based insights.

3.1 Graph-Based Network Representation for Cybersecurity

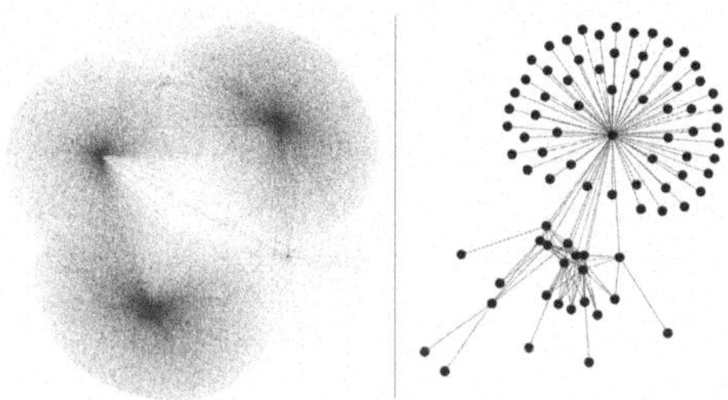

Fig. 1. Graph representations show the TonIot and Botnet datasets respectively from the left to the right.

Graph theory is a crucial tool in cybersecurity, enabling the modeling of networks to detect threats and assess vulnerabilities. By utilizing graph representations Fig. 1, particularly adjacency and Laplacian matrices, network behaviors can be evaluated, and anomalies such abnormal behaviors can be identified. A network is composed of interconnected *nodes*, such as computers and servers, which facilitate data exchange and shared resource access. Essential network characteristics include *interconnectivity*, which enables wired or wireless connections; *data exchange*, which ensures the transfer of files and messages; *shared resources*, such as printers and internet access; and compliance with *protocols* that regulate communication between devices [21]. Networks can be structured as graphs, where nodes represent non-transparent devices -those that can be reliably identified based on traffic log data- and edges signify interactions such as data transmission, communication protocols, or connection rates.

Structural Matrices in Graph Theory. Let G be a graph consisting of n nodes. Its structure is defined by three fundamental matrices. The first is the *Adjacency Matrix* A, an $n \times n$ matrix where each element A_{ij} is 1 if nodes i and j are connected, and 0 otherwise. The second matrix, the *Degree Matrix* D of G, is a diagonal $n \times n$ matrix where D_{ii} is given by $\sum_{j=1}^{n} A_{ij}$, while all off-diagonal elements are set to 0. Finally, the *Laplacian Matrix* L is defined as $L = D - A$ [18]. This matrix plays a crucial role in spectral graph theory, capturing both the spectral properties and structural integrity of the graph.

Laplacian Eigenvalues: The Graph's Spectrum. When a network is modeled as a graph, its topological characteristics can be derived through spectral graph analysis [9] by examining the Laplacian eigenvalues. Various properties emerge from this analysis, such as the number of eigenvalues equal to zero, which corresponds to the count of connected components in the graph, and the largest eigenvalue, which reflect the highest degrees between the interconnected nodes. Additionally, the robustness of this spectral representation to edge rewiring has been demonstrated, meaning that *small alterations in graph weights result in minor variations in the graph spectrum.*

Let Λ be the spectrum of the Laplacian matrix L, $\mathcal{Z}$ is equal to the number of eigenvalues equal to 0 in Λ. We recall that, in linear algebra, any eigenvalue $\lambda \in \Lambda$ of a square matrix $\mathbf{A}$, satisfies the relation $\mathbf{Av} = \lambda\mathbf{v}$, with $\mathbf{v}$ the eigenvector associated to the eigenvalue λ. The Laplacian matrix's spectrum is important as it uncovers critical structural properties of graphs such as the number of components of the graph (sub-networks) and the connectivity of the overall network structure, maintaining its relevance even when the graph's topology is modified [4], thereby emphasizing its utility in exploring and understanding the convolutions of networks.

Singular Value Decomposition (SVD): Theory and Use. Singular Value Decomposition (SVD) is a linear algebra technique that factorizes any real or complex matrix M into three matrices: $M = U\Sigma V^T$, where U and V are orthogonal (or unitary) matrices and Σ is a diagonal matrix containing the singular values of M [1]. The singular values, which are non-negative and sorted in decreasing order, provide insight into the intrinsic geometric and algebraic structure of the matrix. In network analysis, particularly spectral graph methods, SVD is used to extract structural features from matrices representing node interactions, such as adjacency or Laplacian submatrices. Unlike eigenvalue decomposition, which requires square symmetric matrices, SVD applies to any $m \times n$ matrix, making it suitable for analyzing rectangular matrices in bipartite graphs. In this work, we apply SVD to the submatrix of the Laplacian matrix derived from bipartite partitions to compute singular values that reflect connectivity and flow across disjoint node sets. More details will be explained in the upcoming sections.

4 Introducing Spectral Metrics: Indicators

In this context, $\Lambda^{(t)}$ represents the spectrum of L at time t, and for each $i \in [1, n]$, $\lambda_i^{(t)}$ denotes the ith smallest eigenvalue within $\Lambda^{(t)}$. The set $\Sigma^{(t)}$ represents the singular values of L' at time t sorted in increasing order, and for each $j \in [1, n]$, $\sigma_i^{(t)}$ denotes the jth smallest eigenvalue within $\Sigma^{(t)}$.

4.1 Spectral Metrics for Network Analysis

Connectedness. The first metric, *Connectedness*, represented as $\mu_1(t)$, measures the overall connectivity within the network. It is formulated as follows:

$$\mu_1(\mathrm{t}) = \frac{\exp \frac{1}{\mathcal{Z}(\mathrm{t})}}{\exp(1)}$$

where $\mathcal{Z}(t)$ represents the count of zero values in the network's spectrum. This metric helps detect changes in network topology. When μ_1 approaches 1, it may indicate an intrusion, as it suggests the formation of new links between previously disconnected components. As $\mathcal{Z}(t)$ increases towards $+\infty$, μ_1 converges to $\exp^{-1} \approx 0.367$. Conversely, when $\mathcal{Z}(t)$ nears 1, μ_1 tends toward 1.

Flooding. The second metric, *Flooding*, denoted as $\mu_2(t)$, assesses the network's overall connectivity by analyzing the algebraic connectivity of each connected component. It is defined as follows:

$$\mu_2(t) = \left(\frac{1}{\mathcal{N}} \sum_{i=\mathcal{Z}(\mathrm{t})+1}^{\mathcal{Z}(\mathrm{t})+\mathcal{N}} \lambda_i^{(t)} \right) - 1$$

where $\mathcal{N}$ represents the number of central network devices, such as switches and servers. This metric accounts for both the existence of node connections and their associated weights, such as packet count or byte size. A higher number of connections with substantial weights leads to an increase in μ_2.

Wiriness. The third metric, *Wiriness*, represented as $\mu_3(t)$, evaluates the density and strength of network interconnections by considering edge weights. It is given by:

$$\mu_3(\mathrm{t}) = \frac{1}{\mathcal{N}} \sum_{i=n-\mathcal{N}+1}^{n} \lambda_i^{(t)}$$

This metric captures the impact of new connections on the network's structure. It increases with higher packet sizes and a greater number of network requests, making it useful for identifying network flooding or large-scale data transfers.

Asymmetry. The fourth metric, *Asymmetry*, represented as $\mu_4(t)$, quantifies variations in spectral properties across the network:

$$\mu_4(t) = \mathit{Cardinality}\left(\left\{ i \in [2, n] \; ; \; \lambda_i^{(t)} - \lambda_{i-1}^{(t)} > \varepsilon \right\}\right)$$

where $\varepsilon = 10^{-12}$ is included to reduce approximation errors. This metric evaluates the smoothness of the spectral curve. When the network maintains similar structural patterns, μ_4 remains low. However, an increase in μ_4 suggests changes in connectivity patterns, potentially signaling a security threat.

4.2 Introducing BiFlowness Metric

Eigenvector-Based Bipartite Graph Partitioning and Analysis. Partitioning graphs into known topologies is essential for understanding network connectivity and identifying structural patterns like bipartite graphs. Spectral graph analysis provides a mathematical framework for this, using L's eigenvectors to partition node sets efficiently. In cybersecurity, bipartite structures often reflect behaviors from attacks like port scanning and OS fingerprinting. A bipartite graph comprises two disjoint node sets with edges connecting only across sets. Partitioning using spectral methods relies on the *Fiedler vector*, associated with the second-smallest Laplacian eigenvalue, for balanced graph division [11]. The eigenvector $\mathbf{EV}_n$, linked to the maximum eigenvalue, also aids in estimating bipartite sets in our methodology, detailed in the next section.

Bipartite Graph Estimation and Partitioning. In order to apply spectral graph analysis to network topologies, the initial step involves identifying bipartite graphs and estimating their partitions in network traffic data. This is achieved by leveraging the eigenvectors of the Laplacian matrix L, starting with the eigenvector linked to the second-smallest eigenvalue-known as the Fiedler vector [11]-and progressively considering those associated with the third, fourth, and subsequent smallest eigenvalues. Partitioning relies on the sign of each entry in an eigenvector $\mathbf{v}_i$: nodes with positive values are grouped into one set, while those with negative values form the other. As shown in Fig. 2, we examine multiple scenarios on $G_{m,n}$ graphs and select two specific eigenvectors for comparison: the Fiedler vector and the eigenvector corresponding to the maximum eigenvalue, $\boldsymbol{\lambda}_{max}$. To evaluate the quality of the partitions they generate, we use two clustering metrics: the **Adjusted Rand Index (ARI)**, which measures agreement between predicted and actual partitions while correcting for random chance, and the **Normalized Mutual Information (NMI)**, which quantifies shared information between partition sets. Figure 2 begins with a graph of eight nodes and incrementally adds edges until a complete bipartite graph is formed by evolution step fifteen. At each step, we apply both partitioning methods and compute ARI and NMI scores to assess clustering performance. Results in the bottom right of the figure show that the eigenvector of $\boldsymbol{\lambda}_{max}$ consistently achieves better partitioning, especially from step 9 onward, where each node has at least one edge. This indicates that $\boldsymbol{\lambda}_{max}$ provides a more reliable representation of bipartite structure in evolving graph topologies compared to the Fiedler vector.

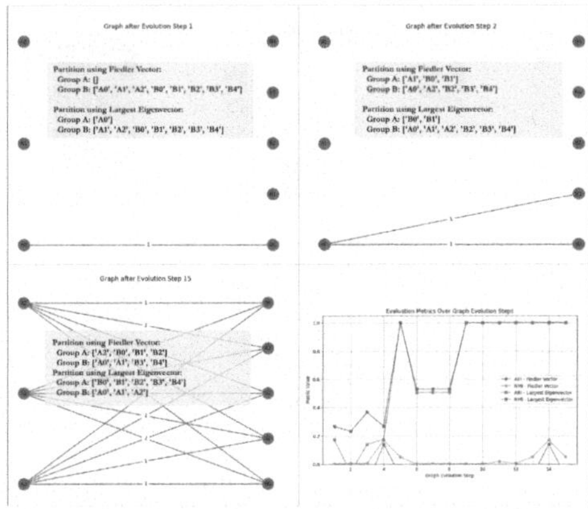

Fig. 2. Graph evolutions representing the procedure used to partition bipartite graph at each step, with comparison metrics.

Sub-matrix Extraction. Let $\mathbf{card}(X)$ and $\mathbf{card}(Y)$ represent the cardinalities of the two sets of X and Y of $G_{m,n}$ respectively after partitioning based on the eigenvector of $\boldsymbol{\lambda}_{max}$. The objective is to extract a submatrix of L, denoted as L', which encapsulates the interconnections between the two sets X and Y.

Consider a bipartite graph $G_{m,n} = (U, V, E)$ consisting of two disjoint node sets, where $X = \{x_1, x_2, \ldots, x_m\}$ and $Y = \{y_1, y_2, \ldots, y_n\}$, and where the edge set E contains only inter-set edges, i.e., edges exclusively between X and Y, with no intra-set connections. The adjacency matrix A of this bipartite graph takes the form:

$$A = \begin{bmatrix} 0 & B \\ B^T & 0 \end{bmatrix}$$

where B is an $m \times n$ incidence matrix such that $B_{ij} = 1$ if x_i is connected to y_j, and B^T represents the transpose of B. The other blocks contain only zeros, ensuring no intra-set edges exist within the same group. A submatrix of A, denoted as $A' = B$, retains information about the edges between the two groups. Consequently, L' is derived from L by extracting its last $\mathbf{card}(Y)$ columns and first $\mathbf{card}(X)$ rows, preserving the spectral, weight distribution and connectivity properties.

An example is provided in Fig. 3, depicting a single evolution window in our methodology. At this stage of evolution, a bipartite graph $K_{3,2}$ is constructed, representing the traffic logs over the network at tw_i. On the right side of the graph, the adjacency matrix A is displayed, with a focused view on its submatrix $A' = B$, which has dimensions $\mathbf{card}(X) \times \mathbf{card}(Y)$. Similarly, the submatrix L' of L is shown, capturing the connectivity structure between the two sets. In the next section (Sect. 4.2), we delve further into singular value decomposition (SVD) and its relevance in $SPECTRA$ methodology.

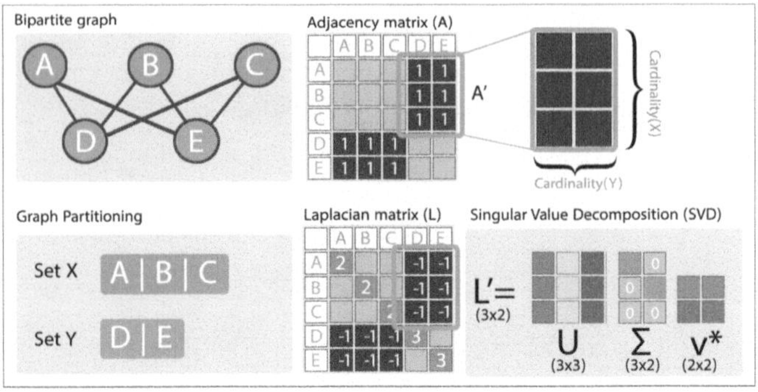

Fig. 3. The figure shows the extraction of L' from $G_{3,2}$, highlighting the partitioned sets X and Y (bottom left) and the construction of matrices A' and L' (top right and bottom middle). SVD is then applied to L' to compute the singular values Σ.

Sigma from SVD. Here is the last step of analyzing the bipartite topologies, in this step SVD decomposition is applied over submatrix L' as shown in Fig. 3:

$$\underset{m\times n}{\overset{\text{Sub-Matrix}}{L'}} = \underset{m\times m}{\overset{U}{\begin{bmatrix} u_1 & u_2 & \cdots & u_m \end{bmatrix}}} \underset{m\times n}{\overset{\Sigma}{\begin{bmatrix} \sigma_1 & 0 & \cdots & 0 \\ 0 & \sigma_2 & \cdots & 0 \\ \vdots & \vdots & \ddots & \vdots \\ 0 & 0 & \cdots & \sigma_r \end{bmatrix}}} \underset{n\times n}{\overset{V^T}{\begin{bmatrix} v_1^T \\ v_2^T \\ \vdots \\ v_n^T \end{bmatrix}}} \tag{1}$$

where $\mathbf{u}_i$ are the left singular vectors (columns of U), $\mathbf{v}_i$ are the right singular vectors (rows of V^T), and σ_i are the singular values in decreasing order. An important observation is that most of the information regarding the graph's topological structure is contained within the set Σ. Notably, Σ singular values reflect the eigenvalues of L, which motivated us to further explore this connection. Consequently, we developed a bipartite spectral metric that leverages the values of Σ, which will be discussed in detail in Sect. 4.

BiFlowness. Here we introduce a new spectral metric, *BiFlowness* denoted as $\mu_5(t)$. *BiFlowness* is a novel contribution that extends spectral analysis by incorporating singular value decomposition (SVD), and since singular values are always non-negative, our metric is guaranteed to be positive. It is defined as follows:

$$\mu_5(\mathrm{t}) = \sum_{j=2}^{n} \sigma j$$

where σj represents the singular values obtained from the decomposition of the sub-matrix L' of L. This metric provides insights into intra-group connections by analyzing eigenvalue structures indirectly. By capturing interconnections between bipartite network groups, μ_5 enhances the understanding of network structural changes over time. It is particularly useful for identifying coordinated attacks that exploit internal network structures. This additional metric strengthens network intrusion detection by complementing the existing four spectral metrics, enabling a more comprehensive evaluation of network behavior especially the interconnections of the two sets of the bipartite topologies.

5 Methodologies and Techniques

In this section, we present a set of methodologies designed to enhance network security analysis. Beginning with fundamental dataset examination and progressing to advanced spectral-time-window classification (with different combination of spectral metrics), each approach incrementally deepens the understanding of network behavior and threat detection. These methods establish a framework for identifying and analyzing anomalies in network traffic, excluding time and sequence data to reduce bias and improve detection accuracy. Next, we will present the methods we used to build our approaches, along with their respective methodologies, beginning with the *OD*, then progressing to *STS* and *STW*, and ending up with *SPECTRA*.

Original Data-logs (*OD*). In this approach, which serves as a baseline for further evaluations, we use datasets without any preprocessing. All dataset features are included, except those related to date, time, and packet arrival sequences. These exclusions prevent bias, as many attacks tend to occur at the same timestamps. The main goal is to establish a reference point for comparing other methods [10]. Our approach consists of two key steps:

- Using the dataset without applying any preprocessing steps.
- Retaining all dataset features except sequence and time-related attributes.

Time-Series for Spectral Analysis (*STS*). This approach extends the previous one by incorporating all dataset features while continuing to exclude time and sequence attributes. The aim behind this approach arises from the observed imbalance between attack and normal traffic within the dataset's log entries. Furthermore, time series analysis offers key advantages in managing dynamic graphs that evolve over time [20]. In this context, each unique timestamp in the dataset corresponds to a specific stage in the graph's evolution. To implement this approach, we follow the methodology illustrated in Fig. 4a, where data is aggregated based on predefined constraints and grouping criteria. First, we define a time constraint t ($t \geq 1sec$) to establish each stage of graph evolution. Then, we iterate through the log set at specific time intervals t, represented as ts.

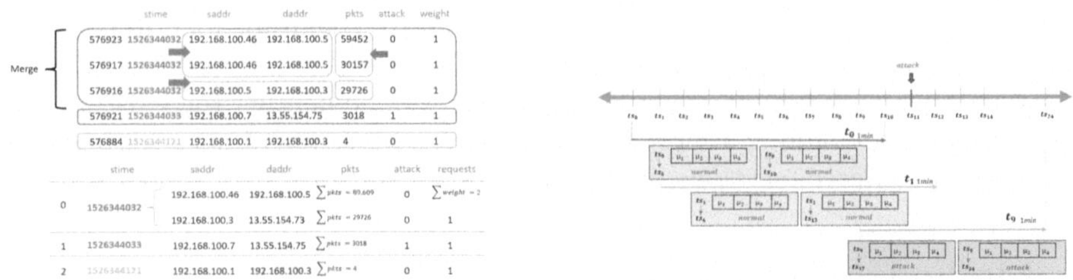

(a) *STS* over Botnet dataset traffic network sample

(b) Times-window methodology over *STS*(*Botnet*)

Fig. 4. Time-series and time-window methodologies over Botnet data sample.

Table 1. Aggregation methods for dataset features

Feature	Aggregation	Feature	Aggregation
Total packets	Sum	Source packets	Sum
Destination packets	Sum	Total bytes	Sum
Source bytes	Sum	Destination bytes	Sum
Sum	Mean	Label	First
Total rate	Mean	Source rate	Mean
Destination rate	Mean	Duration	Mean
Mean	Mean	Sum	Mean
Max	Mean	Min	Mean

Next, traffic logs are grouped based on specific conditions: logs must originate from the same source IP, be directed to the same destination IP, and share the same traffic type (e.g., normal). This grouping strategy is a core component of our approach. If such traffic occurs within the same time window t, it is aggregated. The aggregation process is feature-dependent, with each feature assigned a specific aggregation method. By following this procedure, we transform the dataset into a time-series representation, denoted as $STS_{dataset}$, ensuring well-defined constraints on both time and feature selection.

Time-Windowing for Spectral Analysis (*STW*). In this approach, we model the topological structure of networks in time-series data by applying a sliding time-window technique to the dataset $STS_{dataset}$. This method scans the data using a window of size tw, requiring at least $MIN_{tw} \geq 2$ data logs per window. Each window extracts only quantitative features-such as packet count, byte volume, and rate-related metrics-while excluding topological structure, which is addressed separately in Sect. 5. This exclusion enables a direct comparison with spectral metrics that rely on the same feature set. The sliding window may cover multiple timestamps, and within each window, data is aggregated according to Table 1. Aggregated rows are labeled based on whether attacks occur within the window. The final output consists of two new time-windowed datasets: STW_{bot} for Botnet-IoT and STW_{ton} for Ton-IoT, each representing the aggregated observations over time.

Spectral Methodology Applied on Time-Windowing (*SPECTRA*). In this method, we leverage graph structures within each designated time-window. While this approach shares similarities with the previous *STW* method, it introduces a key difference in the handling of the time-window.

Step 1. Data rows are transformed into a graph and then processed within two distinct sub-time-windows, Sub_{TW_1} and Sub_{TW_2}. Here, Sub_{TW_1} represents the graph constructed from TW_0 to $TW_{n/2}$, while Sub_{TW_2} extends from TW_0 to TW_n, where n denotes the total number of timestamps within the time-window. The division of a time-window into these two sub-windows allows for tracking variations over time, which directly influences the computed metrics.

Step 2. Once graphs are formed over the sub-time-windows, we extract the Laplacian matrix L for Sub_{TW_1} and Sub_{TW_2}, producing L_1 and L_2. Next, we compute the spectra Λ_1 and Λ_2 for each Laplacian matrix. The resulting datasets, $SPECTRA_{bot}$ and $SPECTRA_{ton}$, correspond to the number of sliding windows across the time-series for the Botnet-IoT and Ton-IoT datasets, respectively. With these datasets prepared, we proceed to the application of various classification techniques.

6 Implementation

Table 2. Extracted and eliminated features of Bot-IoT dataset

Features Extracted	Features Eliminated
flgs_number	pkSeqID
dur, proto_number	seq, stime
pkts, bytes	ltime, TnBPSrcIP
state_number, mean	TnBPDstIP, TnP_PerProto
stddev	TnP_Per_Dport
sum, min	AR_P_Proto_P_(SrcIP,DstIP)
max, spkts	N_IN_Conn_P_(SrcIP, DstIP)
sbytes	AR_P_Proto_P_(Sport, Dport)
dbytes	Pkts_P_State_P_Protocol_P_DestIP
rate, srate, drate	Pkts_P_State_P_Protocol_P_SrcIP

Table 3. Extracted and eliminated features of Ton-IoT dataset

Features Extracted	Features Eliminated
duration	ts
src_bytes	ltime
dst_bytes	
src_pkts	
dst_pkts	
label	
type	

We implement the proposed methodology using two Python notebooks that process the *Original Data-logs* (*OD*), *Time-series for Spectral Analysis* (*STS*), *Time-windowing* (*STW*), and *Spectral Analysis on Time-windowed Data* (*SPECTRA*) for both Botnet-IoT and Ton-IoT datasets. The notebooks, available on GitHub[1], enable reproducibility and focus on evaluating μ_5 approaches

[1] https://github.com/majedjaber1996/jaber.2025.eicc.

against the *OD* baseline. Prior implementations comparing *STS* and *STW* are available in another repository[2]. Additionally, the open-source GPML library[3] provides key components for spectral graph analysis.

6.1 Experiments: Practical Applications

This section discusses the applications and case studies conducted as part of our research, building upon the work published in [14]. The focus is on the Botnet-IoT and Ton-IoT datasets 1. The experiments progress through multiple stages, starting with the application of the *OD* methodology as a baseline, followed by the *SPECTRA* methodology incorporating various spectral metrics we introduced. Within *SPECTRA*, traffic logs undergo *STS* and *STW* analysis and processing. Each subsection details the methodology employed. These experiments aim to evaluate the effectiveness of our approaches in detecting network attacks within IoT environments.

Application of *OD*. For the Botnet-IoT and Ton-IoT datasets, the *Baseline* (*OD*) methodology involves selecting a subset of numerical features while excluding time-related, sequence-based, and dataset-specific attributes, as outlined in Tables 2 and 3, and based on prior work [16]. The selected features are those commonly found in the literature across different datasets and attributes present in network traffic which ensures feature consistency and model generalizability. The resulting datasets are denoted as OD_{ton} and OD_{bot}.

Application of *STS*. For the Botnet and TonIoT datasets, the *STS* method builds upon the *OD* approach by incorporating a time-based aggregation. In *STS*, a time window of $t = 1$ s is applied to both OD_{ton} and OD_{bot}. Within this interval, features are aggregated as outlined in Table 1. This process restructures the datasets into a timeseries format, labeled as STS_{bot} for Botnet and STS_{ton} for TonIoT. As a result, the number of data rows typically decreases, aiding in subsequent data analysis and evaluation.

Application of *STW*. To enhance data aggregation, the *STW* method is applied alongside the *STS* approach for the Botnet-IoT and Ton-IoT datasets. In *STW*, different time-window ranges are defined: $tw = 1$ min for STS_{bot} with a minimum threshold of $MIN_{tw} \geq 10$, and $tw = 30$ s for STS_{ton} with $MIN_{tw} \geq 2$. During each t_w, features are aggregated according to Table 1. This process transforms STS_{bot} and STS_{ton} into a new dataset format, denoted as STW_{bot} and STW_{ton}, respectively. In this format, each dataset row represents the aggregated data within the specified tw.

Application of *SPECTRA*. Applying *SPECTRA* requires additional processing for each time window T_i. While *SPECTRA* itself does not impose strict constraints, constructing weighted edges in the network graph depends on selecting a specific feature. For both the Ton-IoT and Botnet-IoT datasets, packet

[2] https://github.com/lre-security-systems-team/jaber.2024.ares.git.
[3] https://gitlab.cri.epita.fr/laboratoires/lse/research-devs/gpml.

count, byte size, and packet rate were chosen as the criteria for edge weighting across three different topologies within the time window. The first topology is weighted by the number of packets transferred, the second by byte size, and the third by packet rate. This selection is based on the observation that these attributes are key indicators for detecting various attack types. For each time window, three sets of ten spectral metrics are computed, corresponding to the three topologies. Metrics from μ_1 to μ_5 are calculated at two intervals: tw (the middle of the time window) and $\frac{tw}{2}$ (the end of the time window). This process generates two datasets, $SPECTRA_{bot}$ and $SPECTRA_{ton}$, where each row consists of spectral metric vectors along with their associated labels.

7 Evaluations: Assessing Techniques and Results

In this section, we evaluate the XGBoost classifier using F1-Score, Balanced Accuracy, and MCC across different methodologies: the *Baseline* (*OD*) and the *SPECTRA*. Building on prior results in [14], we extend the analysis by comparing three *SPECTRA* variants: using only μ_5, using μ_1-μ_4, and using all metrics μ_1-μ_5.

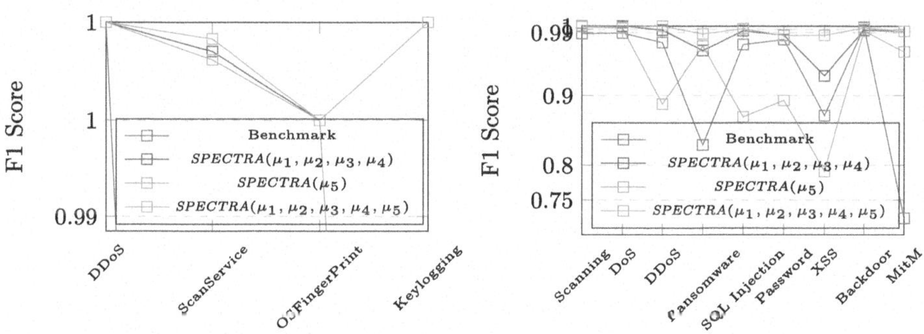

(a) Botnet-IoT Dataset (b) Ton-IoT Dataset

Fig. 5. Comparison of F1 scores across different approaches.

Performance Analysis on Botnet-IoT Dataset: Evaluating the XGBoost classifier's performance on the Botnet-IoT dataset across different approaches reveals substantial variations in classification effectiveness. The Benchmark method achieves a perfect F1-Score of 1.0 for DDoS, but struggles with OS Fingerprint and Keylogging, where F1-Scores drop significantly to 0.2617 and 0.5333, respectively (Fig. 5a). Introducing the *SPECTRA* methodology with different spectral metric combinations significantly enhances classification performance. $SPECTRA(\mu_1, \mu_2, \mu_3, \mu_4)$ achieves an F1-Score of 0.9986 for ScanService, 0.9953 for OS Fingerprint, and 1.0 for Keylogging. Further improvement is observed with $SPECTRA(\mu_1, \mu_2, \mu_3, \mu_4, \mu_5)$, which outperforms all other

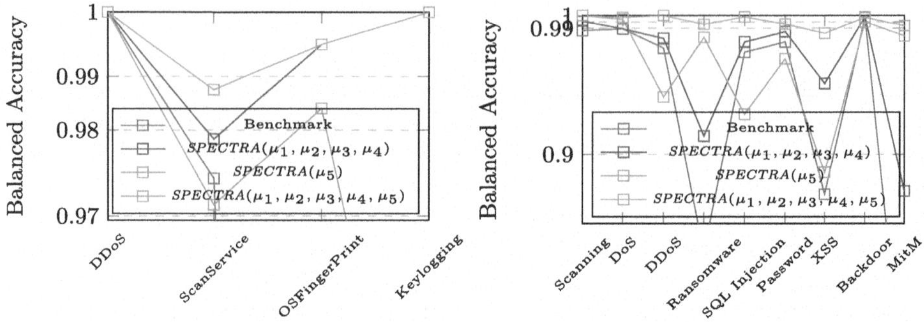

(a) Botnet-IoT Dataset (b) Ton-IoT Dataset

Fig. 6. Comparison of Balanced Accuracy across different approaches.

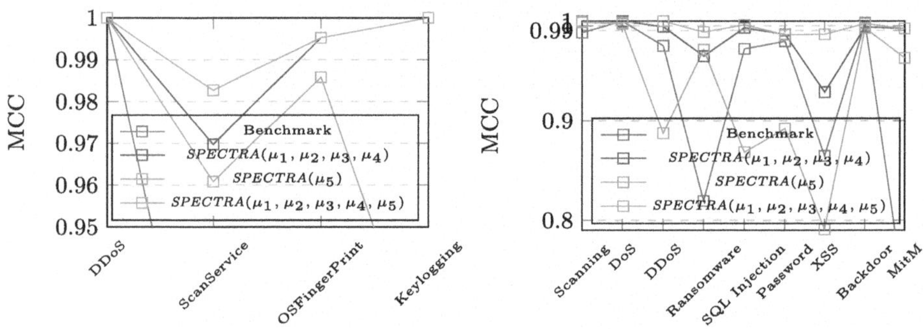

(a) Botnet-IoT Dataset (b) Ton-IoT Dataset

Fig. 7. Comparison of MCC across different approaches.

approaches across all attack types, achieving an F1-Score of 0.9992 for ScanService and maintaining the highest scores for OS Fingerprint and Keylogging (Fig. 5a). Balanced Accuracy and MCC results confirm these trends. The Benchmark approach struggles with OS Fingerprint, achieving a Balanced Accuracy of only 0.5804, while *SPECTRA*($\mu_1, \mu_2, \mu_3, \mu_4, \mu_5$) improves this metric to 0.9953 (Fig. 6a). Similarly, the MCC for OS Fingerprint is lowest in the Benchmark approach (0.3198) but reaches 0.9952 with *SPECTRA* using all spectral metrics (Fig. 7a). These results highlight that incorporating all five spectral metrics optimally refines the classifier's predictive capability, ensuring superior attack detection.

Performance Analysis on Ton-IoT Dataset: Analyzing the Ton-IoT dataset reveals that the XGBoost's performance varies significantly across different attack types and approaches. The Benchmark approach performs well for Scanning and DDoS attacks, achieving F1-Scores of 0.9890 and 0.9764, respectively. However, it struggles with DoS and MitM, where the F1-Scores drop to 0.6175 and 0.7239 (Fig. 5b). The *SPECTRA* substantially enhances the classification performance. *SPECTRA*($\mu_1, \mu_2, \mu_3, \mu_4$) achieves an F1-Score of 0.9944 for DDoS and 0.9650 for Ransomware, demonstrating its effectiveness in detecting complex attacks. The best performance is observed with *SPECTRA*($\mu_1, \mu_2, \mu_3, \mu_4, \mu_5$), which achieves an F1-Score of 1.0 for DDoS and 0.9889 for Ransomware (Fig. 5b), further validating the advantage of incorporating all spectral metrics. The Benchmark approach records a Balanced Accuracy of 0.8290 for Ransomware, while *SPECTRA*($\mu_1, \mu_2, \mu_3, \mu_4, \mu_5$) raises this to 0.9936 (Fig. 6b). MCC values behave similar, where the Benchmark approach achieves only 0.8193 for Ransomware, whereas *SPECTRA* using all five spectral metrics reaches 0.9932 (Fig. 7b). These results confirm that the combination of all spectral metrics significantly improves the classifier's ability to distinguish between various attack types.

8 Conclusion

This work introduces *SPECTRA*, a spectral graph analysis methodology for anomaly detection in dynamic network environments, focusing on medical IoT systems. It uses five spectral metrics-including a BiFlowness metric based on *SVD*-to analyze hybrid graph topologies. Evaluations on IoT datasets show its effectiveness in detecting complex attacks like multistep intrusions. *SPECTRA* supports real-time monitoring of evolving network structures, with future work targeting scalability and broader applications.

Acknowledgments. The authors acknowledge the support of the Région Grand-Est and École Pour l'Informatique et les Techniques Avancées (EPITA) for the joint funding of the XDGMed Project. Their contributions have been invaluable to the research and development efforts.

Disclosure of Interests. The authors have no competing interests to declare that are relevant to the content of this article.

References

1. Abdi, H.: Singular value decomposition (SVD) and generalized singular value decomposition. Encycl. Meas. Stat. **907**(912), 44 (2007)
2. Bilgin, C.C., Yener, B.: Dynamic network evolution: models, clustering, anomaly detection. IEEE Netw. **1** (2006)
3. Bowman, B., Huang, H.H.: Towards next-generation cybersecurity with graph AI. ACM SIGOPS Oper. Syst. Rev. **55**(1), 61–67 (2021)
4. Butler, S., Chung, F., et al.: Spectral graph theory. In: Handbook of Linear Algebra, p. 47 (2006)
5. Cai, L., et al.: Structural temporal graph neural networks for anomaly detection in dynamic graphs. In: Proceedings of the 30th ACM International Conference on Information & Knowledge Management, pp. 3747–3756 (2021)

6. Chen, P.Y., Choudhury, S., Hero, A.O.: Multi-centrality graph spectral decompositions and their application to cyber intrusion detection. In: 2016 IEEE International Conference on Acoustics, Speech and Signal Processing (ICASSP), pp. 4553–4557. IEEE (2016)
7. Chen, Y., Hwang, K.: Collaborative detection and filtering of shrew DDoS attacks using spectral analysis. J. Parallel Distrib. Comput. **66**(9), 1137–1151 (2006)
8. Cheng, C.M., Kung, H., Tan, K.S.: Use of spectral analysis in defense against dos attacks. In: 2002 IEEE Global Telecommunications Conference, GLOBECOM 2002, vol. 3, pp. 2143–2148. IEEE (2002)
9. Chung, F.R.: Spectral Graph Theory, vol. 92. American Mathematical Society (1997)
10. Das, R., Morris, T.H.: Machine learning and cyber security. In: 2017 International Conference on Computer, Electrical & Communication Engineering (ICCECE), pp. 1–7. IEEE (2017)
11. Dutta, S., Metya, N., Mukherjee, S.: Fiedler vector method (2022)
12. Grover, A., Leskovec, J.: node2vec: Scalable feature learning for networks. In: Proceedings of the 22nd ACM SIGKDD International Conference on Knowledge Discovery and Data Mining, pp. 855–864 (2016)
13. Hariharan, A., Gupta, A., Pal, T.: CAMLPAD: cybersecurity autonomous machine learning platform for anomaly detection. In: Future of Information and Communication Conference, pp. 705–720. Springer (2020)
14. Jaber, M., Boutry, N., Parrend, P.: Graph-based spectral analysis for detecting cyber attacks. In: Proceedings of the 19th International Conference on Availability, Reliability and Security, pp. 1–14 (2024)
15. Kipf, T.N., Welling, M.: Semi-supervised classification with graph convolutional networks. arXiv preprint arXiv:1609.02907 (2016)
16. Koroniotis, N., Moustafa, N., Sitnikova, E., Turnbull, B.: Towards the development of realistic botnet dataset in the internet of things for network forensic analytics: BoT-IoT dataset. Futur. Gener. Comput. Syst. **100**, 779–796 (2019)
17. Li, J., Zhou, J., Li, L., Cheng, B., Li, S., He, W.: Robust detection for adversarial attacks-based on a graph neural network for web application security protection system. In: Proceedings of the 4th International Conference on Artificial Intelligence and Computer Engineering, pp. 927–932 (2023)
18. Merris, R.: Laplacian matrices of graphs: a survey. Linear Algebra Appl. **197**, 143–176 (1994)
19. Noble, J., Adams, N.: Real-time dynamic network anomaly detection. IEEE Intell. Syst. **33**(2), 5–18 (2018)
20. Pérez, S.I., Moral-Rubio, S., Criado, R.: A new approach to combine multiplex networks and time series attributes: building intrusion detection systems (IDS) in cybersecurity. Chaos, Solitons Fractals **150**, 111143 (2021)
21. Stallings, W.: Network Security Essentials: Applications and Standards, 4th edn. Pearson Education India (2003)

Hypergraph-Based Model for Adversary Threat Intelligence Analysis

Juan Manuel Matalobos[1,2,3(✉)], Regino Criado[1,2], and Santiago Moral[2]

[1] Departamento de Matemática Aplicada, CC. e Ing. de los Materiales y Tecnología Electrónica, Universidad Rey Juan Carlos, Móstoles, Madrid, Spain
juanmanuel.matalobos@urjc.es

[2] Data, Complex Networks and Cybersecurity Sciences Research Center, Universidad Rey Juan Carlos, Móstoles, Madrid, Spain

[3] Alfonso X el Sabio University, Villanueva de la Cañada, Madrid, Spain

https://www.urjc.es/, https://www.dcncsciences.com/, http://www.uax.es/

Abstract. Threat intelligence is a cybersecurity discipline that focusses on extracting actionable insights from cybersecurity events. This process involves handling large volumes of heterogeneous data with intricate interconnections between different pieces of data. Complex networks is a branch of science closely related to mathematics that study structures made of large amounts of nodes and their relationships, modeled as edges or hyperedges. The modeling of threat intelligence data using complex networks enables the usage of their properties to extract relevant information that can be used for cybersecurity decision making. In this paper we present a model based on a weighted multiplex hypergraph able to select and prioritize the most relevant adversaries for a given organization based on a set of significant characteristics previously selected. This model can be used to support threat intelligence analysts' work, by automating the selection of the short list of relevant adversaries to put focus on, and providing objective measures to the analysis process.

Keywords: Hypergraphs · Higher order networks · Multiplex hypergraphs · Weighted hypergraphs · Sub-hypergraphs · Cybersecurity · Threat Intelligence · Adversary modeling · Threat actor prioritization

1 Introduction

Cybersecurity can be defined as the discipline focused on the protection of digital assets and the information they hold, ensuring their confidentiality, integrity, and availability [1].

With the continuous digitization of the human society, cybersecurity emerges as a cornerstone for the trust of any kind of relationship or activity performed through digital means. As the value of cyberspace for organizations and people has increased, the risk for the owners of digital assets has also grown, as well as the interest of adversaries whose aim is to take advantage of those digital assets.

I. Praça et al. (Eds.): EICC 2025, CCIS 2500, pp. 255–272, 2025.
https://doi.org/10.1007/978-3-031-94855-8_17

In recent years, the number, impact, complexity and diversity of cyberattacks has grown [2] requiring an increasing effort for cybersecurity teams. A major challenge for these teams is identifying and prioritizing the most relevant threats for a specific organization from a vast and dynamic threat landscape.

The Cyber Threat Intelligence field has evolved into the cybersecurity discipline to collect, analyze and deliver information related to potential cyber threats, adversaries, and attack techniques, tactics, and procedures. This information is the basis for decision making regarding the cybersecurity strategy and priorities, leveraging the experience and knowledge of the community [3,4].

Due to the increasing importance of threat intelligence as a basis for sound cybersecurity, there are multiple research lines aiming to improve the processing of cyber threat intelligence, including data collection [5,6], extraction of structured information from sources [7–12], summarization [13,14], analysis [15,16], quality control [17], end to end management of the information pipeline [18,19], sharing [20,21] and protection against malicious interferences [22,23], among many others.

One of the multiple challenges for Cyber Threat Intelligence is to identify adversaries that pose a greater threat to a specific organization, either for the probability that this organization can be targeted by the adversary or the potential impact that a possible attack may cause to the organization.

For an organization, having reliable information about their adversaries can be used for several purposes, such as:

- Understanding the motivation behind those adversaries, the assets of the organization can be prioritized from the cybersecurity standpoint, helping the decision making on where to put the focus on.
- Tailor defenses against the specific techniques, tactics, and procedures of those adversaries, narrowing the threat space to consider.
- Prioritize monitoring and threat hunting with specific IoCs (Indicators of Compromise) and known techniques and tools used by those adversaries.
- Perform hacking and red team exercises that emulate the behaviour of those adversaries, in order to have a more accurate measure of the defense capabilities against them.
- Prioritize the follow up of the adversaries in the threat intelligence activities, in order to promptly detect any evolution, campaign or publication that may be relevant for the organization.

For Cyber Threat Intelligence, Mathematics can provide a formal basis for the modeling of the gathered information and the development of analysis methodologies. So far, several areas of mathematics have been and are being tested for this purpose, like Game Theory, Complex Networks Theory, and many others.

In particular, graphs are powerful mathematics tools that enable the definition of a number of elements, modeled as nodes, and their relationships, modeled as edges. This capacity of graphs to model complex structures of information offers an opportunity to Cyber Threat Intelligence for modeling huge sets of gathered data and make use of the multiple known properties of graphs to obtain results [24].

The development of higher order networks, that allow the definition of hyperedges that model the relationships between more than two nodes, gives an additional opportunity to define models that represent the complexity of current cybersecurity problems in a more accurate and powerful way.

This document explores the feasibility of modeling cyber threat intelligence information related to adversaries and their victimology using higher order networks and the capabilities of those models to provide useful insights from this information.

The requirements for this model to be useful and applicable in the real world are the following:

1. The model must be based on threat intelligence sources, whether open or proprietary. The quality of the results will heavily depend on the quality of the source information.
2. Due to the huge amount of threat intelligence information to process, the ingestion of information must be automated, with low or no human interaction.
3. In order to ensure the objectivity of the results obtained from the model, no expert estimates are required in any step. Any data used by the model must be obtained following a formally defined process.
4. The model must be able to integrate the threat intelligence gathered from the sources with information related to a specific organization that will extract tailored results.

After this introduction, which gives an overview of the context of the problem and the objectives of this work, the structure of the rest of the document is as follows:

- **Section 2 (Preliminaries)** gives the theoretical background about Complex Networks needed for the development of the model.
- **Section 3 (Methodology)** defines the adversary threat intelligence model.
- **Section 4 (Toy model)** explains the definition and the usage of the model with a simplified example.
- **Section 5 (Model implementation and validation)** demonstrates the usage of the model with a real dataset, and compares the results obtained against a real life threat intelligence report.
- **Section 6 (Conclusions and future work)** expose the conclusions obtained in this job and future lines of development of the model.

2 Preliminaries

In this section we give the background of Complex Networks that is used in the development and use of the model.

A **hypergraph**, or **higher order network** is a generalization of the concept of graph where the edges are not limited to link exactly two nodes, but any number of them. Formally, a hypergraph [25, 26] is defined as the pair $\mathcal{H} = (\mathcal{V}, \mathcal{E})$, where:

- $\mathcal{V} = \{v_1, \cdots, v_n\}$ is a set of n nodes, and
- $\mathcal{E} = \{h_1, \cdots, h_m\}$, where $h_k \subset \mathcal{V}$, $\forall k \in \{1, \cdots, m\}$ is a set of hyperedges, each of them linking an arbitrary number of nodes (≥ 0).

A **weighted hypergraph** enables to define the intensity of the relationship defined by each hyperedge. A weighted hypergraph [25,27] is defined as the triplet $\mathcal{H}^{\mathcal{W}} = (\mathcal{V}, \mathcal{E}, \mathcal{W})$, where

- $(\mathcal{V}, \mathcal{E})$ is a hypergraph, and
- $\mathcal{W} : \mathcal{E} \to \mathbb{R}^+ \cup \{0\}$ is a function that define the intensity of the relationship defined by each hyperedge.

In many contexts there are different types of relationships that have to be modeled by a single hypergraph. A **multiplex hypergraph** [28,29] is a multilayer hypergraph where each layer represent a specific type of relationship between the nodes, and every layer contain exactly the same set of nodes. A multiplex hypergraph is defined as the set

$$\mathcal{M} = \{\mathcal{H}_1, \cdots, \mathcal{H}_p\}$$

where

- $\mathcal{H}_\alpha = (\mathcal{V}, \mathcal{E}_\alpha)$, $\forall \alpha \in \{1, \cdots, p\}$ is a hypergraph whose hyperedges define a certain type or relationship.

A **sub-hypergraph induced by a set of nodes** [25,27] of a given hypergraph is the hypergraph whose vertices set is the subset of the given hypergraph' vertices set, and the hyperedges are the intersection between the hyperedges of the given hypergraph and the set of nodes. Formally, given $\mathcal{V}_s \subset \mathcal{V}$, then

$$\mathcal{H}^{\mathcal{V}_s} = (\mathcal{V}_s, \{h_k \cap \mathcal{V}_s | h_k \in \mathcal{E}\})$$

3 Methodology

In this section we provide a formal definition of a multiplex weighted hypergraph-based model that include all relevant information about adversaries and their victimology that may be useful to obtain relevant insights for a specific organization. In this model, the characteristics are identified with the layers, and the attributes are identified with the hyperedges.

Definition 1. *The **characteristics set** is defined as $\mathcal{C} = \{c_1, ..., c_q\}$, with $q \geq 1$. Each element of this set represents one characteristic of the adversaries utilized in the analysis. Each element of this set represent one layer of the multilayer hypergraph.*

Example 1.

$$\mathcal{C} = \{geographic\ scope, targeted\ sector, attack\ aim, victim\ size\}$$

Now, we will go to define the attributes associated to each characteristic. In fact we will identify each attribute with a hyperedge, i.e., the set of adversaries that share that attribute.

Definition 2. *The **adversary model** is defined as follows. Given the set of adversaries $\mathcal{V} = \{v_1, \cdots, v_n\}$, the adversary model for a target organization is the weighted multiplex hypergraph*

$$\mathcal{M}^{\mathcal{W}} = (\mathcal{V}, \{\mathcal{E}_\alpha |\ \alpha \in \{1, \cdots, q\}\}, \{\mathcal{W}_\alpha |\ \alpha \in \{1, \cdots, q\}\})$$

where:

- $\forall \alpha \in \{1, \cdots, q\}$ $\mathcal{E}_\alpha$ *is the layer of the hypergraph that represents the characteristic* α*, and*
- $\forall k \in \{1, \cdots, m_\alpha\}$, h_k^α *is an attribute of the characteristic* α.

Formally, we consider $\mathcal{E}_\alpha = \{h_1^\alpha, \cdots, h_{m_\alpha}^\alpha\}$, *with* $h_k^\alpha \subset \mathcal{V}$, *where each hyperedge (subset of the set of adversaries) represents an attribute, and the elements of that hyperedge are the set of adversaries that share such attribute. This element represent the general knowledge about adversaries that may be useful for any organization.*

So, $\forall \alpha \in \{1, \cdots, q\}$ *the function*

$$\mathcal{W}_\alpha : \mathcal{E}_\alpha \to \mathbb{R}^+ \cup \{0\}$$

assigns a weight to each hyperedge according to the relevance of the attribute it represents for the organization under study such as

$$\forall k \in \{1, \cdots, m_\alpha\}, \quad \mathcal{W}_\alpha(h_j^\alpha) \geq 0 \quad \wedge \quad \sum_{k=1}^{m_\alpha} \mathcal{W}_\alpha(h_k^\alpha) = 1$$

This element enables the model to represent the attributes that are relevant for a specific organization so it can obtain tailored results for it.

Example 2. An organization that has 75% of its operations in Europe and 25% in North America would have the following weights:

$$\mathcal{W}_{Geography}(h_{Europe}^{Geography}) = 0,75$$
$$\mathcal{W}_{Geography}(h_{NorthAmerica}^{Geography}) = 0,25$$
$$\mathcal{W}_{Geography}(h_{Asia}^{Geography}) = 0,0$$

Definition 3. *The model allows **filtering operations** in order to keep the nodes and hyperedges with their weights that are relevant for a certain analysis, based on a set of attributes that the nodes must or must not have.*

- *Given an attribute* h_k^α *represented by the hyperedge* k *that belong to the layer* α*, the set of adversaries that share this attribute are those that belong to this hyperedge.*

- *Given a set* $s = \{h_{k_1}^{\alpha_1}, \cdots, h_{k_r}^{\alpha_r}\}$ *of* r *attributes, that may belong to different layers, this set admits all set operations* $(\cup, \cap, \setminus)$, *where* $\setminus$ *represent the difference of sets, that enables the model to perform complex adversaries selections based on their attributes in order to filter the whole knowledge base to solve a specific question. The result of the application of those operations will be called* $\mathcal{V}_s$.
- *The sub-hypergraph induced by the set* $\mathcal{V}_s$ *of nodes contains the nodes selected according to their attributes plus all the hyperedges they belong to, that represents all their attributes information. Formally: Given* $\mathcal{V}_s \subset \mathcal{V}$, *then*

$$\mathcal{H}^{\mathcal{V}_s} = (\mathcal{V}_s, \{h_k \cap \mathcal{V}_s | h_k \in \mathcal{E}\})$$

Note that if the original hypergraph is weighted, then the sub-hypergraph induced by the set $\mathcal{V}_s$ *can be considered weighted with the same weights of the original hypergraph.*

Definition 4. *The relevance of each adversary for a given organization is measured by an* ***adversary relevance index*** *obtained aggregating the weights of the hyperedges they belong to. Formally:* $\mathcal{R} : \mathcal{V} \to \mathbb{R}^+ \cup \{0\}$ *where*

$$\forall v \in \mathcal{V},\ \mathcal{R}(v) = \frac{\sum_{\alpha=1}^{q} \sum_{k=1}^{m_\alpha} \mathcal{W}_\alpha(h_k^\alpha | v \in h_k^\alpha)}{q}$$

Note that $0 \leq \mathcal{R}(v) \leq 1$.

4 Toy Model Example

4.1 Model Definition

In this section we will explain the definitions in the Methodology section using a simplified model.

We will also consider a reduced set of seven adversaries that will become the nodes of the hypergraph:

$$\mathcal{V}_{Toy} = \{Adv.\ 1,\ Adv.\ 2,\ Adv.\ 3,\ Adv.\ 4,\ Adv.\ 5,\ Adv.\ 6,\ Adv.\ 7\}$$

We will consider a reduced set of three characteristics:

$$\mathcal{C}_{Toy} = \{geographic\ scope,\ targeted\ sector,\ attack\ aim\}$$

For each characteristic, we will define a set of hyperedges linking the nodes that represent adversaries that share a specific attribute.

The **geographic scope** characteristic has the following definition, and is represented in the Fig. 1:

$$\mathcal{E}_{Geographic\ scope}^{Toy} = \{h_{America}^{G.\ scope}, h_{Europe}^{G.\ scope}, h_{Asia}^{G.\ scope}, h_{Africa}^{G.\ scope}\}$$

where:

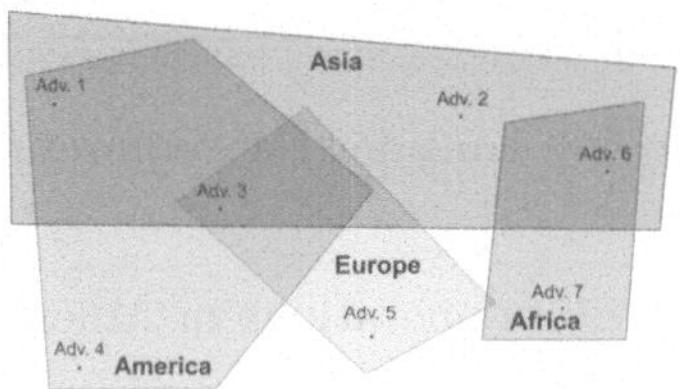

Fig. 1. Hypergraph representing the geographic scope of the adversaries

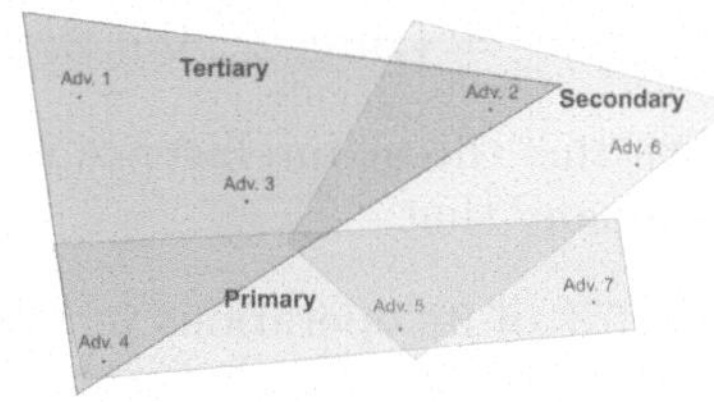

Fig. 2. Hypergraph representing the sectorial scope of the adversaries

$$
\begin{aligned}
h^{Geographic\ scope}_{America} &= \{Adv.\ 1,\ Adv.\ 3,\ Adv.\ 4\} \\
h^{Geographic\ scope}_{Europe} &= \{Adv.\ 3,\ Adv.\ 5\} \\
h^{Geographic\ scope}_{Asia} &= \{Adv.\ 1,\ Adv.\ 2,\ Adv.\ 3,\ Adv.\ 6\} \\
h^{Geographic\ scope}_{Africa} &= \{Adv.\ 6,\ Adv.\ 7\}
\end{aligned}
$$

The **targeted sector** characteristic has the following definition, and is represented in the Fig. 2:

$$
\mathcal{E}^{Toy}_{Targeted\ sector} = \{h^{Targeted\ sector}_{Primary},\ h^{Targeted\ sector}_{Secondary},\ h^{Targeted\ sector}_{Tertiary}\}
$$

where:

$$
\begin{aligned}
h^{Targeted\ sector}_{Primary} &= \{Adv.\ 4,\ Adv.\ 5,\ Adv.\ 7\} \\
h^{Targeted\ sector}_{Secondary} &= \{Adv.\ 2,\ Adv.\ 5,\ Adv.\ 6\} \\
h^{Targeted\ sector}_{Tertiary} &= \{Adv.\ 1,\ Adv.\ 2,\ Adv.\ 3,\ Adv.\ 4\}
\end{aligned}
$$

The **attack aim** characteristic has the following definition, and is represented in the Fig. 3:

$$
\mathcal{E}^{Toy}_{Attack\ aim} = \{h^{Attack\ aim}_{Espionage},\ h^{Attack\ aim}_{Sabotage},\ h^{Attack\ aim}_{Fraud}\}
$$

where:

$$
\begin{aligned}
h^{Attack\ aim}_{Espionage} &= \{Adv.\ 1,\ Adv.\ 3\} \\
h^{Attack\ aim}_{Sabotage} &= \{Adv.\ 5,\ Adv.\ 6,\ Adv.\ 7\}
\end{aligned}
$$

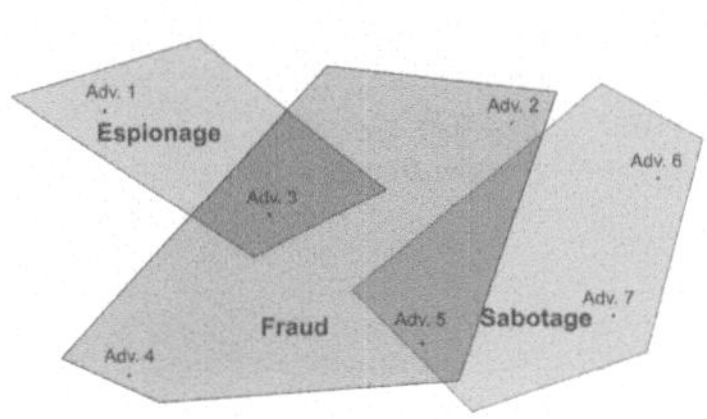

Fig. 3. Hypergraph representing the aim of the adversaries

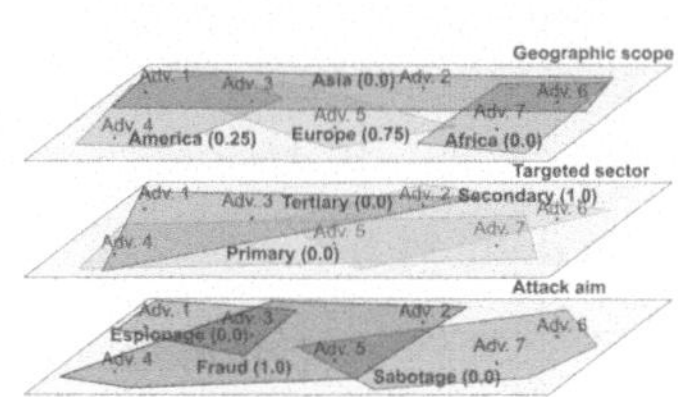

Fig. 4. Weighted multiplex hypergraph representing the whole model

$h_{Fraud}^{Attack\ aim} = \{Adv.\ 2,\ Adv.\ 3,\ Adv.\ 4,\ Adv.\ 5\}$

With this threat intelligence information, we will consider the requirements of a company that:

- Has 75% of its operations in Europe and 25% of its operations in America
- Has 100% of its activity in the Secondary sector
- Considering its activity it is only subject to fraud risk

The description of the company will be modeled using the weights of the corresponding attributes, with the following definition:

$$\mathcal{W}_{\alpha}^{Toy} : \mathcal{E}_{\alpha}{}^{Toy} \rightarrow \mathbb{R}^{+} \cup \{0\}$$

where

$\mathcal{W}_{Geographic\ scope}^{Toy}(h_{Europe}^{Geographic\ scope}) = 0.75$
$\mathcal{W}_{Geographic\ scope}^{Toy}(h_{America}^{Geographic\ scope}) = 0.25$
$\mathcal{W}_{Geographic\ scope}^{Toy}(h_{Asia}^{Geographic\ scope}) = \mathcal{W}_{Geographic\ scope}^{Toy}(h_{Africa}^{Geographic\ s.}) = 0.0$
$\mathcal{W}_{Targeted\ sector}^{Toy}(h_{Secondary}^{Targeted\ sector}) = 1.0$
$\mathcal{W}_{Targeted\ sector}^{Toy}(h_{Primary}^{Targeted\ sector}) = \mathcal{W}_{Targeted\ sector}^{Toy}(h_{Tertiary}^{Targeted\ sector}) = 0.0$
$\mathcal{W}_{Attack\ aim}^{Toy}(h_{Fraud}^{Attack\ aim}) = 1.0$
$\mathcal{W}_{Attack\ aim}^{Toy}(h_{Espionage}^{Attack\ aim}) = \mathcal{W}_{Attack\ aim}^{Toy}(h_{Sabotage}^{Attack\ aim}) = 0.0$

The entire weighted multiplex hypergraph is defined as follows, and is represented in Fig. 4:

$$\mathcal{M}^{Toy} = (\mathcal{V}_{Toy}, \{\mathcal{E}_{Geographic\ scope}^{Toy}, \mathcal{E}_{Targeted\ sector}^{Toy}, \mathcal{E}_{Attack\ aim}^{Toy}\}, \{\mathcal{W}_{Geographic\ scope}^{Toy}, \mathcal{W}_{Targeted\ sector}^{Toy}, \mathcal{W}_{Attack\ aim}^{Toy}\})$$

4.2 Model Filtering

In this example we will suppose that for any reason we are interested in the adversaries that operates in the primary or secondary sectors and don't operate in Africa. We want to obtain a sub-model that contains only the relevant information (nodes and hyperedges) from the original model.

For doing so, we will start with a set of attributes:

$$\mathcal{S} = \{h_{Africa}^{Geographic\ Scope}, h_{Primary}^{Target\ Sector}, h_{Secondary}^{Target\ Sector}\}$$

where, as defined before:

$h_{Primary}^{Targeted\ sector} = \{Adv.\ 4,\ Adv.\ 5,\ Adv.\ 7\}$
$h_{Secondary}^{Targeted\ sector} = \{Adv.\ 2,\ Adv.\ 5,\ Adv.\ 6\}$
$h_{Africa}^{Geographic\ scope} = \{Adv.\ 6,\ Adv.\ 7\}$

We are interested in the nodes that satisfy the condition:

$$\mathcal{V}_S^{Toy} = (h_{Primary}^{Targeted\ sector} \cup h_{Secondary}^{Targeted\ sector}) \setminus h_{Africa}^{Geographic\ scope}$$

$$= (\{Adv.\ 4,\ Adv.\ 5,\ Adv.\ 7\} \cup \{Adv.\ 2,\ Adv.\ 5,\ Adv.\ 6\}) \setminus \{Adv.\ 6,\ Adv.\ 7\}$$
$$= \{Adv.\ 2,\ Adv.\ 4,\ Adv.\ 5\}$$

The sub-model we are looking for is the sub-hypergraph induced by this set of nodes, as represented in Fig. 5, such as:

$$\mathcal{M}_S^{Toy} = (\mathcal{V}_S^{Toy},\ \{h_k^{\alpha} \cap \mathcal{V}_S^{Toy}\},\ \mathcal{W}_{\alpha}^{Toy})$$

$$h_{America}^{Geographic\ scope} \cap \mathcal{V}_S^{Toy} = \{Adv.\ 4\}$$
$$h_{Europe}^{Geographic\ scope} \cap \mathcal{V}_S^{Toy} = \{Adv.\ 5\}$$
$$h_{Asia}^{Geographic\ scope} \cap \mathcal{V}_S^{Toy} = \{Adv.\ 2\}$$
$$h_{Africa}^{Geographic\ scope} \cap \mathcal{V}_S^{Toy} = \varnothing$$

$$h_{Primary}^{Targeted\ sector} \cap \mathcal{V}_S^{Toy} = \{Adv.\ 4,\ Adv.\ 5\}$$
$$h_{Secondary}^{Targeted\ sector} \cap \mathcal{V}_S^{Toy} = \{Adv.\ 2,\ Adv.\ 5\}$$
$$h_{Tertiary}^{Targeted\ sector} \cap \mathcal{V}_S^{Toy} = \{Adv.\ 2,\ Adv.\ 4\}$$

$$h_{Espionage}^{Attack\ aim} \cap \mathcal{V}_S^{Toy} = \varnothing$$
$$h_{Sabotage}^{Attack\ aim} \cap \mathcal{V}_S^{Toy} = \{Adv.\ 5\}$$
$$h_{Fraud}^{Attack\ aim} \cap \mathcal{V}_S^{Toy} = \{Adv.\ 2,\ Adv.\ 4,\ Adv.\ 5\}$$

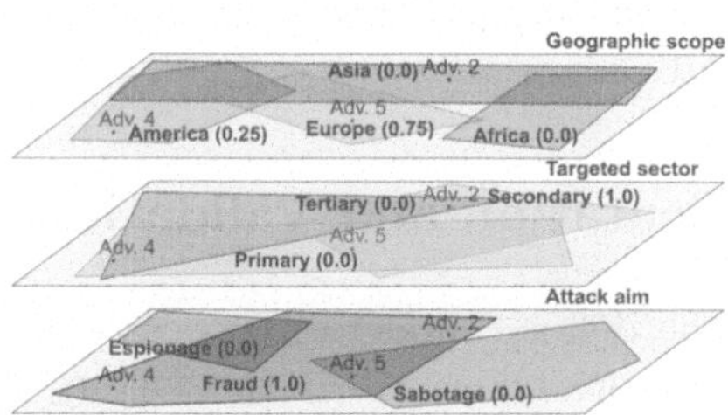

Fig. 5. Filtered model

4.3 Adversary Relevance Index

Adding the weight of the attributes of each adversary, we can calculate the index that enables prioritization, such as:

$$\mathcal{R}(Adv.\,2) = \frac{\mathcal{W}^{Toy}_{G.\,scope}(h^{G.\,scope}_{Asia}) + \mathcal{W}^{Toy}_{T.\,sector}(h^{T.\,sector}_{Secondary}) + \mathcal{W}^{Toy}_{T.\,sector}(h^{T.\,sector}_{Tertiary}) + \mathcal{W}^{Toy}_{A.\,aim}(h^{A.\,aim}_{Fraud})}{3}$$

$$= \frac{0.0 + 1.0 + 0.0 + 1.0}{3} = 0.67$$

$$\mathcal{R}(Adv.\,4) = \frac{\mathcal{W}^{Toy}_{G.\,scope}(h^{G.\,scope}_{America}) + \mathcal{W}^{Toy}_{T.\,sector}(h^{T.\,sector}_{Primary}) + \mathcal{W}^{Toy}_{T.\,sector}(h^{T.\,sector}_{Tertiary}) + \mathcal{W}^{Toy}_{A.\,aim}(h^{A.\,aim}_{Fraud})}{3}$$

$$= \frac{0.25 + 0.0 + 0.0 + 1.0}{3} = 0.33$$

$$\mathcal{R}(Adv.\,5) =$$

$$= \frac{\mathcal{W}^{Toy}_{G.\,s.}(h^{G.\,scope}_{Europe}) + \mathcal{W}^{Toy}_{T.\,s.}(h^{T.\,sector}_{Primary}) + \mathcal{W}^{Toy}_{T.\,s.}(h^{T.\,sector}_{Secondary}) + \mathcal{W}^{Toy}_{A.\,a.}(h^{A.\,aim}_{Fraud}) + \mathcal{W}^{Toy}_{A.\,a.}(h^{A.\,aim}_{Sabotage})}{3}$$

$$= \frac{0.75 + 0.0 + 1.0 + 1.0 + 0.0}{3} = 0.92$$

The obtained results are summarized in Table 1.

Table 1. Weight aggregation for each adversary

Adversary	Relevance index
Adv. 5	0.92
Adv. 2	0.67
Adv. 4	0.33

Adversary 5 is the most relevant adversary for our company, followed by Adversary 2 and then Adversary 4.

5 Model Implementation and Validation

In this section we use the defined model to process real life threat intelligence data and prioritize adversaries relevant to a specific organization.

5.1 Dataset Preparation

Data sharing [30] is one of the pillars of threat intelligence, and there are multiple organizations and forums that aim to encourage the sharing of threat data.

Structured Threat Information Expression (STIX) [31] is a formal language to express all types of cyber threat intelligence data. It has been defined by Organization for the Advancement of Structured Information Standards (OASIS) [32], a non-profit consortium that aims to resolve technical issues in order to enable and improve collaboration and open software development. Threat intelligence information in STIX can be obtained from multiple sources, open and proprietary, free and paid.

STIX define several properties that may be used as characteristics of the threat actors considered in this model, including:

- **Threat actor type**, that defines the profile of the threat actor, considering its aim, support, resources, etc.
- **Threat actor role**, that classifies the threat actors according to their role in the adversaries ecosystem.
- **Goals**, that define the objectives the adversary usually tries to accomplish.
- **Sophistication**, that define the level of capabilities available to the threat actor.
- **Resource level**, that define the material resources available to the threat actor.
- **Primary and secondary and personal motivation**, that define the motivations behind the actions of the threat actor.

MITRE ATT&CK (Adversarial Tactics, Techniques, and Common Knowledge) [33] is a free and publicly accessible knowledge base of threat intelligence information that can be accessed through STIX. At the time of this writing the last available version is 16.1 [34], that contains references of 159 adversaries. ATT&CK represents the information related to Groups using attack-pattern objects, instead of threat-actor objects. As a consequence, all characteristics and attributes are included in free text fields that have to be analyzed in order to extract this information [35]. This information has been normalized using the STIX 2.1 open vocabularies, as follows:

- **Adversary type** using threat-actor-type-ov
- **Geographic scope** using region-ov, that is based on the United Nations geoscheme.
- **Target sector** using industry-sector-ov
- **Goals** using attack-motivation-ov
- **Sophistication level** using threat-actor-sophistication-ov
- **Resource level** using attack-resource-level-ov
- **Origins** using ISO 3166-1:2013 Country Codes plus cybercrime

The raw data from MITRE ATT&CK has been processed using ChatGPT, version GPT-4-turbo. The prompts developed for the conversion defined:

- the adversaries list taken from MITRE ATT&CK,
- the characteristic to process, and
- the valid values for the attributes, taken from the STIX 2.1 vocabularies

Hence, the information from MITRE ATT&CK has been enriched by ChatGPT using other open sources in order to have a set of information as complete as possible.

5.2 Model Implementation

The software for data processing has been developed using Python and the HypergraphX 1.7.5 library [36].

The dataset has been loaded to a HypergraphX multiplex hypergraph, as shown in Fig. 6.

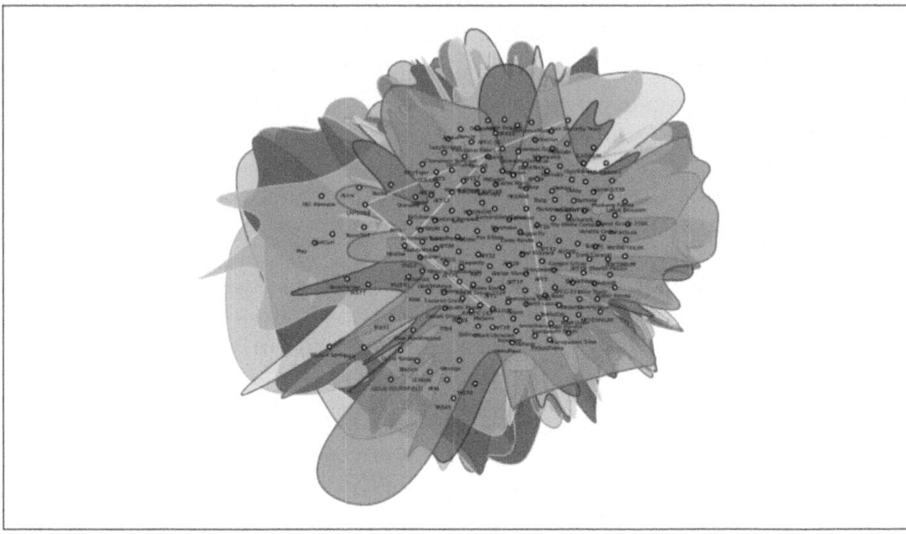

Fig. 6. Weighted multiplex hypergraph with adversaries information

The order of each hyperedge gives information about the size of the knowledge base and also information about the relevance of each hyperedge in terms of number of adversaries that have this attribute. This information can be found in Table 2.

Table 2. Order of each hyperedge of the hypergraph

Characteristic	Attribute	# of adv.
Adversary type	Crime-syndicate	32
	Criminal	2
	Hacker	1
	Hacktivist	1
	Nation-state	118
	Spy	3
	Unknown	2
Geographic scope	Central-Asia	2
	Eastern-Africa	1
	Eastern-Asia	53
	Eastern-Europe	26
	Latin-America-Caribbean	2
	Northern-Africa	1
	Northern-America	28
	Northern-Europe	3
	South-America	4
	South-Eastern-Asia	9
	Southern-Asia	8
	Unknown	7
	Western-Asia	39
	Western-Europe	25
Target sector	Aerospace	4
	Defense	96
	Education	4
	Energy	20
	Gaming	1
	Government	101
	Hospitality	3
	Manufacture	2
	Retail	13
	Health	1
	Finance	33
	Public services	1
	Technology	35
	Telecommunications	35
	Transport	6

Characteristic	Attribute	# of adv.
Goals	Dominance	79
	Ideology	41
	Notoriety	2
	Organizational-gain	133
	Personal-gain	41
Sophistication level	Advanced	37
	Expert	16
	Innovator	1
	Intermediate	67
	Minimal	38
Resource level	Government	75
	Organization	75
	Team	9
Origin	Brasil	2
	Cambodia	1
	China	48
	Colombia	1
	India	5
	International	2
	Iran	16
	Israel	7
	Lebanon	2
	Middle East	3
	Nigeria	1
	North Korea	8
	Pakistan	2
	Palestine	1
	Philippines	1
	Russia	31
	South America	2
	Taiwan	1
	UAE	1
	Unknown	8
	USA	14
	Venezuela	1
	Vietnam	1

5.3 Model Usage and Validation

In order to demonstrate the usage of the model and to validate the results obtained, we have simulated and checked our results against a real threat intelligence report [37] issued by the Spanish public administration analyzing the key cyberthreats for the country.

This report analyzes the most relevant threat actors focusing on three types of actors: state sponsored, hacktivists and cybercriminals.

Regarding the state sponsored threat actors, the report is structured according to the origin of the threat actors: Russia, China, North Korea and Iran.

In order to test the results of our model we have simulated the selection of the most relevant cyber threat actors with origin in each country, compared the results and analyzed the differences.

For each originating country the following process has been followed:

1. The adversaries data set has been filtered to analyze only the threat actors from the source country under analysis.
2. A profile has been defined as approximate as possible to the conditions aligned to the report. This profile has considered the following characteristics and attributes:
   - **Type:** State sponsored, because it is the scope we want to simulate from the reference report.
   - **Resources:** Government, because it is the resource level that state sponsored actors usually have.
   - **Sophistication:** Innovator, expert and advanced, because those are the competence levels that can lead to a high level incident, with 0.6, 0.3 and 0.1 weight respectively, to give more relevance to the more sophisticated. There are no adversaries classified as "strategic" in the data set.
   - **Geographic scope:** Southern Europe, which is the region that contains Spain.
3. For each actor we have calculated the relevance index.
4. We have sorted the resulting in decreasing order, so we have a prioritized listing of threat actors.
5. We have compared the simulation's results with the report and analyzed the differences.

Russia. The results obtained for Russian adversaries by the model are summarized in Table 3.

Table 3. Relevance indexes for Russian adversaries (top 10)

Adversary	Relevance index
APT28	0.83
APT29	0.83
Sandworm Team	0.83
Turla	0.83
Dragonfly	0.78
Strider	0.78
TEMP.Veles	0.78
Ember Bear	0.75
Gamaredon Group	0.75
Moonstone Sleet	0.75

Table 4. Relevance indexes for Chinese adversaries (top 10)

Adversary	Relevance index
APT41	0.83
Axiom	0.83
Deep Panda	0.83
Elderwood	0.83
HAFNIUM	0.83
PLATINUM	0.83
Volt Typhoon	0.83
Winnti Group	0.83
Mustang Panda	0.78
admin@338	0.78

APT28 and APT29 are the adversaries highlighted in the real report [37]. In the model's results they belong to the top 4, with a relevance index of 0.83 both of them.

For example, the calculations for the relevance index of APT28 are the following:

$$\mathcal{W}^{APT28}_{Type}(h^{Type}_{StateSponsored}) = 1.0$$
$$\mathcal{W}^{APT28}_{Resources}(h^{Resources}_{Government}) = 1.0$$
$$\mathcal{W}^{APT28}_{Sophistication}(h^{Sophistication}_{Expert}) = 0.3$$
$$\mathcal{W}^{APT28}_{Geographic\ scope}(h^{Geographic\ scope}_{SouthernEurope}) = 1.0$$

$$\mathcal{R}(APT28) = \frac{1.0 + 1.0 + 0.3 + 1.0}{4} = 0.825$$

China. The results obtained for Chinese adversaries by the model are summarized in Table 4.

In this case, Mustang Panda, which is the only threat actor highlighted in the threat intelligence report [37] ranks in the top 22, in a second group of adversaries with a relevance index of 0.78. The main difference of Mustang Panda's attributes with those of the top tier in the model's results is the consideration of the sophistication level as Advanced instead of Expert.

North Korea. The results obtained for North Korean adversaries by the model are summarized in Table 5.

Table 5. Relevance indexes for North Korean adversaries

Adversary	Relevance index
Lazarus Group	0.83
APT37	0.75
Kimsuky	0.75
APT38	0.53
Darkhotel	0.53
Andariel	0.50
BRONZE BUTLER	0.50
Higaisa	0.50

Table 6. Relevance indexes for Iranian adversaries (top 10)

Adversary	Relevance index
OilRig	0.83
APT39	0.78
Cleaver	0.78
MuddyWater	0.78
NEODYMIUM	0.78
APT33	0.75
Agrius	0.50
Ferocious Kitten	0.50
Fox Kitten	0.50
HEXANE	0.50

In this case, the two adversaries highlighted in the threat intelligence report [37], Lazarous Group and Kimsuky, are ranked first and second (together with APT37), with a relevance index of 0.83 and 0.75, respectively.

Iran. The results obtained for Iranian adversaries by the model are summarized in Table 6.

In this case, the adversary highlighted in the threat intelligence report [37], MuddyWater, is ranked in the top 5 of the model's result, with a relevance index of 0.78, on a second tier. The difference with the tier one is the consideration of advanced in our data set instead of expert as attribute of the sophistication characteristic.

6 Conclusions and Future Work

6.1 Conclusions

We study the weighted multiplex hypergraph as a new approach to the threat intelligence data analysis. This model has been able to process the information related to adversaries from an open source and obtain actionable results, comparable with those obtained by an analysts' research. A key advantage of this approach is its explainability, as it allows for a clear understanding of how specific attributes influence the results. Even though the expert criteria is valuable in any threat intelligence analysis, the usage of this model can bring efficiency in the process, allowing to automate certain tasks, like the elaboration of a short list of relevant adversaries where the analysts must put the focus for a more fine grained analysis.

6.2 Future Work

Starting with this work, there are multiple investigative lines that can be followed.

- The data set can be improved in several ways, like adding additional sources, additional characteristics or treating attributes with a higher granularity.
- The model's capabilities can be enhanced with different additions, like weighting the assignation of attributes to adversaries, weighting the characteristics or developing a cost function for the addition of new attributes to an adversary.
- The usage of hypergraphs as the base of this model bring the opportunity to use additional hypergraphs properties, like similarity measures, centrality measures, projections or community detection, to extract more relevant information from the source data.
- The model can also grow to analyze additional threat intelligence data, like Techniques, Tactics and Procedures (TTPs) or countermeasures.

Acknowledgments. This work has been supported by INCIBE/URJC Agreement M3386/2024/0031/001 within the framework of the Recovery, Transformation and Resilience Plan funds of the European Union (Next Generation EU).

Disclosure of Interests. The authors have no competing interests to declare that are relevant to the content of this article.

References

1. NIST cybersecurity glossary: Cybersecurity. https://csrc.nist.gov/glossary/term/cybersecurity. Accessed 27 Dec 2024
2. FBI Internet Crime Complaint Center: 2023 Internet Crime Report (2023). https://www.ic3.gov/AnnualReport/Reports/2023_IC3Report.pdf
3. Rahman, M.R., Basak, S.K., Hezaveh, R.M., Williams, L.: Attackers reveal their arsenal: an investigation of adversarial techniques in CTI reports (2024). https://arxiv.org/abs/2401.01865
4. Bermudez, J.D., Castro, J.J., Peralta, D.A., Guacaneme, P.A.: Tecnicas Avanzadas de Ciberseguridad: Integracion y Evolucion de la Kill Chain en Diversos Escenarios (2023). https://arxiv.org/abs/2306.09242
5. Hasan, S.M., Alotaibi, A.M., Talukder, S., Shahid, A.R.: Distributed threat intelligence at the edge devices: a large language model-driven approach. In: 2024 IEEE 48th Annual Computers, Software, and Applications Conference (COMPSAC). IEEE (2024)
6. Nunes, E., et al.: Darknet and DeepNet mining for proactive cybersecurity threat intelligence. CoRR, volume abs/1607.08583 (2016). http://arxiv.org/abs/1607.08583
7. Wang, Y., Ren, Y., Cui, Z., Yu, H.: Proactive security defense: cyber threat intelligence modeling for connected autonomous vehicles (2024). https://arxiv.org/abs/2410.16016
8. Rani, N., Saha, B., Maurya, V., Shukla, S.K.: TTPXHunter: actionable threat intelligence extraction as TTPs from finished cyber threat reports. In: Digital Threats: Research and Practice, no. 4, pp. 1–19. Association for Computing Machinery (ACM) (2024). http://dx.doi.org/10.1145/3696427
9. Arazzi, M., et al.: NLP-based techniques for cyber threat intelligence (2023). https://arxiv.org/abs/2311.08807
10. Bayer, M., Frey, T., Reuter, C.: Multi-level fine-tuning, data augmentation, and few-shot learning for specialized cyber threat intelligence (2022). https://arxiv.org/abs/2207.11076
11. Rahman, M.R., Hezaveh, R.M., Williams, L.: What are the attackers doing now? Automating cyberthreat intelligence extraction from text on pace with the changing threat landscape: a survey. ACM Comput. Surv. **55**(12), 1–36 (2023). https://doi.org/10.1145/3571726
12. Hossen, M.I., Islam, A., Anowar, F., Ahmed, E., Rahman, M.M.: Generating cyber threat intelligence to discover potential security threats using classification and topic modeling (2022). https://arxiv.org/abs/2108.06862
13. Peng, W., et al.: CTISum: a new benchmark dataset for cyber threat intelligence summarization (2024). https://arxiv.org/abs/2408.06576
14. Alves, F., Bettini, A., Ferreira, P.M., Bessani, A.: Processing tweets for cybersecurity threat awareness (2019). https://arxiv.org/abs/1904.02072
15. Angelelli, M., Arima, S., Catalano, C., Ciavolino, E.: A robust statistical framework for cyber-vulnerability prioritisation under partial information in threat intelligence. Expert Syst. Appl. **255**, 124572 (2024). https://doi.org/10.1016/j.eswa.2024.124572
16. Almashor, M., et al.: Unraveling threat intelligence through the lens of malicious URL campaigns (2023). https://arxiv.org/abs/2208.12449
17. Alam, M.T., Bhusal, D., Nguyen, L., Rastogi, N.: CTIBench: a benchmark for evaluating LLMs in cyber threat intelligence (2024). https://arxiv.org/abs/2406.07599

18. Alevizos, L., Dekker, M.: Towards an AI-enhanced cyber threat intelligence processing pipeline. Electronics **13**(11), 2021 (2024). https://doi.org/10.3390/electronics13112021
19. Mittal, S., Joshi, A., Finin, T.: Cyber-all-intel: an AI for security related threat intelligence (2019). https://arxiv.org/abs/1905.02895
20. Ali, H., Papadopoulos, P., Ahmad, J., Pitropakis, N., Jaroucheh, Z., Buchanan, W.J.: Privacy-preserving and trusted threat intelligence sharing using distributed ledgers (2021). https://arxiv.org/abs/2112.10092
21. Mavroeidis, V., Eis, P., Zadnik, M., Caselli, M., Jordan, B.: On the integration of course of action playbooks into shareable cyber threat intelligence (2021). https://arxiv.org/abs/2110.10540
22. Ranade, P., Piplai, A., Mittal, S., Joshi, A., Finin, T.: Generating fake cyber threat intelligence using transformer-based models (2021). https://arxiv.org/abs/2102.04351
23. Khurana, N., Mittal, S., Joshi, A.: Preventing poisoning attacks on AI based threat intelligence systems (2018). https://arxiv.org/abs/1807.07418
24. Chapela, V., Criado, R., Moral, S., Romance, M.: Intentional Risk Management through Complex Networks Analysis. Springer (2015)
25. Berge, C.: Hypergraphs: Combinatorics of Finite Sets. Elsevier (1984)
26. Boccaletti, S., et al.: The structure and dynamics of networks with higher order interactions. Phys. Rep. **1018**, 1–64 (2023)
27. Dai, Q., Gao, Y.: Mathematical foundations of hypergraph. In: Hypergraph Computation. Artificial Intelligence: Foundations, Theory, and Algorithms. Springer, Singapore (2023). https://doi.org/10.1007/978-981-99-0185-2_2
28. Kanawati, R.: Multiplex network mining: a brief survey. IEEE Intell. Informatics Bull. **16**(1), 24–27 (2015)
29. Lotito, Q.F., Montresor, A., Battiston, F.: Multiplex measures for higher-order networks. Appl. Netw. Sci. **9**, 55 (2024). https://doi.org/10.1007/s41109-024-00665-9
30. Johnson, C., Feldman, L., Witte, G.: Cyber threat intelligence and information sharing. In: ITL Bulletin. National Institute of Standards and Technology, Gaithersburg, MD (2017). https://csrc.nist.gov/pubs/itlb/2017/05/cyberthreat-intelligence-and-information-sharing/final. Accessed 03 Jan 2025
31. Jordan, B., Piazza, R., Darley, T.: STIX™ Version 2.1: OASIS Committee Specification 02 (2021). https://docs.oasis-open.org/cti/stix/v2.1/cs02/stix-v2.1-cs02.html
32. OASIS Homepage. https://www.oasis-open.org/. Accessed 23 Dec 2024
33. Strom, B.E., Applebaum, A., Miller, D.P., Nickels, K.C., Pennington, A.G., Thomas, C.B.: MITRE ATT&CK, design and philosophy. Technical report. The MITRE Corporation (2018)
34. MITRE ATT&CK version 16.1. https://attack.mitre.org/versions/v16/. Accessed 23 Dec 2024
35. Zych, M., Mavroeidis, V.: Enhancing the STIX representation of MITRE ATT&CK for group filtering and technique prioritization (2022). https://arxiv.org/abs/2204.11368
36. Lotito, Q.F. et al.: Hypergraphx: a library for higher-order network analysis. J.Complex Netw. **11**(3) (2023). http://dx.doi.org/10.1093/comnet/cnad019. ISSN 2051-1329
37. Centro Criptológico Nacional: Cyberthreats and trends, 2024 edn. https://www.ccn-cert.cni.es/es/informes/informes-ccn-cert-publicos/7274-ccn-cert-ia-04-24-ciberamenazas-y-tendencias-edicion-2024/file.html

Medical Device Security and Privacy - MeDSec 2025 Special Session

IoT-Driven Security Solutions for Resource-Constrained Wearable Medical Devices

Emmanouil Bakiris[1](✉) and George Oikonomakis[2]

[1] Space Hellas, 312 Messogion Avenue, 15341 Athens, Greece
ebakiris@space.gr

[2] Zortenet, 87 Makris Street, 12241 Aigaleo, Greece
goikonomakis@zortenet.eu

Abstract. The rapid adoption of wearable medical devices has transformed modern healthcare, enabling continuous patient monitoring and real-time health management. However, these resource-constrained devices are highly susceptible to cyber threats due to their limited computational power, reliance on wireless communication, and the challenges associated with secure software execution. This paper examines the unique security challenges faced by wearable medical devices and presents an IoT-driven approach to mitigating software-based attacks. Specifically, we propose the integration of crypto-authentication chips and compiler polymorphism techniques to enhance runtime security and disrupt large-scale exploitability. By generating structurally unique binaries for each device, our approach prevents code reuse attacks, such as Return-Oriented Programming (ROP), while maintaining system efficiency. Experimental validation demonstrates the effectiveness of compiler polymorphism in preventing unauthorized code execution and reinforcing software integrity. While our work primarily focuses on secure software execution, we acknowledge that additional security mechanisms, including secure boot, encrypted communication, and real-time threat detection, can further enhance overall device security. This study highlights the need for lightweight, adaptive security solutions tailored to the constraints of wearable medical technology, paving the way for more resilient and trustworthy healthcare systems.

Keywords: Wearable medical devices · IoT security · secure software execution · compiler polymorphism · runtime protection · return-oriented programming (ROP) · crypto-authentication

1 Introduction

The rapid adoption of medical devices in modern healthcare has revolutionized the monitoring and treatment of various health conditions. From implantable cardiac defibrillators to wearable insulin pumps, these devices have enhanced

I. Praça et al. (Eds.): EICC 2025, CCIS 2500, pp. 275–287, 2025.
https://doi.org/10.1007/978-3-031-94855-8_18

patient care and quality of life. However, this increasing reliance on medical devices has also exposed a critical vulnerability: their susceptibility to cyberattacks. In recent years, the number of reported incidents involving medical device breaches has grown significantly. For instance, a 2023 report revealed that cyberattacks targeting healthcare systems increased by 55% compared to the previous year, with a substantial proportion involving connected medical devices [1].

Wearable medical devices, a subset of IoT (Internet of Things) technology, are particularly vulnerable due to their limited computational resources, wireless communication capabilities, and often inadequate security measures [2]. These devices collect sensitive patient data, monitor vital signs, and even deliver life-saving treatments, making them an attractive target for malicious actors. Threats such as unauthorized data access, tampering, and denial-of-service attacks can compromise patient safety and privacy, potentially leading to catastrophic outcomes.

To address these challenges, it is imperative to implement robust security mechanisms that ensure secure software execution and provide real-time identification and prevention of potential attacks. By focusing on lightweight and resource-efficient solutions, developers can mitigate risks without compromising device performance. This paper explores the critical need for IoT-driven security measures tailored to resource-constrained wearable medical devices.

2 Current Challenges on Securing Medical Devices

The protection of wearable medical devices presents unique challenges, particularly in ensuring secure software execution. These devices are inherently resource-constrained, designed to be lightweight, energy-efficient, and portable [3]. Limited processing power and battery life make traditional security approaches infeasible, requiring innovative methods that balance security and performance.

One primary challenge is preventing the execution of malicious software and detecting vulnerabilities that could be exploited for attacks. Wearable devices often operate on simplified operating systems or firmware that may lack advanced safeguards against malware [4]. In addition, these devices must execute software updates or patches seamlessly without introducing vulnerabilities or causing operational disruptions.

Another critical issue lies in the ability to identify and mitigate run-time threats dynamically. Attackers may exploit software flaws, such as buffer overflows or privilege-escalation vulnerabilities, to gain unauthorized control over the device. Implementing lightweight run-time monitoring and anomaly detection mechanisms can help address these risks but requires careful optimization to avoid impacting device performance [5].

Ensuring the integrity of software execution is further complicated by the devices' reliance on wireless communication for updates and data sharing. Interception or tampering during these processes can compromise the device's functionality or expose it to further exploitation [2]. Thus, secure boot mechanisms, encrypted communications, and trusted execution environments are crucial components in protecting portable medical devices.

Finally, the fragmented nature of regulations and standards in the wearable medical device industry adds to the challenge. Developers must navigate diverse guidelines and compliance requirements, which can lead to inconsistencies in security implementations and hinder the adoption of comprehensive solutions [3].

Addressing these challenges requires a focused effort to develop tools and methodologies that prioritize secure software execution. By integrating run-time protection mechanisms, secure update protocols, and real-time threat detection, developers can enhance the resilience of wearable medical devices against software-based threats while maintaining their operational efficiency.

3 Security Mechanisms and Secure Software Execution in Wearable Medical Devices

3.1 The Role of the Crypto-Authentication Chip

Crypto-authentication chips are specialized hardware components designed to provide secure key storage and cryptographic processing, ensuring the confidentiality and integrity of sensitive operations. One such example is the ATECC608 secure element module [6]. This hardware module is specifically engineered to address the security needs of IoT devices, including wearable medical devices.

The ATECC608 offers a range of functionalities that can significantly enhance the security posture of wearable medical devices. Provides secure key storage to protect encryption keys. These keys are used to ensure that sensitive data remain inaccessible to unauthorized users. Additionally, the module supports hardware-accelerated cryptographic operations, such as hashing functions and key generation functions.

For wearable medical devices, the integration of a low-cost secure element chip like the ATECC608 can provide robust protection against software vulnerabilities and malware. Its' ability to execute critical functions associated with cryptographic operations can aid the secure software execution, ensuring that only authorized code runs on the device, preventing potential exploits. Furthermore, the low power consumption and compact size of the module make it ideal for resource-constrained environments, aligning with the design constraints of wearable medical devices. Figure 1 shows the crypto-authentication chip.

Fig. 1. The ATECC608 crypto-auth chip from Microchip.

3.2 Generating Polymorphic Binaries

Compiler polymorphism is a cybersecurity technique that enhances the resilience of IoT devices by generating unique executable binaries at compile time (Fig. 2). This approach disrupts the scalability of cyberattacks by ensuring that each device runs a structurally distinct version of the software, making code reuse attacks, such as Return-Oriented Programming (ROP), significantly more difficult to execute.

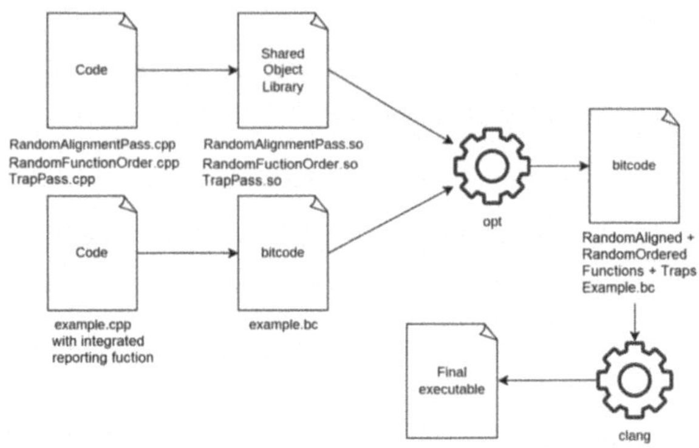

Fig. 2. Compiler Polymorphism Workflow: Function Randomization, Alignment, and Trap Insertion.

Return-Oriented Programming (ROP) attacks exploit memory vulnerabilities by assembling existing instruction sequences, or gadgets, within a program to execute malicious operations [7]. Unlike traditional exploits that inject new code, ROP attacks reuse existing code, allowing attackers to bypass modern defenses such as Data Execution Prevention (DEP) and write-xor-execute policies. By carefully manipulating memory using techniques such as buffer overflows, adversaries overwrite return addresses and hijack a program's control flow [8]. This makes ROP-based exploits highly scalable, as attackers can craft a single exploit that remains effective across multiple devices running identical binaries. The threat is particularly concerning for resource-constrained medical IoT devices, where security mechanisms are often minimal, increasing the risk of large-scale attacks.

The compiler polymorphism randomizes the order of the functions, the alignment of the memory, and introduces security traps, making sure that even if an attacker successfully exploits one device, the same exploit is unlikely to succeed on another. This fundamentally weakens attack scalability, as adversaries cannot rely on static memory layouts to craft universal exploits. As a result, compiler polymorphism provides a proactive security layer, increasing the unpredictability of software execution and fortifying IoT medical devices against advanced cyber threats.

Randomizing Function Order. Function Order Randomization (FOR) is a key compiler polymorphism technique designed to disrupt Return-Oriented Programming (ROP) attacks by altering the order of functions within an executable. Attackers rely on predictable function layouts to craft exploits that chain together existing instruction sequences (gadgets). By shuffling function placement at compile-time, FOR ensures that even if an attacker reverse-engineers one instance of the software, the same exploit will not work on another device with a different function layout.

To implement this technique, the LLVM (Low-Level Virtual Machine) compiler infrastructure is used. Custom LLVM passes are developed to transform the Intermediate Representation (IR) code before the final compilation. An LLVM pass is a specialized C++ module that modifies or analyzes IR code to apply function randomization. During compilation, the LLVM framework loads a shared object file containing the FOR implementation, executing the transformation logic on the LLVM IR. This process dynamically shuffles function placements for each compiled binary (Fig. 3), ensuring that the memory locations of functions vary across devices.

Original Compilation

00000000000011a9	Function1
00000000000011dc	Function2
000000000000120f	Function3
0000000000001242	Function4
0000000000001275	Function5
00000000000012a8	main

Random order Compilation

00000000004011d0	Function5
0000000000401210	main
0000000000401240	Function3
0000000000401280	Function2
00000000004012c0	Function4
0000000000401300	Function1

Fig. 3. Comparison g++ compilation with custom LLVM plugin showcasing random function order of polymorphic executable.

To further enhance randomness, true random numbers generated by the ATECC608 crypto chip serve as a seed for the function shuffling process. The chip provides entropy-based randomness, ensuring that each compiled executable has a unique function order, even when compiled from the same source code. This eliminates predictability, making ROP-based exploits impractical, as attackers can no longer rely on static function locations to construct malicious control flows. By randomizing function placement at compile-time, function order randomization introduces an additional security layer to resource-constrained wearable medical devices, preventing large-scale exploitation and improving overall system resilience.

Randomizing Memory Address Alignment. Function Address Alignment Randomization is designed to disrupt memory-based attacks by introducing randomness in function placement within a binary's memory layout. This technique modifies function alignments, ensuring that functions are placed at random memory addresses within specified boundaries. By doing so, it obfuscates the memory structure, making it significantly more difficult for attackers to exploit Return-Oriented Programming (ROP) vulnerabilities, which rely on predictable memory layouts to chain existing instruction sequences into malicious payloads.

Original Compilation	Addresses	Offset
Function1 – Function2	0x11dc - 0x11a9	0x33 (51 bytes)
Function2 – Function3	0x120f - 0x11dc	0x33 (51 bytes)
Function3 – Function4	0x1242 - 0x120f	0x33 (51 bytes)
Function4 – Function5	0x1275 - 0x1242	0x33 (51 bytes)

Custom Alignment	Addresses	Offset
Function1 – Function2	0x1300 - 0x1280	0x80 (128 bytes)
Function2 – Function3	0x1380 - 0x1300	0x80 (128 bytes)
Function3 – Function4	0x1400 - 0x1380	0x80 (128 bytes)
Function4 – Function5	0x1480 - 0x1400	0x80 (128 bytes)

Fig. 4. Comparison of original g++ compilation with custom LLVM plugin's random function alignment.

This technique leverages LLVM's compiler infrastructure to randomize function alignments (Fig. 4) at compile-time, enhancing security without incurring runtime overhead. By modifying function placement within the LLVM Intermediate Representation (IR), the approach ensures consistent transformation across different compilation stages. Unlike runtime-based defenses such as Address Space Layout Randomization (ASLR), this method eliminates the need for execution-time reordering, making it particularly suitable for resource-constrained medical IoT devices. Additionally, the optimization techniques applied within the Clang compiler maintain system efficiency while integrating security enhancements seamlessly.

Trap Mechanism and Detection. The trap mechanism within compiler polymorphism is a sophisticated security feature designed to detect and report unauthorized access attempts, particularly those exploiting stack-based buffer overflows and return-oriented programming (ROP) techniques. This mechanism is implemented via the EnhanceRandomPass script, which embeds random values at strategic points in compiled binaries, actively monitoring for suspicious execution patterns.

The mechanism works by placing "traps", a known secret value, between the local variables and the control data (such as the return address) on the stack (Fig. 5). Before a function returns, the program verifies that this value remains unchanged. If a buffer overflow has occurred, it would likely alter the trap value, signaling a potential security breach. Upon detecting a modified trap value, the program can terminate execution or take other predefined actions to prevent further exploitation.

Memory stack layout order	Purpose of memory region
Buffer (16 bytes)	User input storage
Polymorphic trap (4 bytes)	Secret random value
r11 register	Frame pointer
pc register (program counter)	Control flow return

Fig. 5. Stack Layout of Polymorphic Executable with embedded trap value.

A key enhancement in this mechanism is the integration of the Z3 theorem solver, which performs lightweight symbolic execution to identify unreachable paths in the code. These unreachable paths serve as ideal locations for inserting traps, ensuring that any attempt to execute unauthorized code paths is promptly detected. Additionally, random padding with NOP (No Operation) instructions increases unpredictability, complicating static analysis, and making it harder for attackers to locate actual operations.

Beyond static security mechanisms, the watchdog verification system plays a crucial role in runtime integrity enforcement. As an independent executable that takes the polymorphic binary as input, it continuously monitors execution by retrieving the embedded trap value from the function prologue and verifying its integrity before control flow transitions occur. Upon detecting a buffer overflow attempt or unauthorized control flow deviation, the watchdog invokes a dedicated response function that promptly reports the attack details to a centralized security monitoring system. The reporting mechanism constructs a JSON-formatted alert that contains critical metadata, including attack type, device identifier, IP address, and timestamp. This alert is transmitted over Bluetooth to a security gateway, enabling real-time threat monitoring and rapid incident response.

However, it is worth mentioning that if an attack attempt is detected via the embedded trap mechanism, the system logs the event and restricts unauthorized execution without halting critical device functions. This ensures continued operation while alerting security personnel, maintaining patient safety and device integrity.

4 Use Case: Securing Wearable/Portable Medical Devices

The increasing reliance on wearable medical devices to monitor and manage patient health underscores the need for robust security mechanisms. This use case demonstrates how the techniques and mechanisms outlined in previous sections can be applied to protect these devices from potential compromises and attacks.

The primary focus of this use case is enabling secure software execution as a cornerstone of device security. By leveraging tools such as crypto-authentication chips and compiler-level polymorphism, we aim to safeguard the integrity of the software running on these devices. Specifically, the compiler polymorphism component plays a critical role by generating functionally equivalent but structurally unique binaries for each device. This ensures that even if an attacker discovers and exploits a vulnerability on one device, the same attack cannot be replicated on another device of the same type, regardless of whether it belongs to the same patient or a different one.

Our approach is carried out in three distinct phases: (a) the pre-deployment phase, (b) the deployment phase, and (c) the run-time phase.

Pre-Deployment Phase. This phase leverages the True Random Number Generation and key generation capabilities of the crypto-authentication chip to supply the compiler polymorphism component with the inputs needed to create unique binaries. These binaries are designed to resist and identify potential attacks. In addition, the binaries are signed with a unique private key produced by the chip, enhancing security by ensuring that only authenticated and untampered software can be installed and executed on the devices. This cryptographic signing process protects against unauthorized modifications and verifies the source of the binaries. Public and private key pairs are also generated by the crypto-auth chip. The private keys are stored within the secure storage of the chip. Each wearable device is monitored by the system administrator or an IT engineer of the hospital infrastructure. This manual key generation process ensures that the correctness of the software can be validated during installation, providing a direct and controlled method to verify the software's authenticity. However, this approach also introduces potential challenges, such as the reliance on human intervention, which could lead to errors or delays in large-scale deployments. Addressing these challenges may require supplemental automated processes or additional safeguards.

Deployment Phase. Each binary is automatically installed on separate portable medical devices. This architecure offers the ability to a development server within a hospital system to incorporate secure over-the-air (OTA) updates.

By leveraging the crypto-authentication chip's secure key storage, software updates could be cryptographically signed and verified before installation, eliminating the need for manual intervention. This approach ensures security while improving real-world applicability. The Elliptic-curve Diffie-Hellman key agreement protocol is being used to validate the source and authenticity of the binaries, ensuring that only authorized software can be installed and executed on the device. This validation process involves cryptographic signature verification, where the device checks that the software matches a pre-signed, trusted signature generated during the pre-deployment phase. This prevents any unauthorized or malicious software from being executed, safeguarding the integrity of the device against potential attacks. This approach enhances security by leveraging the ATECC608 module's robust hardware protections against unauthorized key access, reducing the risk of key extraction or tampering. Compared to traditional software-based key storage, the ATECC608 module offers a more isolated and tamper-resistant environment, ensuring that cryptographic operations remain secure even if the device is physically compromised. By integrating this module, developers can mitigate risks associated with key theft and improve overall device resilience against attacks.

Run-Time Phase. This phase involves executing the binary on wearable medical devices while actively monitoring for potential attack attempts. Monitoring includes real-time attack detection and logging of security-related events. If an attack is detected, an alert is sent to a medical gateway, notifying the user about the security event and providing relevant diagnostic information. A notification will be sent to a medical gateway, informing the user about the events that occurred on the device.

4.1 Simulation Platform Description

Components. The components used to demonstrate the use case of securing a wearable medical device are as follows:

- **Development server:** A Raspberry Pi set up to compile executables for wearable medical devices. It runs an ARM-32 OS for compatibility and integrates the ATECC608 crypto-authentication chip for secure key generation and binary signing. Compiler polymorphism ensures unique binary generation for each device, enhancing security.
- **Simulated Medical Gateway:** A Raspberry Pi or a Laptop. A Bluetooth Low Energy client receives event notifications and a dashboard visualizes status information related to security events occurring on the wearable medical device.
- **Simulated wearable medical device:** This is a Raspberry Pi microcontroller running a lightweight operating system (e.g. Rasbian, Debian or Ubuntu). Each Raspberry Pi has a crypto-auth ATECC608 chip connected to securely store the private keys of each device.

The technical artifacts developed and utilized as part of the required security implementation include:

- **Compiler polymorphism:** The compiler polymorphism mechanism is implemented using C++ and LLVM. It introduces function order randomization, function alignment randomization (32, 64, or 128 bytes), and instruction layout randomization by inserting NOP instructions. The implementation consists of LLVM passes that modify function order, alignment, instruction layout and inserts trap values. It utilizes standard C++ libraries and LLVM core libraries, including llvm/IR/Module.h, llvm/IR/Function.h, llvm/Support/raw-ostream.h, llvm/Transforms/Utils/cloning.h. A watchdog mechanism is also integrated to monitor stack integrity and detect unauthorized memory access attempts.
- **ATECC608 crypto-authentication chip:** The ATECC608 crypto-auth chip from Microchip has been used for the experiments. We established a wired I2C connection of the chip with a Raspberry Pi 3 micro-controller. In order to utilize the functionalities that the ATECC608 crypto-auth chip offers, we used the driver for Microchip ATECCx08 [9] and the cryptoauthlib, which is the library for interacting with the Crypto Authentication secure elements [10].

These components and security mechanisms are illustrated in Fig. 1 (Fig. 6).

4.2 Validation Methodology

Compiler Polymorphism. The validation process begins with setting up three identical wearable medical devices running a lightweight operating system on Raspberry Pi microcontrollers. One of these devices is equipped with an original vulnerable executable compiled without polymorphic transformations, while the other two execute polymorphic versions of the same software. To ensure that the evaluation isolates the impact of compiler polymorphism from other security mechanisms, Address Space Layout Randomization (ASLR) is explicitly disabled. The goal of the experiment is to demonstrate how an attacker can successfully execute a Return-Oriented Programming (ROP) attack on the vulnerable device and how the same exploit fails against polymorphic binaries, triggering security mechanisms that notify the medical gateway of the intrusion.

The attack exploits a buffer overflow vulnerability present in the vulnerable executable, allowing the attacker to overwrite the saved return address on the stack and take control of the program counter (pc). To construct the malicious payload, the attacker first identifies an appropriate memory location where controlled input can overwrite the return address. A crafted input is designed to exceed the allocated buffer space and overwrite pc with an address pointing to a carefully chosen ROP gadget. In this case, the attacker selects the gadget pop r0, pc, a short instruction sequence already present in the executable's memory. This gadget is useful because it pops values from the stack into the r0 register and pc, thereby controlling both an argument for function calls and the execution flow.

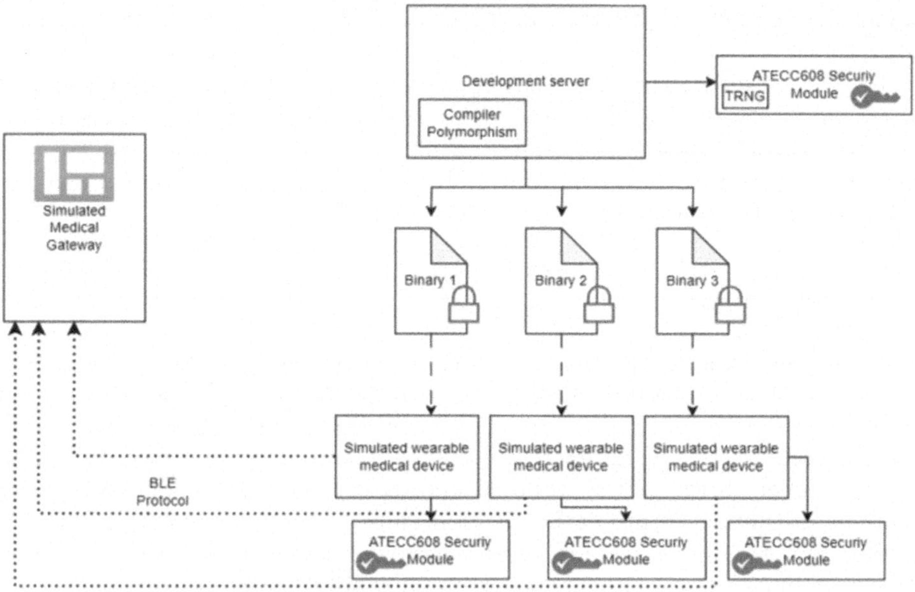

Fig. 6. Simulated platform diagram of the use case and the corresponding components.

To execute an arbitrary shell command, the attacker places the address of a known writable memory location containing the string /bin/sh immediately after the gadget's address in the crafted input. When the gadget executes, it pops this string address into r0, setting it as the first argument to a system call. The next value placed on the stack is the address of the system() function itself. As execution continues, the gadget transfers control to this function, which then executes /bin/sh, effectively spawning a shell. This results in full control over the device, allowing an attacker to manipulate or extract sensitive medical data.

Once the exploit is verified to work on the vulnerable device, the same payload is tested against the two devices running polymorphic executables. Unlike the original binary, these versions have undergone compiler polymorphism transformations, which alter the function order and memory layout. Because the structure of each binary is unique, the offsets and addresses assumed by the payload are no longer valid. The memory locations of the gadgets, stack layout, and system call entry points are randomized, preventing the payload from successfully chaining instructions into a valid execution path. Instead of gaining control over the device, the attack results in execution failure, often causing an immediate crash or unexpected behavior that disrupts the exploit. More importantly, compiler polymorphism integrates security traps within the binary to detect unauthorized execution patterns. When the malformed execution flow is detected, the embedded security mechanisms trigger an alert, logging the attack attempt, and notifying the medical gateway in real time. The system logs relevant details, including timestamp, device identifier, attack vector classification, and extracted payload signature, enabling forensic analysis and further security actions.

While a detailed evaluation of the impact of compiler polymorphism on device operations is planned for future research, preliminary expectations suggest that function randomization may introduce a minor increase in binary size without significantly affecting execution time or battery life. Further empirical analysis will be conducted to quantify these effects accurately.

5 Conclusion

Ensuring the security of resource-constrained wearable medical devices is a critical challenge in modern healthcare, as these devices play a vital role in patient monitoring and treatment. This paper has examined key vulnerabilities and emphasized the necessity of secure software execution mechanisms to mitigate cyber threats. Our proposed approach leverages IoT-driven security solutions, including crypto-authentication chips, compiler polymorphism, and runtime monitoring, to enhance resilience against sophisticated software-based attacks, such as Return-Oriented Programming (ROP) exploits.

By integrating hardware-assisted security with compiler-level transformations, we ensure that each device runs a structurally unique binary, significantly reducing the feasibility of large-scale attacks. Our validation results demonstrate that compiler polymorphism effectively disrupts exploitability while maintaining system efficiency.

While our work specifically focuses on secure software execution, we acknowledge that additional security measures such as secure boot mechanisms can further strengthen overall device and software security. Although these techniques are beyond the scope of this study, their integration would contribute to a more comprehensive, multi-layered defense against potential threats.

Future research should explore further optimizations to these security mechanisms, ensuring minimal impact on power consumption and computational performance. Additionally, the establishment of standardized security frameworks and regulatory guidelines will be crucial in promoting the widespread adoption of these protective measures. Advancing security solutions for wearable medical devices is essential to safeguarding patient data, preventing unauthorized access, and reinforcing the reliability of next-generation healthcare technologies.

Acknowledgments. The work described in this paper has received funding from the European Union's Horizon Europe research and innovation programme registered with grant agreement No 101094901 (www.septon-project.eu) under the call HORIZON-HLTH-2022-IND-13-01: Enhancing cybersecurity of connected medical devices.

References

1. Blaze Information Security: Cybersecurity Risks in Digital Health Apps. Blaze Information Security (2023). https://www.blazeinfosec.com/post/cybersecurity-risks-digital-health-apps/. Accessed 27 Jan 2025
2. Health Tech World: The Growing Cybersecurity Risks of Wearable Medical Devices. Health Tech World (2023). https://www.htworld.co.uk/insight/opinion/the-growing-cybersecurity-risks-of-wearable-medical-devices/. Accessed 27 Jan 2025
3. U.S. Department of Homeland Security: Securing Consumer Mobile Healthcare Devices and Applications (2017). https://www.dhs.gov/sites/default/files/publications/phase_iii_-_securing_consumer_mobile_healthcare_devices.pdf. Accessed 27 Jan 2025
4. National Center for Biotechnology Information (NCBI): Challenges in Securing Wearable Devices: A Review of Current Solutions. PubMed Central (PMC) (2021). https://www.ncbi.nlm.nih.gov/pmc/articles/PMC10758361/. Accessed 27 Jan 2025
5. MoldStud Solutions: Securing Wearable Devices: Challenges for System Engineers. MoldStud Solutions (2022). https://moldstud.com/articles/p-securing-wearable-devices-challenges-for-system-engineers. Accessed 27 Jan 2025
6. Adafruit: ATECC608 Secure Element Breakout Board. Adafruit Learning System. https://learn.adafruit.com/adafruit-atecc608-breakout. Accessed 27 Jan 2025
7. Sustainable Computing: Informatics and Systems: Return-oriented programming on a resource constrained device. https://www.sciencedirect.com/science/article/pii/S2210537917303931?via%3Dihub. Accessed 30 Jan 2025
8. IEEE Access: RIO: Return Instruction Obfuscation for Bare-Metal IoT Devices. https://ieeexplore.ieee.org/stamp/stamp.jsp?arnumber=10177731. Accessed 30 Jan 2025
9. Driver for Microchip ATECCx08 cryptographic co-processors with secure hardware-based key storage. https://github.com/adafruit/Adafruit_CircuitPython_ATECC/tree/main. Accessed 11 Feb 2025
10. Library for interacting with the Crypto Authentication secure elements. https://github.com/MicrochipTech/cryptoauthlib/tree/master. Accessed 11 Feb 2025

Cyber-Risk Indicators for Connected Medical Devices

Simeon Tverdal[1(✉)], Gencer Erdogan[1], Andrea Neverdal Skytterholm[2], Samuel M. Senior[3], Steve Taylor[3], and Laura Carmichael[3]

[1] Sustainable Communication Technologies, SINTEF Digital, Oslo, Norway
simeon.tverdal@sintef.no
[2] Software Engineering, Safety and Security, SINTEF Digital, Trondheim, Norway
[3] IT Innovation Centre, University of Southampton, Southampton, UK

Abstract. Connected Medical Devices (CMDs) play a crucial role in modern healthcare, enabling real-time monitoring, automated decision making, and remote patient care. However, increasing reliance on digital connectivity introduces significant cybersecurity risks. As CMDs are used in increasingly diverse contexts, traditional risk assessment approaches may fail to capture the nuances of each use case. The increased attack surface and additional lack of cybersecurity experience among daily users further highlight the need for accurate and comprehensible risk management information. Cyber-risk indicators are additional pieces of information that can connect the risk assessment to its dynamic context, enabling risk assessors to estimate risk values more precisely and accurately. In this article, we define and present domain-specific cyber-risk indicators to facilitate dynamic risk assessment in the CMD domain and demonstrate how they can be used in the risk assessment process. Preliminary results from four real-world industry case studies are promising and validate the feasibility of our approach.

Keywords: cyber risk · indicator · cybersecurity · privacy · risk assessment · connected medical devices · case study

1 Introduction

Connected Medical Devices (CMDs) have become indispensable in contemporary healthcare, revolutionizing the way medical professionals monitor patients, make decisions, and provide care remotely. These devices facilitate the collection and analysis of real-time data, allowing timely interventions and improved patient outcomes. Additionally, CMDs support automated decision-making processes, which can enhance the accuracy and efficiency of medical treatments.

However, the growing dependence on digital connectivity in CMDs also presents substantial cybersecurity challenges [3]. As these devices become more interconnected, they become potential targets for cyberattacks, which can compromise patient data, disrupt medical services, and even endanger patient lives.

I. Praça et al. (Eds.): EICC 2025, CCIS 2500, pp. 288–306, 2025.
https://doi.org/10.1007/978-3-031-94855-8_19

Therefore, it is imperative to carefully balance the benefits of CMDs with the associated cybersecurity risks.

As Connected Medical Devices are increasingly utilized in diverse settings, including home patient care, traditional approaches to risk assessment often fall short in addressing the unique context of each use case. The expanded attack surface, coupled with the general lack of cybersecurity expertise among everyday users, underscores the critical need for precise, current and easily understandable risk management information.

In this paper, we further explore the concept of cyber-risk indicators by utilizing four industry case studies all centred on CMDs. An indicator is defined as *a piece of information relevant for assessing risk levels.* They can help facilitate a way to bridge the gap between a risk model and the system under evaluation, providing the cyber-risk analyst with a more dynamic risk assessment process. Indicators can further inform the context of the system under evaluation by providing a more dynamic window into the realities of the system, allowing the risk model and accompanying risk evaluation to reflect the current reality with more accuracy. Ideally, these indicators should be sourced from a diverse range of inputs, including documentation and real-time monitoring tools.

This article first outlines the background of the indicator identification process (Sect. 2), before providing related work (Sect. 3), followed by detailing the methodology and presenting several example indicators (Sect. 4). In addition, we will discuss how to translate the source data into indicator values, offering an initial overview of the methodology. We will then introduce four industry case studies to demonstrate the results of the indicator identification process (Sect. 5). Finally, we conclude the paper including future work and threats to validity (Sect. 6).

2 Background

2.1 The CORAS Method

CORAS is a risk management approach that consists of a risk assessment method in line with ISO27005, a modelling language and a tool. It falls under the umbrella of defensive risk management, meaning it concerns itself with preventing risk in contrast to weighing it against possible benefit. CORAS is developed for the express purpose of being understood by laypeople without sacrificing complexity, and includes both models showing complete attack paths, as well as simplified representations of such for the purpose of stakeholder presentation and information dissemination. In line with this concept, the risk assessment is performed by expert risk assessors in concert with stakeholders. This is done through collaborative meetings or brainstorming sessions, allowing the stakeholders to provide input and otherwise contribute to the modelling process during the entire process.

The CORAS method uses five types of models, representing different stages in the assessment process. These are *Asset models*, *Threat models*, *Risk models*, *Treatment models* and *Treatment overview models.* Asset models present the assets included in the scope of the risk assessment, Threat models contain the full threat paths from threat source to asset, and include vulnerabilities and indicators. This is where most of the modelling is performed. A CORAS Risk model is an overview of the Threat model, with most detail omitted for the purpose of readability. A Treatment model is a threat model with appended treatments or mitigations, while a Treatment overview model is an overview of risks and treatment with most detail omitted. In this paper we use the more generic term *risk model* when referring to risk models in general, but the term *threat model* when referring to CORAS threat models specifically.

A typical full-scale CORAS risk assessment is carried out in eight steps [8]:

1. **Preparation for the analysis**: Define the scope, objectives, and context of the analysis, gather preliminary information, and set up the analysis team.
2. **Customer presentation of target**: Engage with stakeholders to align on the objectives, scope, and context of the risk analysis, ensuring mutual understanding and agreement.
3. **Refining the target description**: Collect detailed information about the system, context, and assets, and identify the main security objectives.
4. **Approval of target description**: Document the system and its environment thoroughly, and have stakeholders review and validate this documentation.
5. **Risk identification using threat diagrams**: Conduct brainstorming sessions and workshops with stakeholders to identify potential threats, vulnerabilities, and unwanted incidents.
6. **Risk estimation using threat diagrams**: Assess the likelihood and impact of identified risks using qualitative or quantitative methods.
7. **Risk evaluation using risk diagrams or risk evaluation matrices**: Compare estimated risks against predefined risk criteria to determine their acceptability.
8. **Risk treatment using treatment diagrams**: Develop and evaluate options for mitigating, transferring, avoiding, or accepting risks, and plan for implementing the chosen risk treatments.

The terminology used in CORAS and in this paper is explained below [8].

1. A **threat source** is a potential cause of an unwanted incident. We distinguish between deliberate human threat, accidental human threat, and non-human threat such as malware or failure.
2. A **threat scenario** is a chain or series of events that is initiated by a threat and that may lead to an unwanted incident.
3. A **vulnerability** is a weakness, flaw or deficiency that opens for, or may be exploited by, a threat to cause harm to or reduce the value of an asset.
4. An **unwanted incident** is an event that harms or reduces the value of an asset.
5. An **asset** is something to which a party assigns value and hence for which the party requires protection.
6. A **party** is an organisation, company, person, group or other body on whose behalf the risk assessment is conducted.

7. An **indicator** is a piece of information that is relevant for assessing the risk level. An indicator may be assigned to any element in a CORAS threat model.

These elements are presented in the simple model shown in Fig. 1, with the numbering matching the above list. The party element is only present in asset models, and so is not displayed in this example threat model.

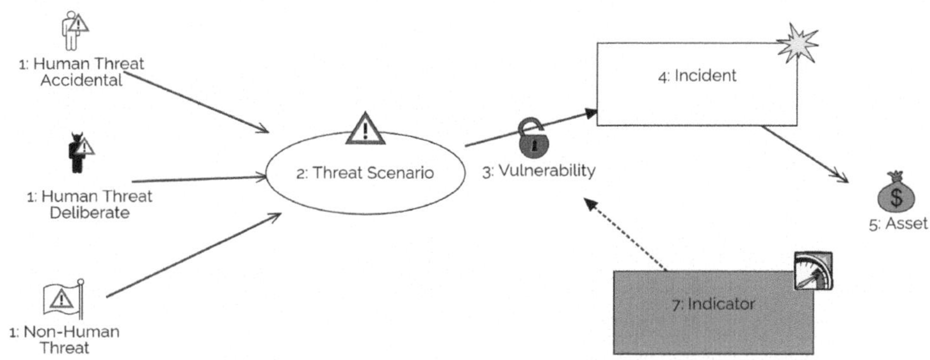

Fig. 1. CORAS threat diagram elements.

2.2 Indicators

From the risk management perspective, indicators can be defined as any piece of information that can be used to assess risk levels. This encompasses information gathered during context establishment concerning the overall security of the system, as well as signals of events or situations relevant to the cybersecurity of the operation of CMDs in the context they are deployed. When discussing indicators in this paper, we are referring the *model element* representing this information, as described in Table 1. The indicator value and indicator source are referred to explicitly where relevant. Indicator source refers to where and how the indicator is obtained, while the indicator value is a metric received from the source, either directly or after processing, which is used to modify the risk value of the model.

We distinguish between four types of indicators: expert knowledge, test results, network-layer monitoring, and application-layer monitoring. Tables 2 through 5 explain these categories in more detail (Tables 3, 4 and 5).

These indicator types represent potential sources of information that can be used to feed the risk assessment script with information about the likelihood or consequence of elements leading up to the unwanted incident of a risk. Moreover, the implementation of indicators in the assessed system will help facilitate a more dynamic assessment of risks.

Table 1. Indicator Fields and Descriptions

Field	Description
ID	Uniquely identifies each indicator. Naming convention: IN-<Number>, where IN stands for Indicator
Question	Defines the indicator in terms of a question that needs to be answered to obtain a value for risk assessment
Data type	Specifies the output type of the indicator, e.g., binary (yes/no) or numerical value
Motivation	Describes why the indicator is useful
Indicator type	Shows the type of the indicator. The author marks the type with an "X"
Means of obtaining indicator	Describes how the indicator may be obtained from a technical perspective

Table 2. Expert knowledge

Indicator	Expert knowledge
Description	Indicators derived from the insights, experience, and specialized knowledge of domain experts, typically used to identify risks or vulnerabilities that may not be immediately apparent through automated tools or predefined methods
Purpose	To incorporate human expertise and contextual understanding into risk assessment, especially for complex or novel scenarios that cannot be fully captured by standard tests or monitoring
Examples	– An expert's identification of emerging cybersecurity threats based on trends in the medical device industry – Recommendations from a security specialist on potential vulnerabilities in custom-built systems – Insights from clinical professionals about how device malfunctions might lead to patient harm

3 Related Work

This article builds on the foundational work done by Erdogan et al. (2024) [4], where a more in-depth explanation of the CORAS methodology in the context of CMDs is presented. The concept of indicators, as used in this paper, was also introduced in that earlier work.

Indicators as a concept have been discussed in the literature from various perspectives. One of the contributions most relevant to this paper is the article by Meland et al. (2021) [9], which presents a systematic literature review of indicator data in the cybersecurity domain. Meland et al. found that the literature predominantly focuses on technically sourced indicators, particularly those

Table 3. Test result

Indicator	Test result
Description	Indicators derived from the outcomes of testing procedures, such as security tests or functional tests, to assess the state of the system or identify specific weaknesses or vulnerabilities
Purpose	They provide information about potential vulnerabilities or failures identified through testing
Examples	– Calibration tests for In Vitro Medical Device (IVD) glucose sensors reveal inaccuracies, risking improper diagnoses or treatments – Latency tests for IVD devices highlight delays, risking data loss and delayed decisions – Firmware integrity checks detect tampering, risking device functionality and security

Table 4. Network-layer monitoring

Indicator	Network-layer monitoring
Description	Indicators focused on network-level activities and configurations, providing insights into network-related risks and vulnerabilities
Purpose	Monitor and detect risks associated with network communications, such as unauthorized access, data interception, or weak encryption
Examples	– Abnormal traffic patterns indicating potential Distributed Denial of Service (DDoS) attacks – Unencrypted data transmissions between medical devices and servers – Weaknesses in network segmentation or firewall configurations

related to network and system resources. This category of indicators has seen the most significant increase in attention and utilization.

In addition to technically sourced indicators, Meland et al. highlighted a gap in the use of domain-specific data and expert opinions. Despite their potential value, these sources of indicator data are less frequently utilized in the literature. This finding underscores the need for a more balanced approach that incorporates both technical and domain-specific indicators, as well as expert insights, to provide a more holistic view of the risk landscape.

Erdogan et al. propose a method for quantitative cyber-risk assessment using CORAS models and Bayesian Networks [5]. Indicators, derived from expert input and monitoring data, are used to dynamically adjust likelihood and impact estimates, enabling continuous and adaptive risk evaluation. The authors also

Table 5. Application-layer monitoring

Indicator	Application-layer monitoring
Description	Indicators related to the security and functionality of applications interacting with CMDs
Purpose	Identify vulnerabilities or threats specific to software applications and their interfaces
Examples	– Use of outdated libraries or dependencies with known vulnerabilities – Unvalidated input fields leading to potential injection attacks – Lack of secure APIs exposing sensitive patient data

present a method for developing qualitative cyber-risk assessment algorithms, combining CORAS with DEXi, a multi-criteria decision-making tool [6]. The method helps produce executable algorithms that assign qualitative risk levels based on the dynamic state of systems. It is targeted at practitioners who lack advanced programming skills but need transparent and easy-to-maintain tools. Our work is closely related to the work presented in these papers, but the usage domains are different. Our work is closely related to these contributions, but targets a different domain. While Erdogan et al. [5,6] provide generic methods that enable dynamic risk assessment through the use of indicators and threat patterns, our focus is on systematically identifying threat scenarios and domain-specific cyber-risk indicators tailored to connected medical devices (CMDs).

Asiri et al. [2] provide a comprehensive survey of Indicators of Compromise (IOCs) in Industrial Control Systems (ICS), focusing on their role in post-incident cyber forensics and threat detection. The authors highlight that while traditional IT networks have well-established IOC practices, ICS environments pose unique challenges due to differences in architecture, limited resources, and safety-critical operations. Moreover, they analyse standards for representing IOCs, existing tools and techniques, and challenges in ICS environments, and propose a list of potential IOCs tailored to ICS attacks. The paper also maps these IOCs to common attack types and discusses how they can support incident detection and response. The authors classify IOCs into four main types: 1) Atomic Indicators (basic data points like IP addresses, URLs, filenames, etc.), 2) Computed Indicators (data derived through computation, e.g., file hashes, certificate hashes), 3) Behavioral Indicators (observable patterns of adversary behaviour (e.g., spear phishing attempts, repeated login failures), 4) Physical Measurement Indicators specific to ICS (anomalies in physical parameters, such as unexpected voltage levels and actuator states). Our approach and the one proposed by the authors are complementary: while they leverage indicators primarily to detect and map observed information to known attack types, we use the information obtained through indicators to assess the risk level of potential attack paths.

Adaros-Boye et al. [1] introduce an Indicator of Risk (IoR) Library aimed at enabling continuous cyber-risk monitoring in Industrial Control Systems (ICS). These indicators are similar to the indicators we refer to as network-layer monitoring and application-layer monitoring. The authors seek to identify what information is necessary to monitor cyber risks and how this information can be derived from observable, measurable variables. To achieve this, they developed a library of 95 IoRs mapped to adversary techniques from the MITRE ATT&CK [10] for ICS framework.

Silvestri et al. [13] propose a method for identifying and prioritising cyber threats in healthcare ecosystems by using Natural Language Processing to extract threat information related to specific assets from widely available security-related information on the Internet, and assess risk levels based on frequency of occurrence, supporting informed mitigation actions. Our approach and theirs are complementary, as the output of their method can serve as input to the indicators in our approach, enabling dynamic risk assessment.

4 Methodology

This section explains the general methodology for identifying indicators, possible sources of indicator data, and how to translate these data into indicators of various types. Finally, the method used with the case studies will be presented.

4.1 Indicator Identification

Indicator identification is done as part of an extended CORAS risk assessment. First, we describe the target of analysis and its assets. As part of this step, we also describe high-level potential threat sources, threat scenarios, vulnerabilities, and unwanted incidents. The output of this step is a high-level risk table that captures the aforementioned information.

We base ourselves on the descriptions provided by the case study partners to identify the assets that need protection and to identify the main components in the target system which need to be considered during a risk assessment. In addition, we hold regular meetings with the case study partners to align the objectives, scope, and context of the risk analysis, ensuring mutual understanding and agreement.

Next, based on high-level risk tables, we create CORAS threat models that include threat sources, threat scenarios caused by threat sources, vulnerabilities exploited by the threat sources, and unwanted incidents caused by the threat sources as a result of exploiting vulnerabilities. The specific assets that are harmed by the unwanted incidents are also captured by the CORAS threat models. This corresponds to Step 5 of the CORAS method. Having created CORAS threat models, we then identify as many relevant indicators as possible for each element in the threat model. Identifying indicators is an extension to the original CORAS method and is performed to facilitate a more dynamic risk assessment.

Indicators can be identified during all eight stages of the CORAS process. As the creation of CORAS models is based on collaboration, new information and, as such, possible indicators can be discovered at any stage. However, some stages are more relevant. Context establishment in addition to risk identification are particularly useful, as this is where system information is gathered and the threat models are created. After having completed the risk identification, most possible indicators will likely have been identified. The formal modelling of indicators is done after the risk identification is completed, which means in betweens steps 5 and 6 of the CORAS method, and is performed in a similar manner to risk identification. With a complete threat diagram and an overall view of the system, the risk assessor will be aware of the likely threat paths, components and vulnerabilities in the system, and can target more specific parts of it to identify possible indicators.

Some care must be taken to separate vulnerabilities and indicators, as overlap may occur. In many cases, a vulnerability can be more usefully presented with some uncertainty or probability attached. Here, an indicator may better represent inexplicit vulnerabilities than the vulnerability element. However, in other cases, indicators can be phrased so that the answer may be represented by the presence or absence of a vulnerability. In such cases, the risk assessor must exhibit judgment as to how likely that is to change. Change in the state of a system is more easily modelled with indicators than with more static vulnerability elements. For visual clarity, there is also the possibility of adding a vulnerability with an attached indicator. This will make the possible vulnerability explicit to the reader while still benefiting from the risk calculation provided by the indicators (Fig. 2).

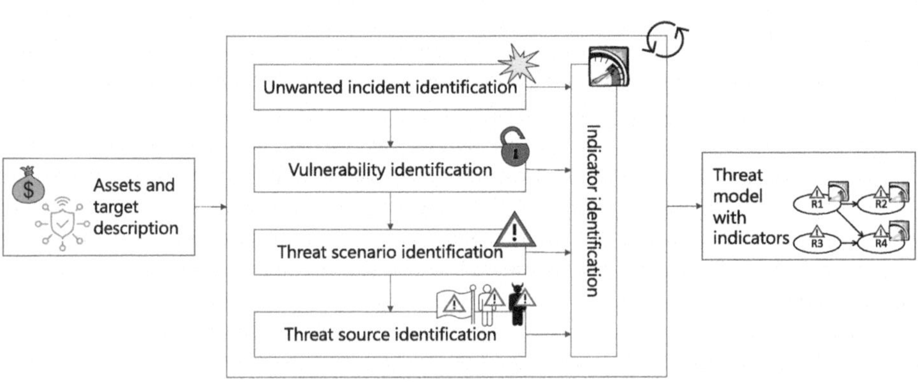

Fig. 2. Indicator identification method high level.

4.2 Indicator Translation

Indicator translation is the process of converting data from various indicator sources into a value that can modify the final risk value. The CORAS calculus explains how the final risk value is calculated. Provided the model is complete, the likelihood of an element can be determined based on its incoming relationships or links. There are established rules for working with both probabilities and frequencies, based on the user's preference. In this context, frequencies refer to how often an event occurs or is expected to occur.

The calculation is designed to work with value ranges, meaning the risk assessor can, for instance, set the category 'Rare' to a range of frequencies that do not overlap with the next level on the scale. These ranges can be defined according to the context of the case and do not need to be uniform. In other words, one range might cover years, while another might be measured in days.

The rules for determining likelihood differ based on whether the incoming links are from mutually exclusive or statistically independent elements. In most cases, in a regular diagram, the majority of elements stemming from a single preceding element can be considered statistically independent. These calculations are performed across the entire model to determine the likelihood of the unwanted incident, and thus the final risk value.

The calculus will be extended to account for the indicators. This means defining the allowed types of indicator values and how they impact the likelihood, conditional probability, or consequence of the elements they affect. How multiple indicators interact with one another will also be defined, to account for instances where there are many indicators affecting a single element and for instances where some indicators may be mutually exclusive.

Indicator sources present data in a variety of formats, ranging from simple booleans indicating presence or absence or integers counting detected events to more ambiguous data where a domain expert estimates policy quality. Certain tools may also present large amounts of data that must be parsed to represent a quantifiable risk-modifying value. For this reason, guidelines must be established to translate these heterogeneous sources into quantified metrics.

At this point, three valid formats have been identified: booleans representing the presence or absence of an indicator, integers representing the number of occurrences of an event, and integers representing the indicator data as a level on a scale.

Boolean indicators are the most common format because many questions can be phrased this way. Technical indicators can be on or off, indicating whether an occurrence or event is registered. In addition, for more social indicators, the risk assessor can remove ambiguity by rephrasing the indicator question. For example, instead of asking the case study owner to grade their access control policy, they could ask if it meets a certain standard. Boolean indicators are also well suited for categories related to test results and monitoring, where a positive test result or an event from monitoring software can be easily represented in this manner.

For indicators represented by the number of occurrences, the most common use case is application or network monitoring that records and logs events. A certain number of occurrences could indicate unusual or malicious behaviour, but in some cases, this threshold is not well represented by a Boolean yes or no. Increased or decreased activity can be used as indicators to gradually and subtly adjust the level of risk faced by the system. A system administrator could present a numerical metric that can be implemented as an indicator without defining an absolute threshold where an attack would be suspected.

Finally, some indicators cannot be represented as a boolean or a raw numeric metric. For example, an indicator representing the privacy risk inherent in a machine learning model should be rated based on how much data is leaked. In this case, there is no reasonable way to measure the number of occurrences, and setting a Boolean value would omit valuable information. For these cases, a scale system can be used in which the case study owner ranks severity on a scale. These scales can be of any granularity. For example, a risk assessment tool can provide a numerical metric from one to one hundred, which can be used to inform an indicator. Note that some effort must be made to ensure that the scale remains relatively linear to reduce ambiguity.

For all indicators, a certain level of user control is necessary. Attempting to capture the complexities of a system with simple, universally applicable quantifiable standards is unrealistic at this level of abstraction. Therefore, both the translation and calculation must be modifiable at the implementation level. This means that while reasonable defaults are set, users should be able to modify them if necessary. This could involve changing the number of levels on a scale, adjusting how much each level affects the calculation, or modifying the impact of an indicator value being on or off. Often the presence of an event carries more weight than its absence, though the opposite could also be the case.

For Boolean indicators, the user can determine how much the likelihood or consequence of the connected element is altered by a positive or negative result. This is represented by a float, by which the value is then multiplied. For numerical indicators, the user must set the maximum expected impact themselves, while the default remains as no change for no detected occurrences. In the case of scales, defaults are set for common scales, with the user able to modify the minimum and maximum impact.

Finally, it is noteworthy that most indicators can change format by rephrasing the question. Instead of implementing an indicator as a number of occurrences, it can simply ask if the number is above or below a certain threshold. Care should therefore be taken to ensure that the indicator question can be answered in the format most beneficial to the creator of the risk model.

4.3 Case Studies

Our method has been evaluated in four case studies carried out within the context of the EU project NEMECYS [12], covering both the deployment and operation phases of connected medical devices. These case studies span diverse healthcare scenarios, including home dialysis treatment, wearable devices for continuous monitoring of movement disorders, software as a medical device, and hospital-based point-of-care testing.

Due to time constraints, the full CORAS risk assessment process was condensed. The context establishment phase relied primarily on interview transcripts and supporting text documentation provided by each case study partner. Three meetings were held for each case study: (1) to confirm the initial high-level risk table, (2) to co-develop the base threat model, and (3) to identify relevant cyber-risk indicators. Preliminary models and tables were prepared in advance and refined collaboratively in each session. While the estimation, evaluation, and treatment steps are not yet completed, this phase delivered essential insight into relevant threats, vulnerabilities, and indicators in each context.

For each case study, we created a high-level risk table capturing potential threat sources, threat scenarios, vulnerabilities, unwanted incidents, assets at risk, and relevant components in the system under analysis. Section 5 provides a summary of the case studies, including the risks identified in each.

5 Results

In this section we first provide a summary of the case studies, then it shows an example CORAS threat model with indicators, followed by an overview of the number of indicators we identified for each indicator type. The reader is referred to our public deliverable [11] for a more detailed explanation of each risk indicator.

Case Study 1 - home dialysis treatment: This case study explores the cybersecurity implications of using a wearable sensor patch developed for real-time and continuous hydration monitoring in home dialysis scenarios. The technology aims to expand access to home dialysis for older patients by enabling remote clinical decision support. The primary cybersecurity risks identified include unauthorized modification or extraction of sensitive sensor data by hackers or malicious insiders, through methods such as tampering with firmware, exploiting debug ports, or performing adversary-in-the-middle (AiTM) attacks. These incidents could compromise the confidentiality, integrity, or availability of patient data, potentially leading to incorrect clinical decisions and health deterioration due to fluid imbalance. Additional risks include the theft of proprietary algorithms, which could result in financial loss and reputation damage. The main vulnerabilities are related to insufficient physical security, weak encryption, lack of secure communication protocols, and inadequate detection mechanisms.

Case Study 2 - wearable devices for continuous monitoring of movement disorders: This case study focuses on a wearable device system designed to provide continuous and objective monitoring of motor symptoms in patients with Parkinson's disease, improving treatment beyond traditional clinical exams. The system collects sensitive personal and clinical data using sensors and cloud-based infrastructure. The primary cybersecurity risks identified include unauthorized access to raw or processed patient data via insecure cloud services, physical access to local storage, or side-channel attacks on networked components. Potential threat actors include hackers, insiders, and employees, who may exploit vulnerabilities such as a lack of encryption, poor access controls, static passwords, and insufficient physical security. Critical risks include breaches of patient confidentiality, loss of data integrity and availability due to configuration errors or malicious tampering, and extraction of proprietary algorithms, all of which could result in regulatory penalties and reduced trust in the system. The main components at risk include the monitoring devices, a smartbox, a physician portal, and Azure cloud services.

Case Study 3 - software as a medical device: This case study investigates mobile applications used by diabetes patients for therapy management, such as calculating carbohydrate intake and insulin dosage. Classified under the EU Medical Device Regulation (2017/745) [7] as Software as a Medical Device (SaMD), these apps must comply with stringent risk assessment requirements. The primary cybersecurity risks stem from their operation within general-purpose mobile environments, where vulnerabilities may be introduced through interactions with other applications, insecure device configurations, or weak communication protocols. Identified threats include adversary-in-the-middle (AiTM) attacks, data modification or extraction by hackers, technical malfunctions of external services, malicious code injection, and user error. These risks threaten patient data confidentiality, integrity, and availability, and in some cases directly harm patient health. Case Study 3 also highlights risks to the manufacturer's finances and reputation due to potential data breaches and non-compliance with data protection regulations. Key components at risk include the SaMD mobile app, NFC interfaces, mobile operating environment, external services, and web-based physician interfaces.

Case Study 4 - hospital-based point-of-care testing: This case study examines cybersecurity risks associated with self-testing using in-vitro diagnostic (IVD) medical devices, particularly for patients managing long-term conditions from home. While self-testing improves patient autonomy, reduces healthcare costs, and alleviates pressure on hospital resources, it also introduces new vulnerabilities. The primary risks identified involve unauthorized access to patient data across various components, including mobile apps, clinical portals, and the devices themselves (e.g., readers and sensors). Threat actors may exploit weak session management, improper authorization, insecure physical environments, and limited user security awareness to access, extract, or manipulate sensitive patient information. Additional concerns include malware injection through compromised systems, sensor data manipulation affecting clinical decisions, and broader reputation and financial risks to healthcare providers. The main assets at risk include patient data, sensor integrity, hospital systems, and the overall trustworthiness of the self-testing infrastructure.

5.1 Threat Model

Figure 3 shows an excerpt from the model derived from case study 2. The scenario considers the risk of patient data becoming unavailable due to the edge device, called a SmartBox, malfunctioning. The Smartbox is responsible for collecting, processing and transferring data to the system cloud servers. A lack availability of data can cause harm to the patient via improper or delayed treatment. In this part of the model three possible paths were identified, namely intentional and unintentional misconfiguration as well as an adversary getting access to the configuration web portal and altering the settings. The three adversaries are respectively *TS4: Employee TS5: Insider* and *T6S: Hacker*, and the asset in question is *A1: Availability of data.*

For the unintentional misconfiguration path, we see it initiated by *TS4: Employee*, with the vulnerability *Lack of training* affecting the likelihood of initiation. To inform this vulnerability we use the indicator *IN47: How comprehensive is employee training?*. This leads to the threat scenario *T4: Unintentionally misconfigured Smartbox*, which propagates to threat scenario *T7: Basic device settings deleted or corrupted.*

For intentional misconfiguration the threat source is *TS5: Insider*, which initiates the threat scenario *T5: Intentionally misconfigured Smartbox.* The likelihood of this threat scenario is informed by the indicator *IN15: Are there screening procedures in place?* From here we propagate to threat scenario *T7: Basic device settings deleted or corrupted.*

The final path representing an external adversary is initiated by threat source *TS6: Hacker.* The web portal in question is on the local network, and requires close proximity to the device, so we inform the vulnerability *Lacking awareness training* with the indicators *IN26: How comprehensive is the awareness training?* and *IN10: Is the access policy too broad allowing access from unauthorized personnel?*, in addition to *IN34: How many employees clicked on a phishing link?* to approximate a value for this vulnerability. In addition, we add the vulnerability *CWE-308: Use of Single-factor authentication* which is informed by the indicator *IN48: Does the website use multi-factor authentication?* This leads to the threat scenario *T6: Adversary gets access to the web portal* where the likelihood is affected by the indicator *IN7: How many connection attempts does the portal receive per minute?*. Finally, this propagates to threat scenario *T7: Basic device settings deleted or corrupted.*

All these lead to the threat scenario *T7: Basic device settings deleted or corrupted*, where the edge device will attempt an online backup of its device settings. At this point, we note that a poor internet connectivity could interfere with the planned backup. To cover part of this, we include the vulnerability *Poor internet connectivity* and attach the indicators *IN6: Are there frequent blackouts due to weather and infrastructure?* and *IN5: How reliable is the internet connectivity?* to inform this vulnerability. Should the backup fail we denote this with threat scenario *T8: Device backup fails*, which results in the unwanted incident *U2: Device malfunction/shutdown* harming the asset *A1: Availability of data.*

The indicators in Fig. 1 fall into the following categories: *IN6*, *IN10*, *IN15*, *IN25*, *IN26*, *IN47* and*IN48* are of type **Expert knowledge**, *IN5* and *IN34* are **Test results**, while *IN7* and *IN4* are of type **Application monitoring** and **Network monitoring** respectively.

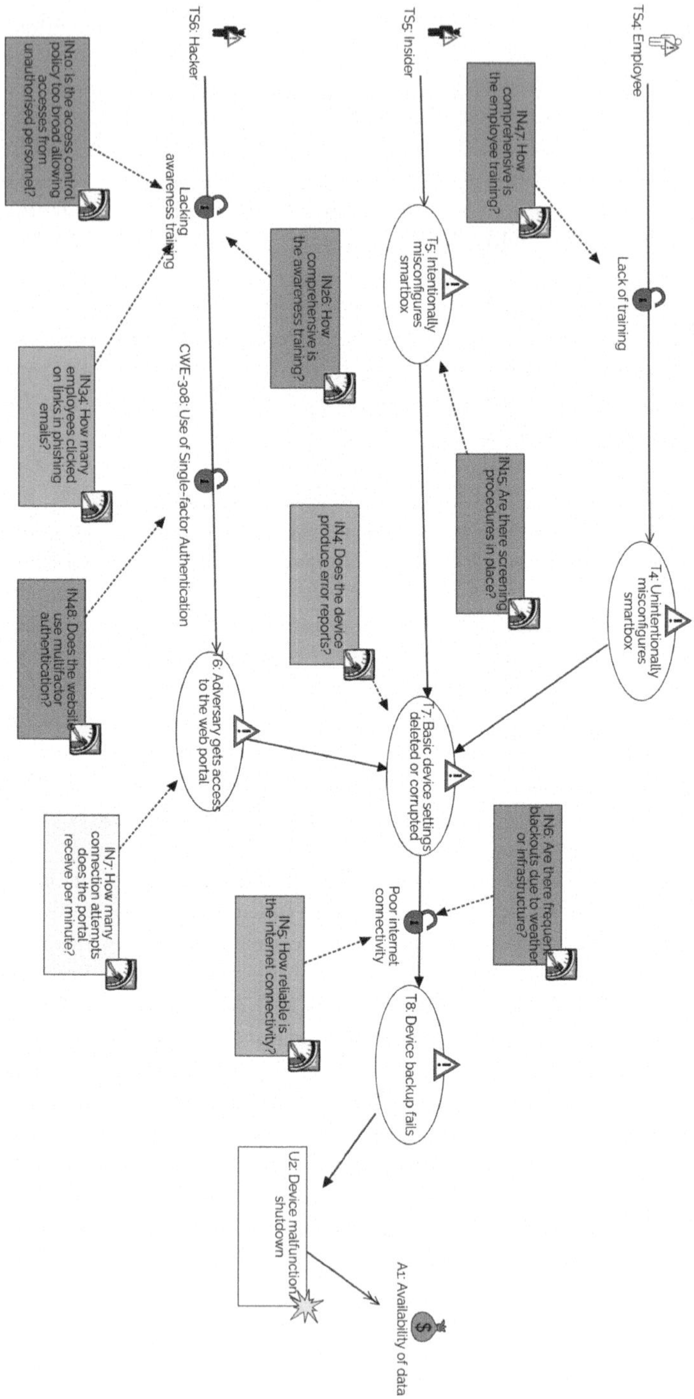

Fig. 3. Threat model

5.2 Indicators

The aggregated results of the risk identification for all case studies are presented in Table 6. The final count of indicator instances identified during the project totals 140, of which the majority are of type *expert knowledge* and *test result* compared to the more technical categories. Some of these are repeated, and the final count of *unique* indicators is 106. The overlap is present mostly in the expert knowledge category, as several indicators are sourced from more general knowledge of a system, though the test results category have some overlap as well. Due to the difference between the case studies there is no overlap in the network and application monitoring categories.

Table 6. Indicator count per type

Case study	Expert knowledge	Test Result	Network monitoring	Application monitoring
1	12	13	2	10
2	17	11	2	6
3	24	10	3	3
4	16	7	0	4
Total	69	41	7	23

To show how the format presented in Table 1 is used, we present another identified example indicator relevant to case study 4, which concerns itself with the transfer of data from the patient device to the physician system. A risk was identified were either a malicious insider or an adversary received access to and inserted a compromised reader or a device falsified to look as such into the hospital system. In this case a possible indicator was identified concerning the physician's knowledge of legitimate devices. This indicator is presented below in Table 7.

5.3 Discussion

Due to the range of defined indicators and the breadth in each category, there are many possible sources of indicators. One such source, as presented in Sect. 2, is the output of technical tools, which could be application or network monitoring tools, or tools used as part of a penetration test. While those indicators are less populous in the findings, they are still present. Most prevalent are the risks related to expert knowledge. One example is the indicator presented in Table 7. Another is *Access control policy* which has a substantial impact on physical security and can vary drastically within each use case. The distribution of identified indicators is affected by several factors. One of these is the lack of a functioning testbed from which to extract information. As a result, some possible technical indicators were omitted due to a lack of data, and what could otherwise have

Table 7. IN66: Staff Awareness of Permitted Readers

ID:	IN 66	
Question:	IN66: Are the staff aware of all permitted readers?	
Data type:	Boolean	
Motivation:	The blood glucose readers transfer data to the system's computer via a USB connection, which can be exploited to infect the system with malware. To mitigate this risk, the staff responsible should be made aware of all permitted readers to avoid inserting a malicious device into the hospital system	
Indicator type:	Expert knowledge	**X**
	Test results	
	Network-layer monitoring	
	Application-layer monitoring	
Means of obtaining indicator value:	Consult policy if possible and interview staff regarding adherence and awareness	

been several separate indicators were aggregated into one indicator at a higher level of abstraction. In addition, as the process was performed in less time than recommended and with less case partner interaction, more generic indicators were often preferred, and thus the resulting models and indicators can still be expanded upon.

5.4 Lessons Learned

All case study owners found the risk assessment process useful, with reports that the threat models and indicators contributed to a clearer and more significant understanding of the risk picture they are exposed to. Indicators were found to be a relevant addition to the models for their respective use cases, and in some cases simplified stakeholder communication. As indicators are phrased as simple questions, it can make it easier for stakeholders with a less technical background to contribute to the discussion. Asking specific yes or no questions allowed for direct answers, or if none were available, pointed out areas of the system where more attention was needed.

It can be challenging to choose the right level of technical abstraction for each indicator. General indicators can be more easily reused across case studies, and will often be easier to understand when reading the model. On the other hand they will often be less useful to each specific case, in particular in the case of technical indicators concerning bespoke system components. For the majority of the indicators, we made them as specific as information allowed, while restricting them to two or three per relation to avoid information overload. This approach resulted in many indicators which were reusable across all case studies, and the enumeration of indicators listed in deliverable [11] simplified the process of indicator identification for the following case studies. It can with benefit be extended and refined further to facilitate future risk assessments.

6 Conclusion

In conclusion, indicators show promise in providing a flexible and adaptable means of accounting for current contextual information when modelling actual risk values. Building on the process of CORAS modelling, our established method for deriving these indicators has shown to be effective, particularly through collaborative brainstorming sessions with our industry case study partners. These sessions have enabled us to generate useful and relevant indicators that can be applied in real-world scenarios.

Moreover, we have developed practical guidelines to translate data from heterogeneous sources into simple risk value modifiers. This capability allows us to handle diverse data inputs and convert them into meaningful risk assessments. However, it is important to note that further work is required to refine these translation processes and to verify the overall usefulness and accuracy of the indicators we derive.

6.1 Future Work

Future work will include further elaboration on the various indicator formats and data types, in addition to specification of the indicator calculus. Besides, more indicators and indicator types should be sourced from real tools to determine their feasibility and usefulness. By incorporating a wider range of indicators from practical applications, we can better assess their relevance and effectiveness.

6.2 Threats to Validity

The indicators have been identified during a shortened version of the CORAS method, with less time spent with the case partners. This will inevitably affect the number and detail of the indicators identified. In addition, the lack of an accessible testbed results in an increased proportion of less technical indicator types, in addition to the fact that technical indicators identified are presented at a higher level of abstraction.

References

1. Adaros-Boye, C., Kearney, P., Josephs, M., Ulmer, H.: An indicators-of-risk library for industrial network security. In: Proceedings of the 16th International Conference on Availability, Reliability and Security. ARES 2021. Association for Computing Machinery, New York (2021). https://doi.org/10.1145/3465481.3470023
2. Asiri, M., Saxena, N., Gjomemo, R., Burnap, P.: Understanding indicators of compromise against cyber-attacks in industrial control systems: a security perspective. ACM Trans. Cyber-Phys. Syst. **7**(2) (2023). https://doi.org/10.1145/3587255

3. Bernsmed, K., Jaatun, M.G.: Security-by-design challenges for medical device manufacturers. In: Proceedings of the 2024 European Interdisciplinary Cybersecurity Conference, EICC '24, pp. 155–160. Association for Computing Machinery, New York (2024). https://doi.org/10.1145/3655693.3661297
4. Erdogan, G., Carmichael, L., Taylor, S., Tverdal, S., Skytterholm, A.N.: Dynamic cyber risk assessment for connected medical devices: the nemecys approach. In: Joint Proceedings of RCIS 2024 Workshops and Research Projects Track Co-located with the 18th International Conference on Research Challenges in Information Science (RCIS 2024), p. 3674 (2024). https://ceur-ws.org/Vol-3674/RP-paper6.pdf
5. Erdogan, G., Gonzalez, A., Refsdal, A., Seehusen, F.: A method for developing algorithms for assessing cyber-risk cost. In: 2017 IEEE International Conference on Software Quality, Reliability and Security (QRS), pp. 192–199 (2017). https://doi.org/10.1109/QRS.2017.29
6. Erdogan, G., Refsdal, A.: A method for developing qualitative security risk assessment algorithms. In: Cuppens, N., Cuppens, F., Lanet, J.-L., Legay, A., Garcia-Alfaro, J. (eds.) CRiSIS 2017. LNCS, vol. 10694, pp. 244–259. Springer, Cham (2018). https://doi.org/10.1007/978-3-319-76687-4_17
7. European Union: Regulation (eu) 2017/745 of the european parliament and of the council of 5 april 2017 on medical devices, amending directive 2001/83/ec, regulation (ec) no 178/2002 and regulation (ec) no 1223/2009 and repealing council directives 90/385/eec and 93/42/eec (2025). https://eur-lex.europa.eu/eli/reg/2017/745/oj/eng
8. Lund, M.S., Solhaug, B., Stølen, K.: Model-Driven Risk Analysis: The CORAS Approach. Springer, Heidelberg (2010). https://doi.org/10.1007/978-3-642-12323-8
9. Meland, P.H., Tokas, S., Erdogan, G., Bernsmed, K., Omerovic, A.: A systematic mapping study on cyber security indicator data. Electronics **10**(9), 1092 (2021). https://doi.org/10.3390/electronics10091092
10. MITRE: MITRE ATT&CK (2025). https://attack.mitre.org/
11. NEMECYS: Deliverable D2.1 - Risk Benefit Schemes (initial) (2025). https://nemecys.eu/resources/public-deliverables/
12. NEMECYS: NEMECYS - New Medical Cybersecurity Assessment and Design Solutions (2025). https://nemecys.eu/
13. Silvestri, S., Islam, S., Amelin, D., Weiler, G., Papastergiou, S., Ciampi, M.: Cyber threat assessment and management for securing healthcare ecosystems using natural language processing. Int. J. Inf. Secur. **23**(1), 31–50 (2024). https://doi.org/10.1007/s10207-023-00769-w

A Secure Data Transmission Protocol for Cardiac Implants to Mitigate Intrusive Cyber Attacks

Anisha Mitra(✉) and Dipanwita Roy Chowdhury

Department of Computer Science and Engineering, Indian Institute of Technology, Kharagpur, India
mitraanisha.15@gmail.com, drc@cse.iitkgp.ac.in

Abstract. The medical implant communication system is a critical cyber-physical system with significant security concerns due to its crucial role in patient health and safety. Implant devices integrate wireless technology, exposing them to potential cyber threats. This paper presents a secure data transmission protocol incorporating cryptographic primitives with a lightweight authentication scheme. This is specifically designed for secure communication in an Implantable Cardioverter Defibrillators (ICD) environment. We present a set of practical attacks and their countermeasures in the ICD environment using the proposed authentication scheme.

Keywords: Medical Device Security · Resource Constraints · Attacks on ICD · Authentication

1 Introduction

Implantable Medical Devices (IMDs) have transformed healthcare by providing cutting-edge treatments for a wide range of crippling and potentially fatal illnesses. The most widely used medical implants include implantable cardioverter defibrillators (ICDs), deep brain stimulation devices, insulin pumps, cochlear implants etc., crucial in monitoring and treating critical health conditions. Researchers have identified that IMDs are vulnerable to security risks that could have life-threatening repercussions. Wireless telecommunication enables new-generation medical implants to transmit real-time health data to healthcare providers, facilitating remote monitoring and timely interventions using a computer-like device called device programmer. Cybercriminals exploit these facilities in IMDs, launching passive attacks to extract patient-specific information or active attacks like replaying older communication, jamming, and sending malicious commands to the IMD. The prevalence of cardiac diseases worldwide has made ICDs one of the most popular IMDs with advanced connectivity and functions. In this work, we primarily focus on the security requirements in the ICD environment, considering the present wireless interfaces. However, safeguarding communication between the IMD and its authorized programmer is

I. Praça et al. (Eds.): EICC 2025, CCIS 2500, pp. 307–323, 2025.
https://doi.org/10.1007/978-3-031-94855-8_20

challenging. Resource constraints with tight power, limited battery capacity, the absence of any direct access interface after implantation, and inherent tensions between safety and security make using traditional cryptography difficult. Thus, there is an urgent need to provide relevant security solutions suitable for specific implant environments. Considering the restricted environmental conditions, this paper proposes a lightweight security model with an efficient authentication scheme for securing time-critical and life-saving ICD-programmer communication.

The rest of the paper is organized as follows. Section 2 deals with the preliminary discussion about the concerned ICD environment and explains the possible security threats. Section 3 presents a secure data transmission protocol and a lightweight authentication scheme for the ICD-programmer communication environment. The necessary security analysis concerning the observed attack scenarios in the ICD environment is discussed in Sect. 4, and we conclude in Sect. 5.

Our Contribution

- In this paper, confidentiality and integrity requirements for functional model based ICD communication are identified, and an optimized message format is proposed to establish an efficient data transmission structure.
- A lightweight authentication scheme is proposed for the ICD-device programmer communication.
- A set of practical security attacks has been presented in the ICD-device programmer communication environment.
- It has been shown how to mitigate the identified intrusive cyberattacks in the ICD environment using the proposed cryptographic primitive.

2 Background Study

This section briefly describes one of the most widely used IMDs, implantable cardioverter defibrillators (ICDs), highlighting its key functionalities in serving patient needs. Additionally, it identifies the existing security threats in the ICD environment based on current literature, including risks of unauthorized access, data breaches, and potential malicious interference, emphasizing the need for robust security measures to protect patient safety and privacy.

ICDs are battery-powered devices that deliver electric shocks to the patient's heart if the heartbeat is too fast [4].The recent ICD models feature wireless telemetry for remote monitoring.

Medical implants, placed through surgical procedures, face resource constraints. It is difficult to use traditional security approaches in the ICD environment due to valid communication and resource limitations, as discussed below.

Resource Constraints: ICDs have limited computational power, a battery life of 7 to 10 years, and a memory capacity of only a few megabytes. Therefore, ensuring ICD security requires an efficient, lightweight, secure protocol that preserves battery longevity while supporting critical functions such as monitoring heart rhythm and delivering lifesaving shocks.

Communication Constraints: Communication transmission in the IMDs causes radiation and power consumption, threatening patient health. Traditional security techniques do not work efficiently for IMDs as their wireless channels are inherently weak due to body-induced fading and low transmission power [8].

2.1 Possible Security Threats

Our work primarily focuses on security aspects in the ICD environment. These aspects may be vulnerable to relevant cyberattacks. Research progress identifies major cyberattacks primarily targeting wireless interfaces in [4,5,8,9], which are mentioned below.

Data Manipulation Attack: ICD configurations are susceptible to data manipulation attacks, allowing malicious programming due to compromised functional parameter values or modification in the correct medical log history. This may lead to incorrect treatment delivery [5,9].

Resource Depletion Attack: Resource-constrained ICD environments are severely vulnerable to resource depletion attacks. Attackers aim to rapidly deplete ICD resources (battery power, memory) to reduce the device's lifetime from several years to a few weeks [6]. Forced authentication attempts from malicious devices near the target ICD cause unnecessary battery consumption and generate numerous security logs.

Communication Attacks: ICDs over insecure channels are susceptible to eavesdropping, Denial of Service (DoS), Man in the Middle (MITM), and Replay attacks. Adversaries capture patient data and manufacturer details, disrupt communication to deplete resources, replay intercepted messages, and use MITM-type attacks to store or block previously executed commands [3].

Other attacks, such as software injection or side-channel vulnerabilities, can replace device firmware, execute malicious code, and create unusual data leakage scenarios.

2.2 Existing Literature Survey

The research in IMD security has identified relevant attack scenarios, but providing comprehensive solutions in such resource-constrained environments remains challenging, necessitating the development of a suitable security model. Our literature survey focuses on two different types of research in this field, as discussed below:

Security Issues of IMDs: A few decades ago, researchers identified security vulnerabilities in medical implant communication interfaces that require immediate attention from the research community. Research like [9] explored the privacy and security design goals of IMDs in real-time deployment scenarios, contrasting them with safety and utility goals without providing definitive countermeasures. In [7], researchers conducted a risk assessment on cardiac implants using actor-based, scenario-based, and combined analysis, but no countermeasure was proposed to prevent identified attacks. Another significant work in [3] highlighted the emerging IMD technologies and innovative ideas for new designs and implementations to improve the security of IMDs. In a recent work [2], N.M. Istiak et al. discussed an analysis of cardiac implant security issues using the STRIDE threat model concerning foundational security principles. They have also presented some trivial mitigation techniques.

Countermeasures: Over the years, research in the IMD security domain has made promising progress, leading to the proposal of various security models.

Halperin et al. in [1] demonstrated attacks on IMD communication protocols using general-purpose software radios, providing comprehensive solutions using radio frequency power harvesting to keep patients in the loop. Another type of security research in this domain employs an external device as a proxy between IMD and programmer communication like IMDGuard [12]. But, one pertinent concern is how secure that external gadget is. The authors in [4] demonstrated that adversaries can easily reverse-engineer a proprietary IMD communication protocol, proposing short and long-term mitigation measures and a new key agreement protocol. In one of the well-defined research [11], a data transmission protocol was proposed considering resource constraints in IMD. This work also employed a continuous authentication approach, which fits well in the IMD environment.

Most of the proposed security models lack a balance between security and computational overhead. Our work focuses on proposing a relevant security solution that is supposed to fill the gap between security goals and environment-specific constraints.

3 Data Transmission Protocol for ICD-Device Programmer Communication

This section proposes a lightweight, secure data transmission protocol for the ICD and its legitimate device programmer communication. Our extensive research on the resource constraints within the ICD environment identifies the necessity for authenticated communication. ICDs administer appropriate therapy based on correct configuration parameters, and even a minor error in these parameter values can result in the device delivering incorrect and potentially life-threatening treatment. Therefore, secure and authenticated data transmission in the ICD environment is essential. Our protocol ensures comprehensive, secure data communication between the ICD and its programmer, particularly on authenticated, unmodified message delivery. The next sections give a detailed description of the proposed protocol, depicting the proposal of an optimised data transmission structure and a lightweight authentication scheme to ensure secure data transmission in the ICD environment.

3.1 Security Goals

Before depicting the proposed protocol, this section discusses the security requirements in the ICD environment, considering the device's operational model. ICDs' resource-constrained, vulnerable environment necessitates establishing security goals costing fewer resources. Sensitive data such as patient's health records, patient-specific device configuration information, ICD monitoring and audit information must be protected from attackers during transfer and storage. Security requirements vary depending on the data transmitted between an ICD and the device programmer. Based on the pseudo-command set proposed in [5], the integrity and confidentiality requirements for the transmitted information are discussed below:

Confidentiality Requirement: The pseudo command set for functional instructions proposed in [5] includes certain pseudo commands that contain important parameter values and need to be kept confidential. Those patient's health-specific parameter values may compromise the patient's safety if disclosed to unauthorised entities other than a legitimate medical practitioner, patient and the patient's trusted sources. An example includes one pseudo command proposed in [5]: **("VF1 REL dose-v1")**. Here, **'dose-v1'** parameter specifies the quantity of electric shock required to deliver after detecting the status of Ventricular Fibrillation condition in the patient's heart. This parameter value depends on the current health condition of the concerned patient, and if the value is released to adversaries, it may be used to launch a replay or data manipulation attack, endangering the patient's life. Thus, to set certain required ICD configurations, messages containing those parameterized instructions must be transmitted securely using encryption and authentication schemes.

Integrity Requirement: Functional instructions without any parameter values are less likely to be kept confidential. The pseudo command set proposed in [5] includes instructions for changing ICD operating mode, aborting or starting specific therapies, or containing generic information not specific to a patient. An example includes one pseudo command: **("START RSHK")**, which is used for releasing rescue shock to the patient's heart. Such instructions fall in this category and must be transmitted without modification. Transmission of correct instruction ensures adequate configuration change at ICDs end.

Thus, for these types of instructions, data is not required to be transmitted in encrypted form. Only data integrity authentication of the raw message will ensure correct instruction delivery without any modification.

In a later section, we propose a security model for the ICD communication environment by dividing the functional instructions into two groups, one requiring integrity and another seeking confidentiality and integrity. Regarding low-end devices like ICDs, this greatly affects computation timing and costs.

3.2 Adversary Model

We consider adversaries with malicious goals to incorrectly configure the ICD and deliver the wrong therapies to the patient. The adversary succeeds if she/he can reverse engineer the ICD-programmer communication protocol and interact with the ICD without being the trusted party. The following considerations are taken into account while proposing a secure data communication protocol in the ICD-device programmer communication environment.

- An adversary cannot get access to the authorized device programmer specific for a particular ICD communication. However, an adversary with an available commercial ICD programmer does not ensure authorized access.
- The medical personnel registered for access to a particular ICD are considered trustworthy.
- Adversaries can use commercial equipment like software radio, oscilloscope, directional antenna, etc., to eavesdrop and reverse engineer the existing ICD-programmer communication protocol.

3.3 Protocol Description

In this section, we emphasize the data transmission protocol for secure ICD-programmer communication. Here, a single session refers to each medical check-up of the patient carrying an ICD. Furthermore, we describe an optimized data transmission structure, as shown in Fig. 2. The legitimate message to be sent between the ICD and the programmer device is contained in this structure. In a later section, we discuss the optimized 32-bit message format obtained by considering ICDs parameterized functional behaviour. This optimized message structure includes additional components, such as an ID and a nonce, to enhance secure communication. A detailed explanation is provided below.

ID: The proposed data structure includes the registered ID of the medical practitioner performing the implant procedure and responsible for the patient's regular health checkup. The uniqueness of this ID is justified by Indian national medical policy that requires each medical practitioner to register in the National Medical Repository (NMR) by submitting their qualification certificates [10]. Once verified, a unique registration ID is assigned to the practitioner. In our proposed protocol, a 32-bit encoded representation of the unique registration ID is used. After the initial implantation procedure, it is assumed that this ID is stored in the ICD storage. The unique ID of each ICD(ID_{ICD}) is also stored in both the device programmer and ICD storage during implantation.

Cross-Clinic Accessibility: Publicly accessible NMR data includes a doctor's name and qualifications, while sensitive information like unique ID access is restricted to the Ethics and Medical Registration Board (EMRB), state medical councils and doctors. If a patient with an ICD is compelled to visit a different clinic for an emergency, a registered doctor can retrieve the original doctor's registration ID from the NMR to facilitate secure communication between the device programmer (DP) and the ICD.

Nonce: We consider a 32-bit nonce. In the initial stage of the communication, we assume that before the first message transmission takes place, a 32-bit random nonce N_1 is generated. For every message transmission, the encrypted versions of both the doctor's registered ID and nonce are transmitted to the ICD. The decrypted version of N_1 is stored as N_2 at ICD's end. For the next upcoming messages, the programmer increments the nonce N_1 by 1, and at the ICDs end, a message is accepted only when $N_1 > N_2$.

The working principle of the protocol is described below:

- R_{id} is the Registration ID of the medical practitioner linked and registered with the concerned ICD who initially implants it in the patient's body and uses a device programmer (DP) to configure the ICD initially and afterwards.
- For each message, the 32-bit plain text instruction (P) or cipher text instruction (C) is transmitted from the DP's end.
- The communication session between the programmer and the ICD begins with an initial message transmission where a 32-bit random nonce is generated. However, each message contains an encrypted version of R_{id} and nonce to keep the secrecy of both the medical practitioner's registration ID and nonce intact from the adversary.
- For each subsequent message after the initial one, the programmer increments the nonce value by 1.
- ICD receives $((E_k(R_{id})),(E_k(Nonce)),P/C,Tag)$. Tags are generated either by considering cipher text(C) or raw message(P) based on the instructions considered to be transmitted. $E(R_{id})$ and $(E_k(Nonce))$ are the encrypted versions of R_{id} and nonce respectively. From the initial message transmission onwards, the received nonce is decrypted and compared against the previously stored value in the ICD's storage.
- ICD verifies R_{id} with its stored value after decrypting encrypted version $E(R_{id})$. If it matches, a message is accepted only when the decrypted nonce is greater than the previously stored value.
- On receiving the correct nonce value, the tag is re-generated at the ICDs end and compared with the received one. If it matches, accept the message either after decryption using a key or accept the direct plain text attached to the received data, as it is applicable; otherwise discard the message.

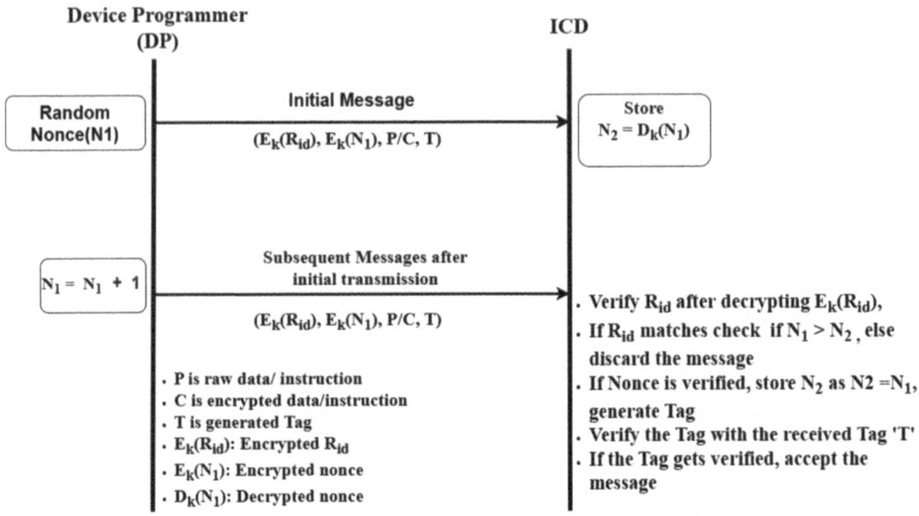

Fig. 1. Data Transmission Protocol for ICD-Programmer Communication

3.4 Data Transmission Structure for ICD-Programmer Communication

ICD is designed to detect potentially harmful ventricular arrhythmias and provide shock treatment. The detection windows of various heart conditions and the treatment sequence can be individually programmed [5]. A pseudo-command set has been proposed in [5] based on identified programmed instructions for market-available ICD manuals. Adversaries may target legitimate programmer devices and use corrupted ones to make incorrect changes to the device configuration by modifying command parameters maliciously. This situation may discontinue correct treatment delivery. Based on the proposed pseudo-command set in [5], we propose a message structure for usage in ICD-programmer communication. An optimized data transmission structure significantly affects communication and computation costs, particularly in a critical environment with limited resources. Figure 2 presents the data encoding procedure utilised by us to propose the 32-bit message structure shown in the same diagram. In this process, we divide the pseudo commands into six parts for bit representation and considered possible parameter values to be used for the parameters present in various pseudo commands. We have shown in the right part of Fig. 2, how the 32-bit message structure is obtained by considering the specific number of bits required to represent different types of values possible for each part of the pseudo commands.

Example: Consider one of the pseudo commands shown in the second row of the table presented in Fig. 2:

Pseudo Command	Instruction Type (3 bits)	Field 1 (5 bits)	Counter (1 bit)	Type (2 bits)	Field 2 (1 bit)	Value (16 bits) Min	Max
SET BT U	SET(000)	BT(00000)	0	01	⊠	60	100
SET VT1 time1	SET(000)	VT1(00001)	0	01	⊠	10	12
SET VT1 timeR	SET(000)	VT1(00001)	0	01	⊠	10	12
SET VT1 timeF	SET(000)	VT1(00001)	0	01	⊠	15	30
SET VT1 time2	SET(000)	VT1(00001)	0	01	⊠	10	12
SET UBT value	SET(000)	UBT(00011)	0	00	⊠	Specific Integer	
SET NBT v	SET(000)	NBT(00100)	0	01	⊠	50	70
SET PBT vp	SET(000)	PBT(00101)	0	00	⊠	50	
SET RH valueH	SET(000)	RH(00110)	0	00	⊠	Specific Integer	
SET MST rate	SET(000)	MST(00111)	0	00	⊠	Specific Integer (depends on 'U')	
SET PCM mode	SET(000)	PCM(01000)	0	00	⊠	Ex:VVI,AAI,VOO.	
SET PA valuep	SET(000)	PA(01001)	0	00	⊠	1.5	
SET LVA valueL	SET(000)	LVA(01010)	0	00	⊠	1.5	
SET RVA valueR	SET(000)	RVA(01011)	0	00	⊠	1.5	
SET PW valuePW	SET(000)	PW(01100)	0	00	⊠	0.5	
SET LVW valueLW	SET(000)	LVW(01101)	0	00	⊠	0.5	
SET RVW valueRW	SET(000)	RVW(01110)	0	00	⊠	0.5	
SET PT valuePT	SET(000)	PT(01111)	0	01	⊠	2.0	5.0
SET VT1 GT value1	SET(000)	VT1(00001)	1	01	1	140	170
SET VT2 GT value2	SET(000)	VT2(00010)	1	01	1	170	200
SET VF GT valueF	SET(000)	VF(10000)	1	01	1	200	250
VF1 REL dose-v1	VF1(001)	REL(10001)	0	01	⊠	10	80
VF2 REL dose-v2	VF2(001)	REL(10001)	0	01	⊠	10	80
VF3n REL dose-vn	VF3n(001)	REL(10001)	0	01	⊠	10	80
ENL DET ON/OFF	ENL(010)	DET(10010)	0	00	⊠	OFF/ON	
DISABLE BEEP	DISABLE (011)	BEEP(10011)	0	10	⊠	⊠	
SELECT MAXS valuesh	SELECT (100)	MAXS(10100)	0	00	⊠	5	
SELECT SHOCK valueS	SELECT (100)	SHOCK(10101)		00	⊠	15	
START RSHK	START (101)	RSHK(10110)	0	10	⊠	⊠	
START DEFT	START (101)	DEFT(10111)	0	10	⊠	⊠	
ABORT RSHK	ABORT (110)	RSHK(10110)	0	10	⊠	⊠	

⊠ denotes don't care bits, which are not considered

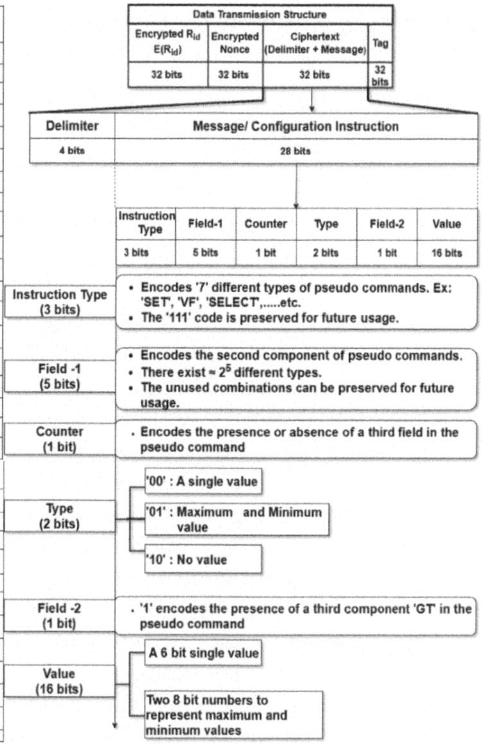

Fig. 2. Data Encoding Process & Data Transmission Structure

Pseudo Instruction: `SET VT1 time1`-This instruction denotes the configuration instruction for an ICD to set the time interval for tachycardia detection in a patient's heart.

Using the data encoding scheme, this command is coded in the following way:

- 'Instruction type' part `SET` is coded as `000`.
- 'Field1' part `VT1` is coded as `00001`.
- 'Counter' part is coded as `0`.
- 'Type' part is coded as `01` as this pseudo instruction consists of parameter value in range.
- Parameter value `time1` is coded as a 16-bit binary number representing its range.

Details about the rest of the pseudo commands can be found in the referred work [5]. In this 28-bit message format, four more bits are added as delimiters. The last bit of the delimiter is reserved to identify whether the received data transmission structure consists of an encrypted message or a raw message. Thus, a 32-bit standard message size is proposed for ICD-programmer communication. The proposed security model is based on two major security primitives: lightweight encryption and authentication. These primitives utilize the specified

32-bit message structure. In this work, considering a 32-bit standard lightweight encryption scheme, our focus is to design a lightweight authentication scheme.

Key Management: The secure communication between ICD and the device programmer(DP) requires two keys, one for authentication and another for encryption. We employ the dynamic key concept, which is efficient in resource utilization and provides a higher level of security than a session or traditional fixed key. Key management comprises three schemes, which are briefly outlined below.

Key Generation: Initially, after the implantation procedure ends, a master key K_M is generated at DP's end and pre-shared with the ICD. Master key generation includes usage of the individual unique device IDs 'ID_{ICD}' and 'ID_{DP}', of ICD and DP, respectively. This key is stored in both the ICD and DP's storage and gets regenerated after a predetermined number of medical check-up sessions.

Key Update: For every communication session between this ICD and the DP, a random dynamic key K_d is generated from the master key on both sides using a lightweight random key update function. K_d is required to get updated after being used for a fixed number of messages (say 'm') in a single communication session. It differs from the traditional fixed key or session key in the granularity of the key's usage. A session key is utilised for a communication session, and a traditional fixed key may be used across multiple communication sessions. However, for a dynamic key, when the predecided threshold for messages is crossed, a new dynamic key must be generated even in a single session.

Key Revocation: Medical implants are crucial for critical health care, requiring secure communication and immediate key revocation in various emergency situations involving patients, practitioners, and hospital premises. The master key must be changed in three scenarios: if it is compromised or stolen, when there is a change in the medical practitioner or DP, or when the ICD-carrying patient relocates or switches their regular healthcare provider.

Our research work suggests that traditional resource-consuming ciphers should be avoided in the ICD environment due to their low-powered, low-memory nature. Traditional lightweight ciphers like NIST standard lightweight ARX-based ciphers are unsuitable for constrained environments. New-generation lightweight ciphers like *Shadow*, which combine Feistel structure and ARX operations, are recommended for ICD environments.

However, in a functional model and parameter-based, ICD's operational environment must emphasize more on protecting data integrity and authenticity. We introduce a lightweight tag generation scheme relevant to resource-constrained ICD environments and provide the required security analysis in subsequent sections.

3.5 A Lightweight Authentication Scheme

Authentication of secret messages prevents revelation, deception, and modification of message content. Through ICD-programmer communication, medical practitioners generally intend to provide the correct therapy to the patient. It is necessary to authenticate the message to prevent any alterations to the valid configuration instructions transmitted through transmission content. At the programmer's end, before transmission begins, the authentication tag is computed on the previously mentioned data transmission structure utilizing a lightweight scheme, which is discussed in this section. The tag is received and recomputed at the receiver (ICD) end. If the recomputed tag matches the one sent, the received message is accepted, and the instructed change is incorporated into the current ICD configuration. Thus, if any adversarial modification is made to the transmitted data, it will lead to immediate rejection of the received message. This prevents any illegitimate, unnecessary configuration change to the ICD settings.

We present a novel approach to generate an authentication tag for the proposed protocol based on simple ARX operations.

Tag Generation

To generate an authentication tag, we define a new lightweight scheme based on bitwise and bytewise circular shifts, as well as bitwise exclusive or(XOR) operation. These resource-efficient operations are applied to various components of the proposed data transmission structure to generate the authentication tag. We define some trivial operations utilised during tag generation as depicted below:

Circular Left Shift: Define **LCR(X,Y)** as left rotate string X by Y bit positions in a circular way.

Bytewise Circular Left Shift: Define **BLCR(X,Y)** as left rotate string X by Y byte positions in a circular way.

Algorithm 1 depicts the proposed authentication tag generation scheme.

Algorithm 1 Tag Generation Scheme

Input:
- N: Nonce
- C: Cipher Text, P: Plain Text
- K_d: Dynamic Key
- ID_{ICD}: least significant 32-bits of the unique ID of the ICD

Output: Authentication Tag T_A

// T_1, T_2:16-bit Registers

// T_3: A 32-bit Register

$T_1 \leftarrow$ Most Significant 16-bits of ($\mathbf{LCR}(ID_{ICD}, 1)$)

$T_2 \leftarrow$ Least Significant 16-bits of ($\mathbf{BLCR}(Nonce, 1)$)

$T_1 \leftarrow$ (Most Significant 16-bits of K_d) $\bigoplus T_1$

$T_2 \leftarrow$ (Least Significant 16-bits of K_d) $\bigoplus T_2$

$T_3 \leftarrow T_1$ Append T_2

$\mathbf{T_A} \leftarrow T_3 \bigoplus \mathbf{M}$

// $\bigoplus$: Bitwise XOR

// M is either C or P, depending on the type of instruction to be transmitted

---

Performance Evaluation of the Proposed Authentication Scheme
The tag generation algorithm discussed previously is proposed for the ICD-device programmer communication environment, which is extremely resource-constrained. The algorithm should consume less resources in terms of power and memory to be implemented in an ICD environment. We have simulated our proposed algorithm in a low-power microcontroller environment, which is quite similar to the ones used in ICDs. Below, we present experimental results on the implementation of the proposed scheme, focusing on cryptographic security and resource consumption aspects.

Avalanche Effect: In cryptography, the avalanche effect ensures that a minor change in input, such as flipping a single bit, results in a significant and unpredictable change in the output. This property enhances security by making it difficult for attackers to predict how alterations in plaintext or keys affect the ciphertext or authentication tag, thereby safeguarding against cryptanalysis. A good cryptographic algorithm should exhibit a high degree of avalanche effect, which means that any alteration in the input should cause a significant change in the output. Table 1 illustrates the results of the avalanche effect for the proposed authentication scheme. It shows experimental results for ten random plaintexts and presents the number of bits changed in the generated tag if one, two and three bits are changed in the plaintext, respectively. We observe from the presented result that, for the majority of plaintexts, alterations in one, two, or three bits result in modifications to over half of the bits in the produced tag.

Table 1. Avalanche Effect

Plaintext (Hex)	Number of bits changed in Tag		
	1-bit flip	2-bit flip	3-bit flip
4334192D	17	16	16
41A90C37	18	20	14
62685024	18	18	16
6F6B0E52	17	19	14
418E3352	14	18	17
462E5B0F	16	19	16
79F6113D	16	17	14
1BEC1880	17	16	16
667E7C74	16	17	18
44442542	17	21	16

Microcontroller Simulation: ICDs are patient-specific personal medical devices that are extremely vulnerable to any unauthorized access. To preserve patients' health security, thus manufacturers disclose only trivial information regarding ICD resources like processors, memory capacity, etc. We choose a representative mid-range microcontroller of **MSP430** series, similar to the one used in resource constraint medical implants to simulate our proposed authentication scheme. This model is an ultra-low-power microcontroller that incorporates 64KB FRAM (Ferroelectric RAM) for energy-efficient, high-speed, and non-volatile data storage, complemented by a 16-bit RISC architecture, and different low-power modes are designed for effective processing in resource-limited settings. Its low-power modes and integrated peripherals render it suitable for battery-operated and energy-harvesting applications such as IoT, medical devices, and industrial sensing.

Experimental Set up: The experimental setup comprises a Texas Instruments (TI) microcontroller model and a few peripherals. Table 2 gives detailed information about the implementation set up. Code Composer Studio (CCS), TI's integrated development environment (IDE) tailored for their microcontrollers, is used. CCS offers comprehensive support for MSP430 devices, including simulation capabilities and extensive debugging tools.

Table 2. Experimental Set up Details

Microcontroller Model	Software Development Environment	Connection Interface
MSP430FR5969 Launchpad	Code Composer Studio v12.8.0	USB Cable

Table 3 depicts the implementation details of our proposed tag generation algorithm in the considered microcontroller environment. The considered microcontroller is capable of running at a frequency range of 16 MHz. However, to simulate a resource constraint environment, we set up the microcontroller frequency at 1 MHz. CCS provides various code optimization levels to balance code size, execution speed, and power efficiency. We simulated our algorithm for three different optimization levels, namely, Opt1(Global), Opt2(Register level) and Opt3(Whole program), respectively, to measure the FRAM consumptions. EnergyTrace, provided by TI with CCS, is a real-time power analysis tool that monitors energy consumption in MSP430 microcontrollers, and it must be run in free-run mode to capture accurate power data without debugger-induced interruptions. The simulation results indicate minimal memory usage, power consumption, and lower CPU cycle utilization.

It is evident that the proposed authentication tag generation scheme is efficient in terms of resource consumption(memory, power, etc.) and will be suitable for the target resource-constrained environment.

Table 3. Implementation on MSP430FR5969 Microcontroller

Operation	Cycles	FRAM (Bytes)			EnergyTrace Measurement (Sampling Time: 1 Sec)		
		Opt1	Opt2	Opt3	Energy (mJ)	Mean Power (mW)	Mean Current (mA)
Tag Generation	158	1454	1482	1464	1.285	1.2283	0.3415

4 Security Against Relevant Attacks

Section 2.1 depicts possible security vulnerabilities identified in the concerned ICD environment. A security model should prevent identified attacks and disallow any unauthorised access.

Design Rationale: The proposed protocol incorporates certain design considerations to satisfy the security and resource constraint criteria of the ICD environment, as discussed below:

- Authentication tag generation scheme do not include R_{id} usage. Even if the attacker gets access to the valid R_{id}, the secrecy of the authentication tag is not violated.
- Use of a random 32-bit number as initial nonce and transmission of its encrypted version reduces the possibility of correct nonce prediction.
- The simulation of the proposed scheme in a similar microcontroller environment, like the ones used in ICDs, ensures the relevance of the scheme.

We have considered the primary security attacks relevant to the ICD environment as identified in [5] and other research findings. Detailed discussion about the prevention of identified attacks utilising the proposed method is depicted below:

Data Manipulation Attack: One of the most dangerous security threats in the ICD environment is described in [5].

Attack Scenario 1: Unauthorized alteration to health-sensitive therapeutic parameter values.

Prevention: Authentication tag is computed on the transmitted data. The data transmission structure consists of parametrized instructions as part of the message. Any modification to parameter values during transmission will result in a failed tag verification at the ICDs end, as shown in Fig. 3.

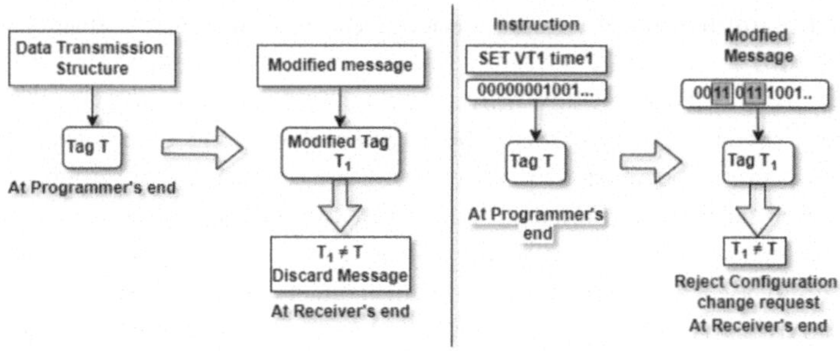

Fig. 3. Authentication scheme thwarting Data Manipulation Attack

Attack Scenario 2: Repetitive execution of Critical Commands.
Ex: **VF1 REL dose-v1**(Shock dose release instruction)
Prevention:

- A 1-bit flag is maintained for every critical instruction of ICD configuration.
- Every time a message containing a critical instruction execution request is received, set the corresponding flag to '1'.
- Keep each set flag value same for a pre-decided 't' time unit.
- A received message with critical instruction whose flag is set, gets discarded. Critical instruction is not executed.
- Flag is reset to '0' after 't' unit time.

Using a flag for each critical instruction prevents the possibility of resource depletion risk involved in the continuous execution of critical instruction.

Eavesdropping Attack
The Security of the encryption and authentication schemes in this proposed protocol depends on the secrecy of the used key. Attackers cannot decode the eavesdropped message content without knowing the secret key. Use of dynamic key K_d which gets updated after every 'm' message transmission prevents key guessing attempt.

Replay Attack
The proposed protocol uses a random nonce for initial message communication. It acts as a timestamp. This nonce gets updated for every individual message during ICD-programmer communication. Thus, for every replayed old message, nonce verification will fail at the ICDs end, leading to message rejection.

5 Conclusion

This paper presents a lightweight authentication tag generation scheme for the resource-constrained ICD environment. It discusses defence against identified attacks. A secure communication protocol for critical ICD-device programmer communication environment must incorporate a lightweight and efficient key management and encryption scheme. Future progress aims to demonstrate the implementation of optimized key management and encryption primitives within the proposed security framework. The proposed scheme currently aligns with the underlying national medical policy. However, due to its reliance on minimal and widely accessible information, the solution can be seamlessly integrated into any existing and emerging frameworks. Future emphasis is placed on the necessity of collaboration with device manufacturers, healthcare institutions, and regulatory boards in adopting the proposed secure framework in the real-time medical world. Real-world integration will facilitate the development of a secure communication environment for not only cardiac implants but also other medical implants equipped with wireless interfaces.

Acknowledgement. This research work is supported by INAE Abdul Kalam Technology Innovation National Fellowship funded by the Department of Science & Technology (DST), Government of India.

References

1. Halperin, D., et al.: Pacemakers and implantable cardiac defibrillators: software radio attacks and zero-power defenses. In: 2008 IEEE Symposium on Security and Privacy (sp 2008), pp. 129–142 (2008). https://doi.org/10.1109/SP.2008.31
2. Istiak Chowdhury, N.M., Hasan, R.: Security analysis of cardiovascular implantable electronic device (cied) using a threat model-based approach. In: SoutheastCon 2023, pp. 81–88 (2023). https://doi.org/10.1109/SoutheastCon51012.2023.10115206
3. Kwarteng, E., Cebe, M.: A survey on security issues in modern implantable devices: solutions and future issues. Smart Health **25**, 100295 (2022). https://doi.org/10.1016/j.smhl.2022.100295. https://www.sciencedirect.com/science/article/pii/S2352648322000307
4. Marin, E., Singelée, D., Garcia, F.D., Chothia, T., Willems, R., Preneel, B.: On the (in)security of the latest generation implantable cardiac defibrillators and how to secure them. In: Proceedings of the 32nd Annual Conference on Computer Security Applications, ACSAC '16, pp. 226–236. Association for Computing Machinery, New York (2016). https://doi.org/10.1145/2991079.2991094
5. Mitra, A., Roy Chowdhury, D.: Unmasking the dominant threat of data manipulation attack on implantable cardioverter defibrillators. In: Proceedings of the 20th Annual International Conference on Privacy, Security and Trust (PST), pp. 1–7. IEEE Xplore (2023). https://doi.org/10.1109/PST58708.2023.10320186
6. Mitra, A., Roy Chowdhury, D.: Guarding the beats by defending resource depletion attacks on implantable cardioverter defibrillators. In: Giri, D., Vaidya, J., Ponnusamy, S., Lin, Z., Joshi, K.P., Yegnanarayanan, V. (eds.) Proceedings of the Tenth International Conference on Mathematics and Computing, pp. 231–243. Springer, Singapore (2024). https://doi.org/10.1007/978-981-97-2069-9_17
7. Ngamboé, M., Berthier, P., Ammari, N., Dyrda, K., Fernandez, J.M.: Risk assessment of cyber-attacks on telemetry-enabled cardiac implantable electronic devices (CIED). Int. J. Inf. Secur. **20**(4), 621–645 (2020). https://doi.org/10.1007/s10207-020-00522-7

8. Rathore, H., Mohamed, A., Al-Ali, A., Du, X., Guizani, M.: A review of security challenges, attacks and resolutions for wireless medical devices. In: 2017 13th International Wireless Communications and Mobile Computing Conference (IWCMC), pp. 1495–1501 (2017). https://doi.org/10.1109/IWCMC.2017.7986505
9. Rushanan, M., Rubin, A.D., Kune, D.F., Swanson, C.M.: Sok: security and privacy in implantable medical devices and body area networks. In: 2014 IEEE Symposium on Security and Privacy, pp. 524–539 (2014). https://doi.org/10.1109/SP.2014.40
10. TOI: "nmc's unique id initiative: Doctors to have it by 2024" (2023). https://timesofindia.indiatimes.com/india/nmcs-unique-id-initiative-doctors-to-have-it-by-2024/articleshow/104868791.cms. Accessed 24 Mar 2025
11. Tutari, V.H., Das, B., Roy Chowdhury, D.: A continuous role-based authentication scheme and data transmission protocol for implantable medical devices. In: 2019 Second International Conference on Advanced Computational and Communication Paradigms (ICACCP), pp. 1–6. IEEE (2019). https://doi.org/10.1109/ICACCP.2019.8883012
12. Xu, F., Qin, Z., Tan, C.C., Wang, B., Li, Q.: Imdguard: securing implantable medical devices with the external wearable guardian. In: 2011 Proceedings IEEE INFOCOM, pp. 1862–1870 (2011). https://doi.org/10.1109/INFCOM.2011.5934987

Strengthening the MDCG Guidance on Cybersecurity for Medical Devices: A Legal Approach

Oguzhan Yesiltuna(✉), Ruoxin Su(✉), Ashwinee Kumar, and Paul Quinn

Health and Ageing Law Lab (HALL), Law, Science, Technology and Society (LSTS) Research Group, Faculty of Law and Criminology, Vrije Universiteit Brussel, Pleinlaan 2, 1050 Elsene, Brussels, Belgium
{oguzhan.yesiltuna,ruoxin.su}@vub.be

Abstract. The increasing digitisation of healthcare, including the integration of connected medical devices (MDs) into healthcare systems, and the growing number of cyber threats necessitate robust cybersecurity measures for MDs to ensure patient safety, data integrity, and system functionality. The MDCG guidance on cybersecurity for MDs plays a pivotal role in aligning industrial regulatory compliance with technical standards across the EU. However, the current guidance lacks currency and consistency with the rapidly evolving EU legal framework, including data protection regulations, cybersecurity regulations, and the AI Act. Adopting a legal approach, this paper identifies the relevant significant shortcomings, examines the implications of legislative developments, and provides targeted recommendations to align the guidance with current cybersecurity frameworks. The analysis highlights the urgency of strengthening the guidance and argues that the strengthened guidance would better support stakeholders, particularly manufacturers, in safeguarding MD cybersecurity through enhancing legal clarity and practical applicability.

Keywords: Connected medical devices · MDCG · Cybersecurity · MDR · IVDR · GDPR · Data Act · EHDS · Cybersecurity Act · NIS2 Directive · Cyber Resilience Act · AI Act

1 Introduction

The Medical Device Regulation (MDR) [1] and the *In Vitro* Diagnostic Regulation (IVDR) [2] establish a robust framework for regulating medical devices (MDs) and *in vitro* diagnostic MDs (IVDMDs) within the EU. Although these regulations do not explicitly refer to *cybersecurity*, their Article 5 requires manufacturers to ensure that MD and IVDMDs meet the general safety and performance requirements set out in Annex I implicitly covering the cybersecurity aspects of such devices. The Medical Device Coordination Group (MDCG), an

I. Praça et al. (Eds.): EICC 2025, CCIS 2500, pp. 324–337, 2025.
https://doi.org/10.1007/978-3-031-94855-8_21

expert body set up by the European Commission (EC) to ensure effective and harmonised implementation of the MDR and IVDR, provides further practical guidance on how these general requirements apply to the specific context of cybersecurity, as demonstrated by the Guidance on Cybersecurity for Medical Devices (MDCG 2019-16) [3]. It delineates the necessary measures that manufacturers must adopt to guarantee the security of their devices against cyber threats throughout the product lifecycle. Despite being an important initiative, the MDCG 2019-16 seems to be falling short of the evolving EU legal frameworks.

This paper, therefore, adopts a legal approach and aims to propose recommendations for a potential revision of MDCG 2019-16. It proceeds as follows: Sect. 2 briefly analyses key reasons for the imperative to revise the guidance. Sections 3–5 identify and analyse the relevant instruments within the evolving EU legal frameworks governing MD cybersecurity, revealing a significant gap in the MDCG 2019-16. Following this analysis, Sect. 6 offers several recommendations designed to improve the effectiveness and timeliness of the guidance. Finally, Sect. 7 presents authors' conclusions.

2 The Imperative to Revise MDCG 2019-16

The MD cybersecurity landscape continues to be a dynamic and evolving area. The growing interconnectedness of MDs expands the cyberattack surface for malicious actors and requires manufacturers to be more vigilant about security measures. Cyber threats such as ransomware attacks, unauthorised access, or vulnerability exploitation can be lethal when they directly target MDs [4,41]. Additionally, as MDs become more integrated into complex hospital networks, their exposure to cyber risks escalates, requiring more robust and adaptive cybersecurity measures to protect both device functionality and sensitive patient data [5]. This has recently been the subject of high-level attention. The WHO Director-General has urged global action on cybersecurity threats to health systems, highlighting the need for coordinated efforts [6]. The EC also unveiled an Action Plan [7] specifically designed to protect hospitals and healthcare providers from cyberattacks. In order to enhance the resilience of healthcare supply chains, Action Plan charges the NIS Cooperation Group, in collaboration with the MDCG, with the execution of a comprehensive security risk assessment addressing both technical and strategic vulnerabilities. Furthermore, it encourages MD manufacturers to *voluntarily* report vulnerabilities, even though it does not address the need to revise the MDCG 2019-16 [7,46].

The MDCG 2019-16, while a soft-law instrument, serves as a crucial document for the industry in the EU, representing the first and only guidance on the regulatory subject of MD cybersecurity [8]. It assists manufacturers on how to fulfil the relevant cybersecurity requirements of the MDR. As one of the most comprehensive guidelines available, it offers advice ranging from secure design and manufacturing principles to post-market surveillance for industry stakeholders [9]. Thus, it is imperative that the guidance be regularly updated to reflect the evolving legal landscape and ensure its practical applicability for manufacturers in addressing new cybersecurity requirements.

Following the publication of MDCG 2019-16 in December 2019 and its subsequent revision in July 2020, the legislative framework governing MD cybersecurity within the EU has undergone substantial changes (*see* Fig. 1). Among the notable developments are new legislative initiatives in the areas of cybersecurity and data regulation, coupled with the nascent regulation of AI. These developments contribute to an overlapping regulatory landscape that is deficient in legislative clarification [10–12]. The MDCG 2019-16 has not been revised to reflect these significant legislative novelties, inevitably resulting in a discrepancy that may impede effective compliance and risk management practices. Moreover, the guidance even failed to sufficiently address certain pre-existing and essential EU laws, including the General Data Protection Regulation (GDPR) [13] and the Network and Information Systems (NIS) Directive [14]. This misalignment calls for a comprehensive review of the MDCG 2019-16 to ensure its consistency with contemporary cybersecurity requirements (Fig. 1).

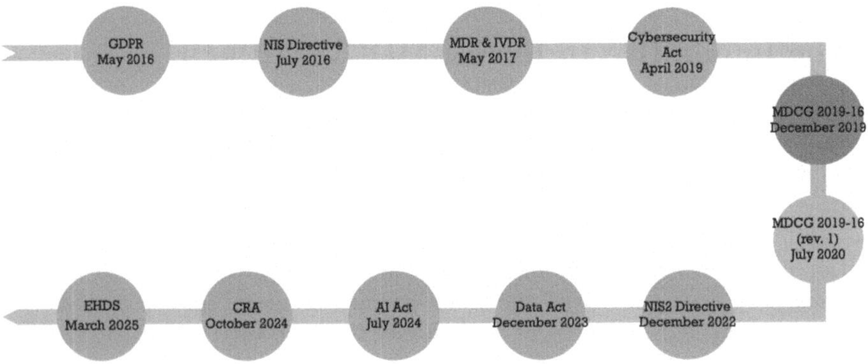

Fig. 1. Legislative developments related to MD cybersecurity (according to publication date in the Official Journal) and the MDCG 2019-16.

Recent scholarship and industry feedback consistently highlight the limitations of current guidance in adequately addressing the emerging regulatory challenges related to MD cybersecurity [12,15–17]. Academic studies have identified inconsistencies in definitions, a lack of clarity regarding stakeholder responsibilities, and insufficient consideration of the interplay among relevant EU laws within the guidance [12,17]. Appropriate guidance is crucial for manufacturers as they navigate the complex legal landscape and the proliferation of new legislative initiatives [15]. The convergence of scholarly critique and industry demand further reinforces the imperative for MDCG to update the guidance.

3 Evolving EU Legal Framework on Data

3.1 GDPR

A key pillar of EU data-related legislation is the GDPR, which came into effect on 25 May 2018 [13]. Although GDPR is often viewed primarily as a data protection regulation, its requirements also have significant implications for the overall cybersecurity of connected MDs. By mandating strict protections for personal data stored within these devices, the GDPR effectively integrates data privacy measures into broader system cybersecurity considerations. It also aligns with the formal definition of cybersecurity in the EU law (*see* Sect. 4.1), which also includes the protection of affected individuals. While MDCG 2019-16 acknowledges the applicability of GDPR, its coverage remains limited and lacks comprehensive guidance on its intersection with other EU legislation, notably the MDR and the IVDR.

The GDPR applies to the processing of personal data relating to an identified or identifiable individual, excluding anonymised data and data that cannot be attributed to a specific person [13]. Notably, data revealing one's health status, genetic data and biometric data are classified as sensitive personal data, subject to strengthened safeguards and stringent processing requirements [13]. This is particularly relevant for connected MDs, which inevitably collect and process sensitive patient data, such as health metrics, vital signs, and wellbeing indicators, to achieve their intended medical purposes. Under the GDPR, data subjects are granted extensive rights, such as the right to access and the right to data portability (only where consent or contract is used as legal basis for personal data processing), reinforcing their autonomy over data processed by controllers. When a MD manufacturer directly manages the processing of data collected by its devices, it becomes a data controller under the GDPR and may need to deal with various data-related requests from users, such as data access. Nevertheless, enabling patient access to device-generated data may inadvertently heighten cybersecurity risks, including potential vulnerabilities to hacking, data interception, or system exploitation [18]. This underscores the urgent need for consistent regulations and interpretive guidance that can help balance individual rights with the security and integrity of MDs.

3.2 Data Act

Following GDPR, another important data regulation applicable to MDs is the Data Act [19]. Entered into force in January 2024 as key component of the EU's digital strategy, the Data Act establishes cross-sectoral rules governing fair access to and use of data, fostering innovation and enabling the EU's digital transformation [20]. It empowers users to access and share IoT-generated data, allows public sector emergency access to private data, protects businesses from unfair contractual terms, facilitates cloud provider switching, and safeguards against non-EU data access requests [20]. Complementing the Data Governance Act (DGA) [21], it promotes interoperability standards and will fully apply from 12 September 2025.

The Data Act provides harmonised rules for fair data access and utilisation across three operational frameworks: business-to-consumer (B2C), business-to-business (B2B), and business-to-government (B2G), with cross-sectoral applicability, including MDs. As a horizontal legislation, it governs data interactions in the healthcare sector while permitting sector-specific vertical regulations to address specialised requirements (e.g., the MDR and IVDR at the EU level) [22]. Recital 14 explicitly integrates MDs into the scope, encompassing health monitoring equipment (e.g. wearables, pacemakers) and diagnostic tools (e.g. MRI scanners), thereby affirming their classification as connected products under the Data Act. The Act delineates dynamic roles, i.e., user, data holder, and data recipient, which may overlap in healthcare settings. For instance, hospitals may act as users *vis-à-vis* device manufacturers while serving as data holders for patient-generated data [23]. Notably, the Data Act's role-based framework diverges from the GDPR's data subject-controller-processor trinity, reflecting the complexity of healthcare ecosystems where institutional users (e.g., healthcare providers) are not necessarily GDPR data subjects. This statutory distinction demonstrates the need for nuanced compliance strategies to address normative overlaps in medical data governance.

3.3 Regulation on the European Health Data Space (EHDS)

The EHDS [24], the most recent addition to the EU data regulations, entered into force on 26 March 2025. Driven by the EC's ambition and commitment to a Health Union, the EHDS builds upon the foundations laid by the GDPR, the DGA, the Data Act, the MDR and the IVDR. It pursues the objectives of enhancing the patient control over their health data and the data exchange for the healthcare delivery (primary use of health data), establishing rules to promote the reliability, security, and interoperability of electronic health record (EHR) systems, and laying down rules for the secondary use of health data [25, 26].

The EHDS is relevant particularly for two reasons: First, certain software components of EHR systems and wellness applications may fall under the scope of MDR and IVDR and thus require certification in accordance with the respective regulations. The EHDS's essential requirements on interoperability set out in Annex II, including the requirements for security and logging, apply to MDs, IVDMDs and wellness applications providing electronic health data to be processed as part of the EHR system, to the extent that their manufacturer claims interoperability with such system. Second, both the minimum categories of electronic health data for secondary use and the purposes for which such data can be processed for secondary use relate to MDs. Personal electronic health data automatically generated through MDs, data from clinical trials, clinical studies, clinical investigations and performance studies subject to the MDR and IVDR, and other health data from MDs are listed among the minimum categories of electronic health data to be made available for secondary use [24]. Electronic health data can be processed for secondary use with the aim of training, testing, and evaluation of algorithms, including in MDs, IVDMDs, AI systems, and digital health applications [24]. Data sharing is permissible only within secure processing environments, the security requirements of which are governed by the EHDS. These two aspects highlight the necessity of integrating the EHDS's interaction with other regulations relevant to MD cybersecurity into the guidance.

4 Evolving EU Legal Framework on Cybersecurity

4.1 Cybersecurity Act (CSA)

The CSA [27], effective since June 2019, aims to enhance the mandate of the European Union Agency for Cybersecurity (ENISA) and establish cybersecurity certification schemes for ICT products (e.g., networked MDs), services (e.g., cloud-based healthcare platforms), and processes (e.g., vulnerability management). The CSA's significance lies in its introduction of a formal, EU-level definition of cybersecurity, encompassing the protection of networks, systems, and individuals from cyber threats. It adopts a rights-based approach distinct from the technical "CIA triad" (confidentiality, integrity, availability) under the NIS Directive, recognising it both as a *praxis* and as a *state*, extending protection beyond systems to include affected users [28]. In April 2023, the EC proposed expanding certification to managed security services, addressing risk management and incident response, which gained provisional approval in March 2024 [29].

A key advancement under the CSA is the establishment of European cybersecurity certification schemes, designed to ensure compliance with security requirements and bolster trust in ICT products. These schemes employ a risk-based approach, classifying assurance levels as basic (documentation review), substantial (security functionality verification), and high (resilience testing against advanced threats) to enhance transparency and consumer confidence. The first major certification is the European Cybersecurity Scheme on Common Criteria (EUCC), which was adopted in January 2024 and became effective from February 2025 [30]. The EUCC scheme will apply on a voluntary basis within the EU and focus on certifying the cybersecurity of all types of ICT products in their lifecycle, such as biometric systems and encrypted storage [31]. Despite these advancements, potential overlaps still exist where the CSA's horizontal scope intersects with sector-specific regulations, such as the MDR. Connected MDs classified as ICT products, for instance, may be required to comply with both the MDR's CE marking, which incorporates cybersecurity risk management provisions, and the CSA's certification framework [32], leading to industry concerns about regulatory fragmentation. Although the MDCG 2019-16 acknowledges the CSA's importance for its EU-wide cybersecurity certification framework for ICT products, the applicability of this framework to MDs and its reconciliation with existing regulations remain unclear, thus complicating implementation.

4.2 NIS and NIS2 Directives

The NIS Directive was the first EU-wide legislative piece focused on improving cybersecurity across the EU. Implemented by EU Member States by 9 May 2018, it targeted operators of essential services and digital service providers. However, its minimum harmonization approach resulted in divergent practices among Member States regarding incident reporting, security requirements, and information sharing [33]. While MDCG 2019-16 recognises the importance of NIS Directive in addressing MD cybersecurity, its engagement with the regulation is minimal, limited to a brief mention of its legislative objectives.

Due to its shortcomings, the EC proposed the NIS2 Directive [34], which entered into force on 16 January 2023. Compared to its predecessor, NIS2 introduces significant changes, including expanded scope, stricter security measures, clearer incident reporting guidelines [35], enhanced Member State cooperation, and reduced Member State discretion [36]. NIS2 aligns with the CSA's definition of *cybersecurity* and replaces *operators of essential services* with *essential and important entities.* Notably, it designates the health sector including healthcare providers and manufacturers of critical MDs during a public health emergency as *high criticality* and the manufacturing sector encompassing MD manufacturers as *critical.* These updated provisions are to be transposed into national law by 17 October 2024, which still entails a degree of minimum harmonisation across the EU.

NIS2, as a horizontal cybersecurity legislative instrument, may regulate the cybersecurity of certain sectors in parallel with sector-specific legislation at the EU level. In this context, NIS2 functions as *lex generalis*, while sector-specific legislation, including MDR, IVDR, and EHDS, acts as *lex specialis* [37]. However, it remains uncertain whether the MDR can be considered functionally equivalent to NIS2, thus raising doubts about its suitability as a *lex specialis* in cybersecurity. Moreover, incident notification requirements overlap between NIS2 and MDR. The potential applicability of the GDPR data breach notification requirements to connected MDs further complicates the regulatory landscape, leading to a proliferation of supervisory authorities [12]. For MD manufacturers, these overlapping regulatory frameworks may lead to divergent notification and reporting requirements, underscoring the need for further clarification and regulatory guidance on this matter.

4.3 Cyber Resilience Act (CRA)

Following the CSA and NIS2, the EC also proposed for a regulation on horizontal cybersecurity requirements for products with digital elements on 15 September 2022. The proposal was finally agreed upon by the EU co-legislators as CRA [38] and it was published in the Official Journal on 23 October 2024. The CRA seeks to improve cybersecurity by setting essential requirements for product design, development, and production, along with vulnerability management obligations for economic operators. It categorises products with digital elements as *important* or *critical*, requiring stricter conformity assessments based on cybersecurity risk. While MDs and IVDMDs are excluded due to existing MDR and IVDR provisions, healthcare-related software and wearable health-monitoring products outside their scope remain subject to the CRA. These digital health products must adhere to essential security requirements, including security-by-design, prevention of unauthorised access, data confidentiality and integrity, and resilience against cyber threats.

5 Nascent EU Legal Framework on AI: AI Act

Entered into force on 1 August 2024, the AI Act [39] is the first comprehensive regulation which sets legally binding standards for the placing on the market, the putting into service, and the use of AI systems. Being a horizontal legislation, it imposes requirements for all AI systems according to their risks, regardless of the sector. As per Article 6, products or safety components of products that are required to undergo a third-party conformity assessment by the Union harmonisation legislation listed in Annex I are classified as high-risk [39]. Annex I covers the MDR and the IVDR. This broadly implies that MDs that incorporate AI to achieve their intended purpose (AI-enabled MDs) with risk class IIa or higher are classified high-risk under the AI Act.

AI Act requires that high-risk AI systems shall be designed and developed in such a way that they achieve an appropriate level of cybersecurity and that they perform consistently in this respect throughout their lifecycle. When compared for AI-enabled MDs, it seems that MDR's general safety and performance requirements, thanks to their technology-neutral formulation, already covers the cybersecurity risks that AI Act aims to address. The AI Act re-imposing requirements that exist under the sector-specific legislation seems to increase the compliance burden for innovators without putting a new layer of protection, at least for AI-enabled MDs. Article 42 of the AI Act introduces a presumption of conformity for high-risk AI systems that have been certified or for which a statement of conformity has been issued under the CSA. However, the lack of a presumption in the MDR similar to that in the AI Act and the voluntariety of the certification mechanism of the CSA do not entirely eliminate the overlap. One solution for the cybersecurity of AI-enabled MDs could have been for the AI Act to adopt the MDR as *lex specialis*, which functions with highly detailed technical standards. Another solution could have been to adopt the model between the MDR and machinery legislation: When a device is covered by both legislation, the requirements of the latter apply to the extent that they are more specific [40].

Further elaboration is required on several other contentious points: This is due to the fact that although both the MDR and the AI Act are part of product safety legislation, they adopt some different terms (e.g. *manufacturer* v. *provider*, *user* (*healthcare professional and lay person*) v. *deployer*; which may be or not the same person) or definitions (e.g. serious incident). For instance, the AI Act's lack of lay person definition may result in users of self-testing devices being subject to cybersecurity obligations [40]. Moreover, the notification mechanisms for a serious incident (e.g. a cyber-attack on an AI-enabled MD) under the two regulations also differ in a way that is likely to create confusion [32].

6 Recommendations for Strengthening the MDCG 2019-16

European medical device regulatory system lacks a centralised authority that oversees the implementation of MDR and IVDR at the EU level. In such absence, the MDCG acts as an expert committee. Both the MDR and the IVDR task the MDCG to contribute to the development of guidance aimed at ensuring effective and harmonised implementation, including the application of the general safety and performance requirements. The MDCG is also tasked with providing advice in the assessment of *any* issue related to the implementation and contributing to harmonised administrative practice with regard to devices in the Member States [1,2]. Nevertheless, the MDCG has only advisory power, which creates the limitation that its guidance is non-binding. It would, therefore, be preferable to clarify the interactions between legal frameworks with legislative revision or the Court of Justice of the European Union's interpretation having binding force. There are some developments related to the former. The EC President von der Leyen's Mission Letter [42] to the Commissioner-designate for Health and Animal Welfare explicitly refers to the Draghi report [43], to stress test the EU legislation and table proposals to eliminate any overlaps and contradictions. Following this, the European Parliament (EP) has also adopted a resolution on the urgent need to revise the MDR [44]. In this context, it should be noted that there are even proposal initiatives, including the establishment of a central office at the EU level [45]. However, as the identified frameworks involve multiple regulations, clarifying the interaction of these regulations may require efforts beyond the MDR revision. Even given these limitations, there is still room to strengthen MDCG 2019-16 by taking into account the existing applicable legal frameworks. Considering these limitations of the MDCG and based on the shortcomings of the MDCG 2019-16 regarding its engagement with the identified legal frameworks, our recommendations are as follows:

Incorporate the Identified Legal Frameworks Governing MD Cybersecurity: The guidance should not merely acknowledge the regulations but should fully integrate them into the MD cybersecurity framework. This should include a detailed explanation of the applicability of the relevant regulation to MDs.

Include or Refer to the Core Concepts Regulated Under the Identified Legal Frameworks: This covers including but not limited to "cybersecurity", "cyber threat", "security-by-design", and "security-by-default" defined under the CSA, "network and information system", "security of network and information systems", and "incident" defined under the NIS2, "data concerning health", "integrity and confidentiality" principle, strengthened protection regime for health data, "data protection by design and by default", and "security of processing" under the GDPR.

Map Cybersecurity Requirements of the Identified Legal Frameworks for MDs and Acknowledge the Overlaps and Inconsistencies Between these Frameworks: MDCG 2019-16 should offer a map of the cybersecurity requirements of all applicable regulations. It should also explicitly acknowledge the overlaps and inconsistencies (i.e. incident reporting requirements) between these regulations. This would serve transparency, which is crucial for building trust and ensuring that stakeholders are aware of the complexities and urge authorities with binding power to take action.

Where Possible, Provide Further Practical Guidance and Contextualised Interpretation in Cooperation with Relevant Actors: Given that the MDCG is established under the MDR and the IVDR, its primary focus should be on providing guidance that supports the implementation of these regulations. However, as MD cybersecurity is a multifaceted concept, it is recommended that MDCG work with other relevant stakeholders, such as the ENISA, the NIS Cooperation Group, the European AI Board, the European Data Protection Board, the European Data Protection Supervisor, and the European Data Innovation Board, to provide an interpretation of horizontal legislation adapted to MD cybersecurity.

Incorporate Dedicated Guidance for Small and Medium-Sized Enterprises (SMEs): To ensure the updated MDCG 2019-16 remains practical and inclusive, it is recommended that dedicated guidance for SMEs should be incorporated, recognizing that these organizations form a substantial part of the MD manufacturing sector yet often operate with limited financial and technical resources. For example, the updated guidance could offer scaled compliance pathways and targeted support (such as accessible technical guidelines, regulatory mapping tools and training materials) for SMEs. Collaborative partnerships among SMEs, industry bodies, and regulators could also be encouraged.

By implementing these recommendations, the MDCG 2019-16 can be strengthened to provide greater clarity for MD manufacturers, promote a more harmonized approach to MD cybersecurity, and ultimately enhance the safety and security of both individuals and MDs in the EU.

7 Conclusion

This paper has demonstrated the critical need for a comprehensive revision of the MDCG 2019-16. The analysis of the evolving EU legal frameworks, encompassing data protection regulations, cybersecurity regulations, and the AI Act, reveals significant gaps and inconsistencies in the current guidance. Such shortcomings may pose a risk to MD security, patienty safety, and innovation economy. Based on this analysis, the paper has proposed several concrete recommendations to strengthen the MDCG 2019-16. These recommendations aim to enhance its legal clarity, currency, and consistency, and to improve its practical utility for manufacturers. By incorporating these recommendations, the MDCG 2019-16 can serve as a more effective tool for promoting robust cybersecurity practices within the MD sector, ultimately enhancing patient safety and public health.

Acknowledgements. This work was supported by the European Commission under the Horizon Europe Program, as part of the project CYMEDSEC (101094218). Views and opinions expressed are, however, those of the author only and do not necessarily reflect those of the European Union. Neither the European Union, nor the granting authorities, can be held responsible for them. Responsibility for the information and views expressed therein lies entirely with the authors.

Disclosure of Interests. The authors have no competing interests to declare.

References

1. EU: Regulation (EU) 2017/745 of the European Parliament and of the Council of 5 April 2017 on medical devices, amending Directive 2001/83/EC, Regulation (EC) No 178/2002 and Regulation (EC) No 1223/2009 and repealing Council Directives 90/385/EEC and 93/42/EEC (Text with EEA relevance) (2017)
2. EU: Regulation (EU) 2017/746 of the European Parliament and of the Council of 5 April 2017 on in vitro diagnostic medical devices and repealing Directive 98/79/EC and Commission Decision 2010/227/EU (Text with EEA relevance) (2017)
3. Medical Device Coordination Group: MDCG 2019-16 Guidance on cybersecurity for medical devices (Rev. 1) (2019)
4. Pycroft, L., Aziz, T. Z.: Security of implantable medical devices with wireless connections: the dangers of cyber-attacks. Expert Rev. Med. Dev. 15(6), 403–406 (2018). https://doi.org/10.1080/17434440.2018.1483235
5. Williams, P.A., Woodward, A.J.: Cybersecurity vulnerabilities in medical devices: a complex environment and multifaceted problem. Med. Dev. Evid. Res. **8**, 305–316 (2015). https://doi.org/10.2147/MDER.S50048
6. WHO: WHO Director-General's remarks at Meeting of the UN Security Council on threats posed by ransomware attacks against hospitals and other health-care facilities and services (2024)
7. EC: Commission unveils action plan to protect the health sector from cyberattacks (2025)
8. Freyer, O., Jahed, F., Ostermann, M., Rosenzweig, C., Werner, P., Gilbert, S.: Consideration of cybersecurity risks in the benefit-risk analysis of medical devices: scoping review. J. Med. Internet Res. **26**, e65528 (2024). https://doi.org/10.2196/65528
9. Biasin, E., Kamenjašević, E.: Regulatory approaches towards AI-based medical device cybersecurity: a transatlantic perspective. Eur. J. Risk Regul., 1–11 (2024). https://doi.org/10.1017/err.2024.23
10. Scherb, C., Hadayah, A., Heitz, L.B.: CyMed: a framework for testing cybersecurity of connected medical devices. http://arxiv.org/abs/2310.03583 (2023)
11. Biasin, E., Yaşar, B., Kamenjašević, E.: New Cybersecurity Requirements for medical devices in the EU: the forthcoming European health data space, data act, and artificial intelligence act. Law Tech Hum. 5, 43–58 (2023). https://doi.org/10.5204/lthj.3068
12. Biasin, E., Kamenjašević, E.: Cybersecurity of medical devices: regulatory challenges in the European union. In: Cohen, I.G., Minssen, T., Price Ii, W.N., Robertson, C., Shachar, C. (eds.) The Future of Medical Device Regulation, pp. 51-62. Cambridge University Press (2022). https://doi.org/10.1017/9781108975452.005

13. EU: Regulation (EU) 2016/679 of the European Parliament and of the Council of 27 April 2016 on the protection of natural persons with regard to the processing of personal data and on the free movement of such data, and repealing Directive 95/46/EC (General Data Protection Regulation) (2016)
14. EU: Directive (EU) 2016/1148 of the European Parliament and of the Council of 6 July 2016 concerning measures for a high common level of security of network and information systems across the Union (NIS Directive) (2016)
15. Taylor, S., et al.: A way forward for the MDCG 2019-16 medical device security guidance. In: Proceedings of the 17th International Conference on PErvasive Technologies Related to Assistive Environments, pp. 593–599. ACM, Crete (2024). https://doi.org/10.1145/3652037.3663894
16. Rak, R., Quinn, P.: Enhancing Digital Health Innovation in the EU with Effective Industrial Strategy Policies. Joint Research Centre of the European Commission (2024)
17. Milojević, D.: Is it time to update the Medical Device Coordination Group's Guidance on Cybersecurity for Medical Devices?. https://www.law.kuleuven.be/citip/blog/is-it-time-to-update-the-medical-device-coordination-groups-guidance-on-cybersecurity-for-medical-devices/
18. Cohen, I.G., Gerke, S., Kramer, D.B.: Ethical and legal implications of remote monitoring of medical devices. Milbank Q. **98**, 1257–1289 (2020). https://doi.org/10.1111/1468-0009.12481
19. EU: Regulation (EU) 2023/2854 of the European Parliament and of the Council of 13 December 2023 on harmonised rules on fair access to and use of data and amending Regulation (EU) 2017/2394 and Directive (EU) 2020/1828 (Data Act) (Text with EEA relevance) (2023)
20. EC: European Data Act enters into force, putting in place new rules for a fair and innovative data economy. https://digital-strategy.ec.europa.eu/en/news/european-data-act-enters-force-putting-place-new-rules-fair-and-innovative-data-economy
21. EU: Regulation (EU) 2022/868 of the European Parliament and of the Council of 30 May 2022 on European data governance and amending Regulation (EU) 2018/1724 (Data Governance Act) (2022)
22. Biasin, E.: The Data Act will concern eHealth apps and Medical Devices. https://www.law.kuleuven.be/citip/blog/the-data-act-will-concern-ehealth-apps-and-medical-devices/
23. De Pauw, V., De Cordier, T., Fonteyn, B., Wiring, R.: Adapting to the new EU Data Act: implications for medical devices and other health devices. https://cms-lawnow.com/en/ealerts/2024/01/adapting-to-the-new-eu-data-act-implications-for-medical-devices-and-other-health-devices
24. EU: Regulation of the European Parliament and of the Council on European Health Data Space and amending Directive 2011/24/EU and Regulation (EU) 2024/2847 (2025)
25. Li, W., Quinn, P.: The European health data space: an expanded right to data portability? Comput. Law Secur. Rev. **52** (2024). https://doi.org/10.1016/j.clsr.2023.105913
26. Quinn, P., Ellyne, E., Yao, C.: Will the GDPR restrain health data access bodies under the European health data space (EHDS)? Comput. Law Secur. Rev. **54** (2024). https://doi.org/10.1016/j.clsr.2024.105993
27. EU: Regulation (EU) 2019/881 of the European Parliament and of the Council of 17 April 2019 on ENISA (the European Union Agency for Cybersecurity) and on

information and communications technology cybersecurity certification and repealing Regulation (EU) No 526/2013 (Cybersecurity Act) (2019)
28. Papakonstantinou, V.: Cybersecurity as praxis and as a state: the EU law path towards acknowledgement of a new right to cybersecurity? Comput. Law Secur. Rev. **44**, 105653 (2022). https://doi.org/10.1016/j.clsr.2022.105653
29. Mammonas, D.: Cyber solidarity package: Council and Parliament strike deals to strengthen cyber security capacities in the EU. https://www.consilium.europa.eu/en/press/press-releases/2024/03/06/cyber-solidarity-package-council-and-parliament-strike-deals-to-strengthen-cyber-security-capacities-in-the-eu/
30. EU: Commission Implementing Regulation (EU) 2024/482 of 31 January 2024 laying down rules for the application of Regulation (EU) 2019/881 of the European Parliament and of the Council as regards the adoption of the European Common Criteria-based cybersecurity certification scheme (EUCC) (C/2024/560) (2024)
31. EC: The EU cybersecurity certification framework. https://digital-strategy.ec.europa.eu/en/policies/cybersecurity-certification-framework
32. Biasin, E., Kamenjašević, E., Rosager, K.: Cybersecurity of AI medical devices: risks, legislation, and challenges. In: Solaiman, B. and Cohen, I.G. (eds.) Research Handbook on Health, AI and the Law. pp. 57-74. Edward Elgar Publishing (2024)
33. EC: Commission Staff Working Document Impact Assessment Report Accompanying the document Proposal for a Directive of the European Parliament and of the Council on measures for a high common level of cybersecurity across the Union, repealing Directive (EU) 2016/1148 (2020)
34. Directive (EU) 2022/2555 of the European Parliament and of the Council of 14 December 2022 on measures for a high common level of cybersecurity across the Union, amending Regulation (EU) No 910/2014 and Directive (EU) 2018/1972, and repealing Directive (EU) 2016/1148 (NIS 2 Directive) (2022)
35. Schmitz-Berndt, S.: Refining the mandatory cybersecurity incident reporting under the NIS directive 2.0: event types and reporting processes. In: Onwubiko, C., et al. (eds.) Proceedings of the International Conference on Cybersecurity, Situational Awareness and Social Media, pp. 343–351. Springer, Singapore (2023)
36. Wanecki, P., Jašek, R., Drofova, I.: The contribution of the European NIS2 directive to the design of the cyber security model. In: 2023 International Conference on Information and Digital Technologies (IDT), pp. 149–154 (2023). https://doi.org/10.1109/IDT59031.2023.10194454
37. Vandezande, N.: Cybersecurity in the EU: how the NIS2-directive stacks up against its predecessor. Comput. Law Secur. Rev. **52**, 105890 (2024). https://doi.org/10.1016/j.clsr.2023.105890
38. EU: Regulation (EU) 2024/2847 of the European Parliament and of the Council of 23 October 2024 on horizontal cybersecurity requirements for products with digital elements and amending Regulations (EU) No 168/2013 and (EU) 2019/1020 and Directive (EU) 2020/1828 (Cyber Resilience Act) (2024)
39. Regulation (EU) 2024/1689 of the European Parliament and of the Council of 13 June 2024 laying down harmonised rules on artificial intelligence and amending Regulations (EC) No 300/2008, (EU) No 167/2013, (EU) No 168/2013, (EU) 2018/858, (EU) 2018/1139 and (EU) 2019/2144 and Directives 2014/90/EU, (EU) 2016/797 and (EU) 2020/1828 (Artificial Intelligence Act) (Text with EEA relevance) (2024)
40. Choi, W., van Eck, M., van der Heijden, C., Hooghiemstra, T., Vollebregt, E.: Legal Analysis: European Legislative Proposal Draft AI Act and MDR/IVDR (2022)

41. Levy-Loboda, T., Sheetrit, E., Liberty, I.F., Haim, A., Nissim, N.: Personalized insulin dose manipulation attack and its detection using interval-based temporal patterns and machine learning algorithms. J. Biomed. Inf. **132** (2022). https://doi.org/10.1016/j.jbi.2022.104129
42. EC: Mission Letter. https://commission.europa.eu/about/commission-2024-2029/commissioners-designate-2024-2029
43. EC: The Draghi report on EU competitiveness. https://commission.europa.eu/topics/eu-competitiveness/draghi-report
44. EP: Resolution of 23 October 2024 on the urgent need to revise the Medical Devices Regulation, P10 TA(2024)0028
45. Liese Initiative. https://peter-liese.de/en/32-english/press-releases-en/4251-epp-revision-of-the-medical-device-regulation-can-save-lives
46. Yesiltuna, O.: A Prescription for Resilience? European Action Plan on the Cybersecurity of Hospitals and Healthcare Providers. https://hall.research.vub.be/a-prescription-for-resilience-european-action-plan-on-the-cybersecurity-of-hospitals-and-healthcare

MDCG 2019-16 Guidelines: Case Study-Based Assessment and Path Forward

Christos Androutsos[1], Steve Taylor[2], Karin Bernsmed[3], Andrea Neverdal Skytterholm[3], Gregory Epiphaniou[4], Nabil Moukafih[4], Theodoros N. Arvanitis[5], Sotiris Messinis[6], Nikos Papadakis[7], Marco Fruscione[8], Andrés Castillo[9], Dusko Milojevic[10], Dimitrios S. Karas[11], Nikolaos Fotos[11], Max Ostermann[12], Oscar Freyer[12], Stephen Gilbert[12], Vasilis Pezoulas[1], Lambros Athanasiou[1], George Gkois[1], and Dimitrios I. Fotiadis[1,13](✉)

[1] University of Ioannina, Ioannina, Greece
dimitris.fotiadis30@uoi.gr
[2] University of Southampton, Southampton, UK
[3] SINTEF Digital, Trondheim, Norway
[4] University of Warwick, Coventry, UK
[5] University of Birmingham, Birmingham, UK
[6] Institute of Communication and Computer Systems, Athens, Greece
[7] SPACE Hellas, Athens, Greece
[8] EBIT, Genoa, Italy
[9] Fundación Para La Investigación Biomédica Hospital Infantil Universitario Niño Jesús, Madrid, Spain
[10] KU Leuven, Leuven, Belgium
[11] UBITECH Ltd., Athens, Greece
[12] Dresden University of Technology, Dresden, Germany
[13] Biomedical Research Institute, FORTH, Ioannina, Greece

Abstract. The Medical Device Coordination Group (MDCG) 2019-16 guidelines provide a structured framework for cybersecurity in Connected Medical Devices (CMDs) throughout their lifecycle, offering guidance on how to fulfil all the relevant essential requirements outlined in Annex I of both the Medical Device Regulation (MDR) and the In Vitro Diagnostic Medical Devices Regulation (IVDR). This paper evaluates the practical applicability and limitations of these guidelines based on feedback from six Horizon Europe (HEU) projects. Each project employed case studies reflecting diverse CMD environments and operational contexts to assess the guidelines' relevance and effectiveness in real-words scenarios. The paper identifies gaps in the practical application of the guidelines and explores their impact on different stages of the CMD lifecycle, from design and development to deployment and post-market activities. Based on these findings, the paper proposes targeted recommendations aimed at enhancing the usability and effectiveness of the MDCG 2019-16 guidelines. The insights contribute to the ongoing evolution of cybersecurity practices in medical technology, ensuring the guidelines are better

I. Praça et al. (Eds.): EICC 2025, CCIS 2500, pp. 338–355, 2025.
https://doi.org/10.1007/978-3-031-94855-8_22

aligned with the needs of CMD stakeholders, including manufacturers, integrators, and operators, while supporting the development of more resilient and secure medical devices.

Keywords: Medical Device Coordination Group · Cybersecurity · Medical Devices

1 Introduction

1.1 Overview of the MDCG 2019-16 Guidelines

The MDCG 2019-16 guidelines [1] provide a structured approach to integrating cybersecurity measures across the lifecycle of CMDs. Recognizing the increasing complexity of CMDs and their vulnerabilities to cyber threats, the guidelines aim to support manufacturers, integrators, and operators in mitigating risks while ensuring compliance with regulatory frameworks, such as the MDR and IVDR. The guidelines emphasize a risk-based approach, requiring stakeholders to assess and address cybersecurity risks at every stage of a device's lifecycle, from design and development to deployment and post-market surveillance. Key measures outlined in the guidelines include preventing unauthorized access to sensitive medical data, safeguarding data from unauthorized modifications and ensuring continuous functionality of medical devices. The guidelines also advocate for the adoption of a defense-in-depth strategy, which involves implementing layered security measures to mitigate potential cybersecurity threats. They provide detailed requirements for device manufacturers, integrators, and operators, making them highly relevant to the diverse CMD scenarios of use.

1.2 Objectives

The primary objective of this paper is to evaluate the practical application of the MDCG 2019-16 guidelines in the context of CMDs by utilizing insights gathered from six collaborative HEU projects: NEMECYS, MEDSECURANCE, SEPTON, CYLCOMED, ENTRUST, and CYMEDSEC, which share the common objective. A first attempt at this collaboration provided some initial recommendations for the MDCG 2019-16 guidelines showcasing existing gaps regarding the applicability and assessment of the guidelines [2]. In this paper, these results are extended by utilizing real-world case studies, to assess the guidelines' relevance, practicality, and alignment with the cybersecurity needs of CMDs. By focusing on stakeholder feedback across diverse device types and operational settings, the paper contributes unique evidence-based recommendations for improving MDCG utility in future revisions. The aim is to identify gaps, particularly in addressing the complexities of cybersecurity within real-world contexts and to provide recommendations, ensuring their clarity and effectiveness for medical device manufacturers, integrators, and operators. Consolidating feedback from the six projects, the paper offers a unified perspective on the strengths and limitations of the guidelines. The paper is structured as follows: Sect. 2 presents the methodology for feedback collection. Sections 3 through 8 describe the case study contexts, identified gaps, and corresponding recommendations from each of the six Horizon Europe projects: NEMECYS, MEDSECURANCE, SEPTON, CYLCOMED, ENTRUST, and CYMEDSEC. Finally, Sect. 9 concludes the paper and outlines directions for future work.

2 Methodologies for Feedback Collection

To systematically evaluate the applicability and relevance of the guidelines in real-world CMD scenarios, a structured and collaborative methodology was employed by the NEMECYS and SEPTON projects. This process ensured a detailed, scenario-driven assessment of the guidelines across the case studies and provided actionable feedback for their refinement. A structured feedback framework was designed to enable a consistent and in-depth review of the MDCG 2019-16 guidelines. Each guideline was assigned a unique identifier to enable traceability and simplify reference across the project's documentation and discussions. To reflect the multi-layered approach required to protect CMDs against cybersecurity risks, the guidelines were categorized based on the principles of defense-in-depth security strategies. The guidelines were further broken down into IT related subcategories, such as basic principles and general security requirements for operating environments. Alignment with the MDR was highlighted, providing stakeholders with specific requirements the guidelines aim to address. Each guideline was outlined in detail, ensuring case studies had the full text for review and analysis. Initial observations and assessments were then gathered to provide actionable insights into the practical application of the guidelines.

The process followed by the MEDSECURANCE project for identifying gaps in the MDCG guidelines involved a structured approach engaging stakeholders and case studies. Feedback was collected through collaborative workshops, and structured templates focusing on practical challenges stakeholders face when striving to comply with these standards. The process also includes testing the tools developed within the project against three case studies. These tests were designed to evaluate the tools' effectiveness in securing medical devices and Internet of Medical Things (IoMT) systems under scenarios that simulate actual healthcare environments. The observational data collected from these tests capture where the tools successfully mitigate risks and where they fall short, specifically in areas not adequately covered by current MDCG guidelines. This direct comparison and data collection help pinpoint precise gaps in the guidelines, highlighting areas for potential enhancement and to better address contemporary cybersecurity challenges in healthcare.

In the CYLCOMED project it was argued whether the guidelines outlined in Sect. 2.6 of the MDCG "Joint Responsibility - Specific expectations from other stakeholders" may eventually be examined to develop a novel method to enhance cybersecurity for a CMD intended for use in hospitals. Active participation from a broad range of stakeholders including clinicians, nurses, IT staff, patients, and administrators facilitated in-depth discussions on the potential implications of implementing the MDCG 2019-16 guidelines, focusing on how the guidelines may affect daily clinical operations and overall patient care. The pilot validation process involved comparing stakeholder experiences before and after testing the CYLCOMED dashboard with real patients, aimed to identify areas for the refinement of practical guidelines applicable to both new and existing cutting-edge technologies. An important part of this process involved semi-structured interviews, conducted in two distinct stages. The first stage gathered insights on stakeholder expectations and concerns prior to system's deployment. The second stage captured feedback on the dashboard's performance highlighting any challenges encountered, and benefits realized. These interviews were conducted by experts including cybersecurity specialists, clinical engineers, and data protection officers with a focus on MDCG guidelines based on cybersecurity, ethics, clinical workflows, and compliance.

The initial set of recommendations from the ENTRUST project focused on addressing the absence of a clearly defined Vulnerability Management process, as outlined in Sect. 2.4 of MDCG 2019-16 guidelines. This was achieved through the design and implementation of a Trust Assessment Framework (TAF) for ensuring that the trustworthiness level of CMDs remains at an acceptable level throughout their operational lifecycle. To this end, a core dimension of the ENTRUST methodology entailed the construction of the threat landscape affecting each medical domain application represented by the use case demonstrators. Thus, a top-to-bottom methodology was applied, beginning with the Top 10 Mobile Risks identified by the Open Worldwide Application Security Project (OWASP) [3] as a generic list of threats and vulnerabilities applicable across domains. To refine this into domain-specific lists, questionnaires were distributed to the case studies to obtain the following information: (i) description of practical, real-world threat scenarios, (ii) correspondence with OWASP threats, and (iii) countermeasures and types of trustworthiness evidence to be monitored to support the safety and security controls for the mitigation of those threats.

As the CYMEDSEC project is still in its early stages the analysis is based on literature reported real-world vulnerabilities and the most impactful Common Vulnerabilities and Exposures (CVE) and Common Weakness Enumeration (CWE) [4], rather than in-project de novo collected data from case studies. This data was analyzed in the context of the requirements of MDCG 2019-16 guidelines and the FDA cybersecurity guidance. These requirements were mapped against the developed baseline to identify areas of alignment, partial coverage, or gaps. The vulnerabilities in the previously identified real-world incidents from the Cybersecurity and Infrastructure Security Agency (CISA) database were analyzed to determine whether adherence to the guidance would mitigate vulnerabilities. The coverage of thematic areas was assessed and inconsistencies identified. This mapping allowed regulatory deficiencies and gaps to be identified and the formulation of recommendations to improve the regulations.

3 NEMECYS

3.1 Case Studies

NEMECYS employs four distinct case studies representing diverse device types, operational settings, and cybersecurity challenges, ensuring a multimodal evaluation across a range of CMD scenarios. The case studies analyzed in this evaluation are as follows:

1. Re:Balans patch sensor [5]: A non-invasive wearable sensor device that measures hydration levels in patients with fluid management-related health conditions, such as patients with end-stage renal disease requiring dialysis. The device collects real-time data and wirelessly transmits them to an external unit such as a smart tablet or IoT gateway for remote patient monitoring.
2. PDMonitor [6]: An IoT wearable medical device designed for the continuous monitoring of Parkinson's Disease (PD) patients both in home and hospital settings, through an approach that spans the device's design, development, and operational phases.
3. Diabetes management mobile application: A Class IIb Software as a Medical Device (SaMD) mobile application that assists people with diabetes to correctly manage their therapy [7]. This application provides optimized advice for bolus and basal insulin doses, adjusted according to entries related to meals, blood glucose and other relevant information provided by the patient.
4. Freestyle Libre 2 Continuous Glucose Monitoring (CGM) kit [8]: An in-vitro diagnostic device consisting of a glucose monitor worn on the patient's upper arm to continuously measure glucose levels, a reader used both at home and hospital to scan the sensor and display glucose values and a web-based application for healthcare professionals to access patient data [9].

3.2 Gaps and Recommendations

The evaluation of the MDCG 2019-16 guidelines in the NEMECYS case studies covered multiple dimensions and identified several gaps that hinder their practical application in CMDs. In response to these gaps, targeted recommendations were provided to enhance the guidelines' effectiveness in addressing real-world cybersecurity needs. Both the identified gaps and corresponding recommendations are summarized in Table 1.

Table 1. Gaps and recommendations across NEMECYS case studies.

Category	Gaps	Case Study	Importance	Recommendations
Guidance Completeness	The guidelines do not effectively provide extensive guidance on fulfilling all the relevant requirements for Annex I of the MDR, leaving stakeholders with incomplete support	1	Without clear MDR alignment, manufacturers face uncertainly compile MDR technical file and get a CE mark	- Restructure the guidelines into a more straightforward, practical document tailored to manufacturers seeking MDR compliance and CE marking - Provide detailed descriptions for labeling, instructions for use (IFU) requirements, training protocols, and usability validation processes
Terminology Clarity	Fail to clearly differentiate between cybersecurity, security, and safety, creating confusion. Key terms such as "layered defense" and "good security hygiene" are undefined	1–4	Ambiguous terminology complicates secure device design, particularly for wearable and IoT devices where layered security is essential	- Draw upon established frameworks, such as the FDA Post market Management of Cybersecurity in Medical Devices, FDA Cybersecurity in Medical Devices Guidance, ISO 14971, and AAMI/TIR57, to enhance the clarity and applicability of the guidelines - Clear terms should be provided regarding tools and standards to be used in threat modeling techniques, such as recommended tools
Risk Management Framework	Reasonably foreseeable misuse is not defined according to ISO 14971 and AAMI TIR57:2016. No specific cybersecurity examples are provided to illustrate risk mitigation strategies	1,3	As part of risk management, manufacturers must identify and assess potential vulnerabilities that could be exploited, including those arising from reasonably foreseeable misuse of the device	- Include requirements for comprehensive risk assessments of use case scenarios, implementation of mitigation measures and controls. - Clearly define key terms, such as "reasonably foreseeable misuse", based on recognized standards like ISO 14971 and AAMI/TIR57:2016 Provide documentation on general best practices in risk management
Stakeholder Responsibilities	The term "integrator" is undefined, and explicit requirements for manufacturers such as the purpose of integration, technical requirements, environmental requirements, verification plan, installation and servicing, and training requirements, are missing. The distinction between "operator" and "user" is also unclear	1–4	The manufacturer is responsible for safety and effectiveness, and these requirements should be explicitly stated by the MDCG	- Clearly delineate responsibilities among manufacturers, integrators, operators, and regulators - Develop a categorized list of common security requirements to assist manufacturers in drafting effective requirements documents

(*continued*)

Table 1. (*continued*)

Category	Gaps	Case Study	Importance	Recommendations
Defense-in-Depth Strategy	Limited and overly generic guidance on integrating, configuring, and maintaining a defense-in-depth strategy	1, 2	Guidance should be provided to address requirements related to technical descriptions, or rational processes/assessments that the manufacturer needs to implement to achieve a secure design	- Incorporate practical examples and scenarios to clarify concepts like "layered defense", "depth approach", and "good security hygiene" within medical device cybersecurity contexts - Provide detailed descriptions, including technical specifications, rational processes, and assessments manufacturers must implement for secure design - Explicitly state requirements for addressing generic security-related threats, including detailed specifications on when specific security capabilities are needed and what devices or scenarios are relevant
Verification & Validation (V & V)	Too brief, lacking actionable recommendations, reporting templates, and guidance on testing methodologies	3	Insufficient V&V guidance hinders the systematic testing and validation of software in medical devices and mobile health applications	- Offer examples of verification and validation testing, along with best practices for addressing various cybersecurity threats - Specific templates for reporting verification and validation results should be emphasized
Safety vs. Security	The guidelines do not clarify the distinction between safety and security in risk management processes, leading to potential overlaps or gaps in mitigation strategies	1, 2, 3	Inadequate distinction may lead to overlooked cybersecurity threats that could compromise patient safety in continuous monitoring devices	- Establish "safety, security, and effectiveness" requirements as concrete outcomes of a systematic design and risk management process - The relationship between safety risk assessments (as addressed by ISO 14971) and security risk assessments (e.g., TIR57) should be clearly explained, including when security risks impact safety and how they must be included in safety risk assessments with clear traceability between security and safety analysis - Ensure the guidelines account for general safety and performance requirements for devices with cybersecurity risks as outlined in Annex I of the MDR and include both functional and non-functional IT security requirements
Risk Re-evaluation Frequency	The guidelines do not specify how often security and safety risks should be reassessed or provide measures to manage evolving risks effectively	3	Without specified risk re-evaluation intervals, devices may remain exposed to emerging threats, especially in rapidly evolving healthcare technology landscapes	–
Post-Market Surveillance	–	1	–	Emphasize post-market cybersecurity maintenance procedures, including protocols for support, maintenance, and anticipated security degradations over time

4 MEDSECURANCE

4.1 Case Studies

MEDSECURANCE focuses on the enhancement of security-for-safety assurances within the IoMT. With an initial focus on healthcare IT systems, the project explores the intersection of clinical benefits and security vulnerabilities. Through the development of innovative methodologies, infrastructures, and technologies, MEDSECURANCE seeks to advance secure system engineering management activities, addressing the evolving complexity of IoMT. To identify gaps in the MDCG guidelines, MEDSECURANCE focuses on three case studies:

1. Remote Patient Monitoring (RPM) system: This case study evaluates the secure transmission of patient data over networks, highlighting the project's capability to ensure data integrity and confidentiality against cyber threats.
2. Portable Polymerase Chain Reaction (PCR) Testing: This case study focuses on security challenges in diverse environmental conditions, particularly in field settings where the risk of unauthorized access and data breaches could compromise both patient confidentiality and integrity of test result.
3. Virtual Ward: This case study simulates a hospital ward environment managed remotely using a network of CMDs and IT systems.

4.2 Gaps and Recommendations

The evaluation of the MDCG 2019-16 guidelines in the MEDSECURANCE case studies covered multiple dimensions and identified several gaps that hinder their practical application in CMDs. In response to these gaps, targeted recommendations were provided to enhance the guidelines' effectiveness in addressing real-world cybersecurity needs. Both the identified gaps and corresponding recommendations are summarized in Table 2.

Table 2. Gaps and recommendations across MEDSECURANCE case studies.

Category	Gaps	Case Study	Importance	Recommendations
Specific IoMT Security Considerations	The guidelines focus predominantly on the software components of medical devices without considering the broader spectrum of IoMT system vulnerabilities	1	These systems feature a diverse array of device types and connectivity options, each introducing distinct security challenges that include heterogeneity in device functionality, varied data privacy requirements, and increased exposure to cyberattacks	Expanding the guidelines to include specific security considerations tailored to IoMT could mitigate these risks by providing clear directives on securing varied device architectures within healthcare ecosystems

(*continued*)

Table 2. (*continued*)

Category	Gaps	Case Study	Importance	Recommendations
IoMT Risk Management	While the guidelines advocate for rigorous risk management, they lack detailed protocols for the unique risk landscape of IoMT systems	1, 2, 3	IoMT devices not only enhance healthcare delivery but also expand the potential attack surface due to their interconnected nature	Implementing a more nuanced risk management framework that includes specific methodologies for assessing and mitigating risks unique to IoMT such as cross-device data breaches or network-based vulnerabilities would greatly enhance the robustness of these critical systems
Secure Software Development Lifecycle (SDLC) for IoMT	The guidelines currently touch on software development practices but do not provide an extensive framework for the secure SDLC applicable to IoMT	2	Given the complexity and the critical nature of IoMT applications, embedding security at each phase of the software development process-from design to deployment and beyond-is vital	The inclusion of IoMT-specific secure coding practices, threat modelling, and continuous testing within the guidelines would help ensure that IoMT software upholds the highest security standards
Data Privacy and Consent in IoMT	–	1, 2, 3	–	– The guidelines should be broadened to include comprehensive strategies for data privacy management, including data consent, data minimisation, and secure data storage and transmission protocols - Aligning these strategies with general data protection regulations (e.g., GDPR) will further ensure that IoMT systems are not only secure but also respect patient privacy
Vulnerability and Incident Response for IoMT	Current guidelines offer little detail on managing vulnerabilities and responding to incidents within IoMT frameworks	1, 2, 3	IoMT systems routinely handle sensitive health data, making them prime targets for breaches	- Clear and detailed process for ongoing vulnerability assessment, regular updates, and patch management - Establish standardised incident response procedures for IoMT can expedite containment and mitigation efforts, minimising the impact of security breaches
Software Validation and Verification	–	2	–	Enhance validation and verification processes for higher-risk software categories

5 SEPTON

5.1 Case Studies

The SEPTON project includes four distinct case studies each addressing critical cybersecurity challenges in CMDs, ensuring robust protection and data security across various medical environments. The case studies analyzed in this evaluation are as follws:

1. Implantable Medical Devices (IMDs): This case study focuses on IMDs designed for real-time seizure suppression utilizing prerecorded electrocorticogram data to detect and stop seizures before they manifest. The IMD

also securely transmits data, including seizure events and treatment information, to a smartphone acting as a reader.
2. Wearable Medical Devices: This case study focuses on wearable medical devices in various configurations to support RPM. These devices collect health metrics and transmit them through gateways, such as smartphones, to hospital or cloud systems.
3. Hospital Infrastructure Security: This case study focuses on the security of critical devices such as Magnetic Resonance Imaging (MRI) scanners, Picture Archiving and Communication System (PACS) servers, and imaging modalities.
4. Distributed Healthcare Infrastructures: This case study explores secure data sharing utilizng blockchain technology to create an immutable and transparent ledger for all interactions, while decentralized storage solutions like IPFS [10] ensure efficient and secure data distribution.

5.2 Gaps and Recommendations

The evaluation of the MDCG 2019-16 guidelines in the SEPTON case studies covered multiple dimensions and identified several gaps that hinder their practical application in CMDs. In response to these gaps, targeted recommendations were provided to enhance the guidelines' effectiveness in addressing real-world cybersecurity needs. Both the identified gaps and corresponding recommendations are summarized in Table 3.

Table 3. Gaps and recommendations Across SEPTON Case Studies.

Category	Gaps	Case Study	Importance	Recommendations
Portable/Remote Medical devices	The guidelines do not explicitly address the unique risks of devices that remain in a patient for extended periods without direct manufacturer access	1, 2	Missing updates could lead to long-term security vulnerabilities affecting patient health. A lack of direct manufacturer access increases risks of cyber intrusions	Introduce specific protocols for secure over-the-air updates to minimize the need for invasive procedures
Third-Party Mobile Platforms & Software Security	The guidelines do not sufficiently address the risks associated with devices that rely on third-party mobile platforms for patching. No structured framework for evaluating and mitigating third-party security risks, despite many hospital devices depending on vendor-supplied or integrated software components	1, 2, 3	Dependence on third-party apps may delay critical security patches	- Establish minimum security requirements for mobile platforms used in conjunction with medical wearables - Ensure that security patches are validated by the manufacturer before distribution to prevent unauthorized tampering

(continued)

Table 3. (*continued*)

Category	Gaps	Case Study	Importance	Recommendations
Healthcare Data Protection	The guidelines do not clearly define practices for protecting data when shared across different healthcare institutions, especially in cases with varying regulations	1–4	Vital for regulatory compliance and patient confidentiality. Weak data protection could expose sensitive medical records. Insecure transmission could lead to unauthorized access	- Introduce standardized cybersecurity measures for securing cross-institutional data sharing - Mandate the use of secure transmission protocols, such as FHIR over TLS 1.3, to protect Electronic Health Records (EHR) and real-time data
International Data Governance	Although GDPR compliance is acknowledged, there is no in-depth discussion on data sovereignty and governance challenges when medical information is exchanged internationally	4	Unclear governance policies may lead to compliance violations, legal challenges, and potential security breaches when transferring medical records between jurisdictions	Expand security requirements to account for personal use environments, recognizing the additional risks associated with compromised personal devices compared to hospital-controlled settings
Interoperability & Secure Data Transmission	Insufficient coverage of interoperability and secure transmission standards such as FHIR, DICOM, or HL7, creating vulnerabilities in data structuring and protection	3–4	Inconsistent interoperability standards risk data loss or corruption. Insecure transmission protocols may expose patient records to cyber threats	Mandate structured vendor security assessments before software integration, ensuring that third-party components comply with strict cybersecurity requirements
Safety vs. Security	The guidelines do not clarify the distinction between safety and security in risk management processes, leading to potential overlaps or gaps in mitigation strategies	1–4	Inadequate distinction may lead to overlooked cybersecurity threats that could compromise patient safety in continuous monitoring devices	- Introduce stronger network security requirements, particularly mandating strict network segmentation to separate critical medical devices from general hospital IT systems - Establish minimum cybersecurity baselines for legacy devices, including firewall-based isolation, Virtual Local Area Network (VLAN) segmentation, and role-based access controls
Post-market Surveillance	–	1–4	–	Enhance post-market surveillance frameworks to include long-term cybersecurity monitoring, incorporating proactive vulnerability assessments and periodic risk analysis to identify emerging threats throughout the device's lifespan

6 CYLCOMED

6.1 Case Studies

The introduction of pilots within the CYLCOMED project is a crucial tactic to actively involve end-users and gain a deeper understanding of the challenges associated with integrating cutting-edge technologies in healthcare scenarios as follows:

1. Hospital Equipment: Although there are no clinical trials planned for the first pilot related to hospital equipment, collaboration with the Hospital Niñ Jesús of Madrid [11] ensures valuable feedback to support the improvement of tools under evaluation.
2. Telemedicine Implementation in Hospital Environments: In this case study end-users from Hospitals Bambino Gesù of Rome [12], Charité of Berlin [13], and Hospital Niñ Jesús of Madrid [11] actively contribute with their experiential insights, which are seamlessly integrated into the definition of the clinical protocol and the formulation of the patient journey, guaranteeing an in-depth approach to technology implementation.

6.2 Gaps and Recommendations

The evaluation of the MDCG 2019-16 guidelines in the CYLCOMED case studies covered multiple dimensions and identified several gaps that hinder their practical application in CMDs. In response to these gaps, targeted recommendations were provided to enhance the guidelines' effectiveness in addressing real-world cybersecurity needs. Both the identified gaps and corresponding recommendations are summarized in Table 4.

Table 4. Gaps and recommendations across CYLCOMED case studies

Category	Gaps	Case Study	Importance	Recommendations
Lack of Terminological Coherence	Does not include explicit definitions or references to terms such as "cybersecurity", "security-by-design", and "security-by-default"	1, 2	Leaving these terms theoretical and undefined present challenges for stakeholders seeking to implement practical cybersecurity measures effectively	Key concepts like cybersecurity, security-by-design, and security-by-default should be explicitly defined, either by referring to legal acts that provide such definitions or following established terminology defined by frameworks such as ISO/IEC standards
Legal ambiguity and uncertainty	Briefly mentions the intersection of the cybersecurity legal frameworks without providing recommendations on addressing the potential simultaneous applicability of these frameworks, such as notification obligations. Moreover, the medical device cybersecurity landscape is a dynamic field that has undergone significant changes since the MDCG guidance endorsement in 2019	1, 2	Proper guidance is essential to facilitate compliance with various legal requirements scattered across various regulations	- The MDCG guidelines could clarify the interplay between the MDR and other cybersecurity frameworks - To enhance the cybersecurity of medical devices, the guidelines should be updated to clarify how different regulatory frameworks, such as GDPR, NIS2, AI Act, and MDR, intersect

(*continued*)

Table 4. (*continued*)

Category	Gaps	Case Study	Importance	Recommendations
Lack of Acknowledgement of Human Factors in Cybersecurity	Hints at addressing cybersecurity challenges to other stakeholders in the medical device supply chains (e.g., integrators and operators). The concept of "joint responsibility" lacks clarity and remains open to different interpretations. MDCG Guidance fails to acknowledge that humans are frequently portrayed as the weakest link in the cybersecurity chain [14]	1, 2	Integrating a human-centric approach and providing more guidance would enhance the capability to prevent and mitigate cyber threats and raise overall awareness	Providing more insights and guidance for other stakeholders would be essential to mitigate cybersecurity risks

7 ENTRUST

7.1 Case Studies

The ENTRUST project evaluates the effectiveness of trust assessment controls across four real-world medical application scenarios:

1. KardinBLU ECG [15] and multi-parameter monitoring system: A Bluetooth Low Energy device that collects and transmits cardiological data, SpO_2 levels, and temperature either to a gateway device, which forwards them to a desktop computer, or to a mobile device, which forwards them to the hospital.
2. Tellu Personal Health Gateway (PHG) [16]: This case study focuses on an RPM system where the gateway receives vital health parameters from CMDS in home settings and transmits them to the Tellucare cloud-based eHealth platform for medical use.
3. Patient Care in Hospital Domains: This case study enhances patient care in hospital domains through two demonstrators, the smart ambulance where CMDs transmit real-time data during patient transport and the emergency care unit, which focuses on maintaining the integrity and trustworthiness of the legacy CMDs, such as the Econet Compact 7 [17] in hospital environments.
4. Feel Emotion Sensor (FES) [18]: This case study explores the precision biomarkers and digital therapeutics in mental health care using this wearable device that collects data, such as sweat, heartbeat, and skin temperature.

7.2 Gaps and Recommendations

The evaluation of the MDCG 2019-16 guidelines in the ENTRUST case studies covered multiple dimensions and identified several gaps that hinder their practical application in CMDs. In response to these gaps, targeted recommendations were provided to enhance the guidelines' effectiveness in addressing real-world cybersecurity needs. Both the identified gaps and corresponding recommendations are summarized in Table 5.

Table 5. Gaps and recommendations across ENTRUST case studies

Category	Gaps	Case Study	Importance	Recommendations
Insufficient guidance on life-cycle security and post-market	The MDCG guidelines state that medical device manufacturers should be responsible for ensuring that risks associated with reasonably foreseeable environmental conditions are removed or minimized, while ENTRUST argues that dynamic runtime monitoring is required at the side of the domain for identifying and mitigating zero-day threats and vulnerabilities	1, 2, 4	It is not realistic to expect manufacturers to be solely responsible for real-time monitoring of devices due to time and resource constraints, as well as domain-specific details	- The definition of reasonably foreseeable misuse should be expanded in order to consider the opinion of both the manufacturer and the deployment domain (e.g., the domain administrator) - Provide frameworks for the continuous and dynamic trust assessment of devices deployed to a medical domain, enabling the identification of new threats and vulnerabilities and the enforcement of the appropriate mitigation measures
Real-world vulnerabilities	There is limited consideration for domain-specific threats and vulnerabilities, as well as prioritization of the most prominent ones that need to be patched or addressed in the context of specific application scenarios	1, 2, 3	Vulnerabilities may arise from the interconnectivity between devices, thus domain-specific considerations are essential	Introduce a top-to-bottom process for the definition of the domain-specific threat landscape, starting from a generic list of most prominent threats and leading to its refinement and expansion based on infrastructure and domain information
Insufficient consideration for addressing real-world and post-market issues	The MDCG guidelines highlight the need for manufacturer side device updates for the mitigation of identified vulnerabilities. However, ENTRUST argues that this may be insufficient (e.g., in cases where manufacturer updates are infrequent)	2, 3	It should be feasible to deploy domain-specific updates in a secure manner to address zero day vulnerabilities in a timely manner	Introduce a process for securely updating devices to mitigate threats and vulnerabilities, considering the involvement of not only the manufacturer, but also the deployment domain
Limited or too generic guidance on security measures	The MDCG guidelines dictate that the manufacturer should be responsible for addressing vulnerabilities that may arise from reasonably foreseeable misuse. However, ENTRUST argues that it is not possible to draw such an unambiguous distinction, thus necessitating the presence of runtime security measures	1, 2, 4	The notion of reasonably foreseeable misuse cannot be solely defined by the manufacturer, as domain input is also needed	- Introduce a process for securely updating devices to mitigate threats and vulnerabilities, considering the involvement of not only the manufacturer, but also the deployment domain - Introduce a process for the identification of domain-specific trust sources to be used in the context of the trust assessment process, such as digital twins, anomaly detection, and attestation mechanisms

8 CYMEDSEC

8.1 Case Studies

The first results of case studies in CYMEDSEC compare the guidance of the MDCG guidelines with the cybersecurity guidance detailed in the FDA. Instead of focusing on individual case studies, CYMEDSEC analyses real-world vulnerabilities of CMDs, such as vitals monitors, medical syringe pumps, medical imaging stations etc. [4]. This analysis determines whether correct adherence to the current state of cybersecurity guidance should have successfully prevented the incidents from occurring and what risks cybersecurity incidents pose in the Hospital-at-Home [19,20]. The theoretical analysis of both guidance documents also included an independent third checklist that we created as a baseline cybersecurity checklist, derived from recognized standards, such as NIST CSF2.0, IEC-81001-5, AAMI TIR 57 & BSI-03161 [21]. This was done to identify different levels of scope in the guidelines and to determine the level of coverage in different areas of cybersecurity.

8.2 Gaps and Recommendations

The analysis of real-world CVEs and CWEs in the CYMEDSEC project revealed gaps in both MDCG 2019-16 guidelines and FDA cybersecurity guidance, while targeted recommendations are proposed as provided in Table 6. The structure of this table differs from those presented for other projects due to the unique approach adopted in the evaluation of the MDCG 2019-16 guidelines. Instead of focusing on direct case study assessments, CYMEDSEC conducted a broader thematic analysis of cybersecurity challenges.

Table 6. Gaps and recommendations across CYMEDSEC case studies.

Category	Gaps	Recommendations
No actionable requirements	MDCG 2019-16 references security principles but does not provide concrete, actionable steps for mitigating specific vulnerabilities [22]	–
Inconsistent level of detail	Across the guidance documents, different levels of depth can be found. This can lead to confusion regarding requirements, potentially limiting the attention given to less detailed requirements	- Clearly defining the depth and level of technical detail in the guidance document prevents confusion across requirements - A consistent level of detail clearly sets expectations to the integrating and validating personnel
SMART Standards	–	Implementation of SMART standards allows for enhanced accessibility of standards, enabling machine-automated checks, better readability and updatability
Clear Nomenclature	–	Remaining consistent in nomenclature and redefining recommended principles prevents confusion and allows for easier adherence to the guidance

Table 7. Identified Gaps in the MDCG 2019-16 Guidelines

Category	Gaps	Recommendations	No. of Projects
Risk-Benefit Analysis	Lack of guidance on risk-benefit analysis, which is essential to support practitioners for trade off analysis in proportional cybersecurity without compromising the safety and effectiveness of CMDs	Necessity for guidance on risk assessment and risk-benefit trade-offs	4
Whole-Lifecycle Security	Lack of guidance in whole-lifecycle security for medical devices, especially at operation time such as monitoring intrusions, vulnerabilities, frequency of risk re-assessment and managing risks	Guidance on full lifecycle support for secure device management, including intrusion detection, monitoring, dynamic risk assessment and controls such as secure patching	4
Terminology and Definitions	Lack of clear and consistent definitions, with discrepancies between the MDCG guidelines and other regulatory standards creating uncertainty in implementation	- Clear definitions of terms and resolution of definitional inconsistencies between MDCG and other relevant regulations - Practical and actionable guidance for manufacturers towards MDR compliance and CE marking	3
Post-Market Cybersecurity Maintenance	Lack of structured post-market monitoring, vulnerability assessments, and response measures	Emphasize post-market cybersecurity maintenance procedures	3
Domain-specific Threats	Lack of guidance on domain-specific threats and vulnerabilities	Provide links to external checklists of common vulnerabilities, threats and mitigations	3
Security vs. Safety Distinction	Lack of differentiation between security and safety in risk management, leading to potential conflicts	Clarify how security and safety should be addressed separately within risk management strategies	2
V & V Processes	Insufficient recommendations for verifying and validating cybersecurity measures in CMDs	Enhance V & V processes offering examples for addressing various cybersecurity threats	1
Regulatory Navigation	Lack of guidance on navigating overlapping regulations and the need for regular MDCG updates to keep pace with evolving legal requirements	Clarify how manufacturers can navigate the complex and evolving regulatory landscape	1
Legacy Device Management	Lack of guidance for secure operation and management of legacy devices	Provide guidance on secure management of legacy unpatched devices	1
Secure Data Exchange	Lack of guidance on secure data exchange between institutions	Establish secure data-sharing protocols and provide guidelines for protecting healthcare data across healthcare institutions	1
Human Factors in Cybersecurity	Insufficient guidance on operational security related to human factors	Provide guidance on how to consider human factors in terms of human vulnerabilities and how to address them	1

9 Conclusions

This study involved an extensive evaluation of the MDCG 2019-16 guidelines, gathering insights from six HEU projects with contributions from medical device manufacturers, integrators, cybersecurity experts, regulatory professionals, healthcare providers, and IT specialists. Beyond the tabulated feedback, several shared themes emerged, such as the lack of clarity in role definitions, challenges in aligning cybersecurity measures with clinical safety, and the need for actionable guidance at implementation level. These insights highlight barriers that stakeholders face when translating the guidelines into practice, indicating opportunities for refinement of terminology and structure. The analysis covered a wide range of CMDs in various real-world applications, including wearable and implantable medical devices, RPM systems, hospital IT infrastructure and secure data-sharing platforms. These devices operate in heterogeneous environments, such as home healthcare, hospital networks, distributed healthcare infrastructures, and telemedicine systems, highlighting the diverse cybersecurity risks that must be addressed across different case studies. The identified gaps along with the recommendations are summarized in Table 7.

While this study provides an assessment of the MDCG 2019-16 guidelines, certain limitations should be acknowledged. The analysis relied on structured expert feedback and scenario-based evaluations, which may not fully capture the complexities of real-world regulatory compliance. Additionally, the findings were derived from specific case studies, and while they cover a broad spectrum of CMDs, expanding the scope to include more diverse environments and stakeholders could further enhance the generalizability of the results. The study recognizes the need to further explore how organizational, management, and operational practices in medical device administration influence cybersecurity. As a next step, this consultation process should continue by collecting additional feedback and may include comparative analysis with other international regulatory frameworks. Overall, the projects concluded that the MDCG guidelines are very useful and beneficial and that the main content of this paper is focused on a gap analysis in a spirit of constructive criticism with an honest intent to suggest potential improvements to make updates to the guidelines more beneficial.

References

1. MDCG 2019-16 - Guidance on Cybersecurity for medical devices. Document date: 06/01/2020 - Created by GROW.R.2.DIR - Last update: 22/06/2020. https://ec.europa.eu/docsroom/documents/41863
2. Taylor, S., et al.: A way forward for the MDCG 2019-16 medical device security guidance. In: 17th International Conference on PErvasive Technologies Related to Assistive Environments, pp. 593–599. Association for Computing Machinery (2024)
3. OWASP. https://owasp.org/www-project-mobile-top-10/2023-risks/. Accessed 06 Feb 2025
4. Mejía-Granda, C.M., Fernández-Alemán, J.L., Carrillo-de-Gea, J.M., García-Berná, J.A.: Security vulnerabilities in healthcare: an analysis of medical devices and software. Med. Biol. Eng. Comput. **62**(1), 257–273 (2024)
5. Mode Sensors AS Homepage. https://www.modesensors.com/. Accessed 06 Feb 2025
6. PD Neurotechnology Homepage. https://www.pdneurotechnology.com. Accessed 06 Feb 2025
7. Debiotech Homepage. https://www.debiotech.com. Accessed 06 Feb 2025

8. FreeStyle Libre Homepage. https://www.freestyle.abbott/us-en/products/freestyle-libre-2.html. Accessed 06 Feb 2025
9. Ospedale San Raffaele Homepage. https://research.hsr.it/en/index.html. Accessed 06 Feb 2025
10. IPFS Homepage. https://ipfs.tech. Accessed 06 Feb 2025
11. Hospital Niñ Jesús Homepage. https://www.comunidad.madrid/hospital/ninojesus/. Accessed 06 Feb 2025
12. Hospitals Bambino Gesù of Rome. https://www.ospedalebambinogesu.it. Accessed 06 Feb 2025
13. Charité of Berlin. https://claim.charite.de/en. Accessed 06 Feb 2025
14. Cartwright, A.J.: The elephant in the room: cybersecurity in healthcare. J. Clin. Monit. Comput. **37**(5), 1123–1132 (2023)
15. KARDINBLU project Homepage. https://www.kardinero.com.tr/english/news/tubitak-grant-for-our-kardinblu-project. Accessed 06 Feb 2025
16. Tellu Homepage. https://www.tellu.no/en/services/remote-patient-monitoring/. Accessed 06 Feb 2025
17. Eukon Homepage. https://www.eukon.it/products/compact-7/. Accessed 06 Feb 2025
18. Feel Homepage. https://rpm.feeltherapeutics.com. Accessed 06 Feb 2025
19. Gilbert, S., Ricciardi, F., Mehrali, T., Patsakis, C.: Can we learn from an imagined ransomware attack on a hospital at home platform? NPJ Digit. Med. **7**(1), 65 (2024)
20. Ostermann, M., Freyer, O., Jahed, F., Rosenzweig, C., Gilbert, S.: Cybersecurity in the hospital at home: assessment of patient risks when using IoMT devices. Zenodo (2024)
21. Ostermann, M., Gilbert, S., Freyer, O.: Cybersecurity requirements for medical devices in the EU and US - a comparison and gap analysis. Zenodo (2024)
22. Freyer, O., Jahed, F., Ostermann, M., Rosenzweig, C., Werner, P., Gilbert, S.: Consideration of cybersecurity risks in the benefit-risk analysis of medical devices: scoping review. J. Med. Internet Res. **26**, e65528 (2024)

Threshold Multiparty Private Set Intersection via Federated Graph Neural Networks

Sotirios Messinis[1](✉), Nicholas E. Protonotarios[2], Dimitris Kalogeras[1], and Nikolaos Doulamis[1]

[1] Institute of Communication and Computer Systems, 15773 Zografou, Greece
smessinis@mail.ntua.gr

[2] Mathematics Research Center, Academy of Athens, 11527 Athens, Greece

Abstract. Multiparty private set intersection (MPSI) protocols allow participants to compute the intersection of their datasets while preventing leakage of non-intersecting elements. However, these methods often enforce exact matching, which can be overly restrictive in scenarios where partial intersections provide more meaningful insights. To address this limitation, we introduce a federated learning-enabled threshold multiparty private set intersection (T-MPSI) framework that ensures elements are included in the intersection only if they appear in at least a certain number of datasets. This approach is particularly relevant in cyber threat intelligence sharing and secure data collaboration. Our proposed framework employs federated graph neural networks to dynamically optimize the intersection threshold based on real-time data distributions. We integrate homomorphic encryption to enhance privacy while minimizing communication and computational costs. To evaluate the effectiveness of our method, we conducted simulations in three client population scenarios, comparing against popular protocols. Our results show that our approach achieves lower computation overhead, better intersection accuracy, and scalable performance, making it a viable solution for large-scale privacy-preserving applications.

Keywords: multiparty private set intersection · federated learning · graph neural networks · cyber threat intelligence

1 Introduction

Multiparty private set intersection (MPSI) is a fundamental cryptographic protocol extensively used across various domains, including contact discovery, private profile matching, privacy-preserving genomics, and collaborative learning [4]. MPSI enables multiple mutually untrusting parties to compute the intersection of their datasets, ensuring that no information is revealed about elements outside the intersection. While MPSI has gained significant attention for its ability to maintain data privacy, most existing protocols focus on achieving communication efficiency and cater to specific adversarial models. These approaches,

I. Praça et al. (Eds.): EICC 2025, CCIS 2500, pp. 356–371, 2025.
https://doi.org/10.1007/978-3-031-94855-8_23

however, are typically limited to exact matches, where an element is included in the intersection only if it is present identically in every participating party's dataset [4]. This exact matching constraint, while valuable in certain applications, limits the broader utility of MPSI in scenarios where approximate matches or thresholds could provide more meaningful results.

In privacy-preserving data mining and machine learning [21], where data is vertically partitioned across parties, collaboration may only proceed if the intersection of their datasets is sufficiently large. Similarly, in privacy-preserving ride-sharing applications [18], users may wish to share rides only if significant portions of their trajectories overlap. In this case, users are only interested in discovering the intersection of their routes if the overlap exceeds a predefined threshold. This problem can be formalized as a so-called "threshold-MPSI" (T-MPSI) problem, where parties only learn about the intersection of their datasets if they differ by at most T elements, where T denotes the threshold. In some scenarios, determining the common elements across all parties may be overly restrictive. To this end, the over-threshold MPSI (OT-MPSI) addresses the problem of multiparty involvement in a private set intersection protocol [13,17]. This notion of T-MPSI naturally extends to federated learning (FL) where collaborative model training relies on securely identifying and leveraging shared data while preserving privacy. This approach is particularly relevant in decentralized learning scenarios, where maintaining both privacy and model utility is essential for effective collaboration. In federated graph neural networks (FedGNNs), where graph-structured data is distributed across multiple parties, such techniques are particularly relevant for secure node matching and subgraph alignment, allowing entities to collaborate only when a sufficient structural or feature overlap exists. This threshold-based approach not only enhances privacy but also ensures that federated graph models capture meaningful relationships without compromising the confidentiality of non-overlapping data. In modern healthcare, the role of cyber threat intelligence (CTI) has become indispensable in protecting hospitals and connected medical devices from evolving cyber threats. CTI facilitates the collection, analysis, and dissemination of threat-related information, empowering organizations to proactively defend against attacks, manage risks, and respond swiftly to incidents [1,25].

In the present work, we introduce a novel approach that utilizes FedGNNs to enhance the efficiency and scalability of T-MPSI. Unlike traditional methods, which often struggle with computational overhead and scalability issues, our approach employs FedGNNs to efficiently model relationships between datasets across multiple parties, improving intersection computation. Unlike fixed-threshold approaches, we introduce an FL-driven dynamically adaptive threshold mechanism. Furthermore, by employing homomorphic encryption, our approach improves communication and computational efficiency. We evaluated our method across three client population scenarios, demonstrating lower computation overhead, improved intersection accuracy, and scalable performance compared to popular protocols, proving its effectiveness for large-scale privacy-preserving applications.

2 Related Work

Existing work on threshold MPSI, federated learning, and federated graph neural networks explores cryptographic and decentralized techniques to enhance privacy-preserving collaboration while addressing computational and communication efficiency.

2.1 Threshold Multiparty Private Set Intersection

In [30] the authors propose two efficient MPSI protocols tailored for scenarios involving a large number of participants with small set sizes. The proposed protocols are proven secure against collision attacks in both semi-honest and malicious models. In [28], the authors introduce a protocol that utilizes Function Secret Sharing (FSS) and enables multiple participants to compute the intersection of their private datasets without revealing any additional information. The authors of [32] introduce a novel paradigm for multiparty Private Set Intersection Cardinality (PSI-CA), enabling n parties to compute the size of the intersection of their datasets without revealing additional information. The authors of [29] address the challenge of inefficiency when dealing with a large number of participants and small set sizes in MPSI protocols. The authors propose two novel MPSI protocols designed to be efficient in scenarios with many participants holding small datasets. In [11] the authors introduce the concept of Delegated Private Set Intersection (D-PSI), wherein users delegate the computational and communication tasks to a cloud server while ensuring that the server learns nothing about the individual datasets or the intersection result. In the standard MPSI, each party holds a private set $\mathcal{S}_i$. The goal is to compute the intersection $I = \bigcap_{i=1}^{m} \mathcal{S}_i$ without revealing any $\mathcal{S}_i$ to the other parties. In threshold MPSI, the intersection generalizes to $I_t = \{x \mid x \in \mathcal{S}_i \text{ for at least } t \text{ out of } m \text{ sets}\}$ where $t \leq m$ is the predefined threshold. It is worth mentioning that although the threshold parameter defines the minimum number of dataset occurrences required for an item to be included, it does not directly represent the size of the intersection. Instead, the actual intersection size varies with the overlap characteristics of the input datasets and decreases as the threshold increases.

For T-MPSI, there are several works [6,10,22,33] that address the functionality of threshold multiparty private set intersection by first computing the intersection's cardinality and then verifying whether it meets the threshold. However, these protocols incur communication costs that scale linearly with the size of the smallest input set. Recently, Ghosh and Simkin [7] revisited the communication complexity of two-party threshold PSI, introducing a novel approach computing the intersection cardinality while verifying if the intersection is sufficiently large without revealing its exact size. The authors of [8] introduce a protocol that achieves a quasi-linear dependency on the threshold using simpler cryptographic primitives, such as additively homomorphic encryption and oblivious transfer. In [15], the authors propose a probabilistic T-MPSI framework, where the intersection is revealed to the parties with a probability proportional to the intersection size. This design introduces a trade-off: a small probability of revealing insufficiently large intersections in exchange for significantly improved efficiency and scalability. The authors of [9] address this gap by presenting a generic composition method that extends any T-MPSI individual protocol into a T-MPSI collective protocol. The proposed extension employs a multiparty private set union to efficiently aggregate outputs, enabling all parties to learn the collective threshold intersection. Recently, the authors of [17] addressed the problem of a specialized class of private set operations that allows multiple parties to securely compute which elements appear in at least t private sets, where t is an adjustable threshold.

2.2 Federated Graph Neural Networks

Federated learning (FL) is a privacy-preserving distributed machine learning framework designed to mitigate data silos while collaboratively constructing a shared model. Its decentralized training approach, combined with a secure aggregation mechanism, makes it well-suited for real-world applications that require stringent privacy protection [31]. New privacy challenges have emerged in FL-based applications as various inference attacks can extract sensitive information about the raw data from local models or gradients, leading to potential privacy breaches [2].

FL operates through several key strategies that define how clients contribute to training and how the central server aggregates model updates. Aggregation techniques, such as federated averaging (FedAvg), ensure that local updates are combined effectively to improve model performance [23] while optimization algorithms are designed to handle non-independent and identically distributed (non-IID) data, addressing the heterogeneity of real-world datasets across different clients [24]. An averaging operation of FedAvg can be formulated as:

$$w^r = \sum_{c=1}^{C} \frac{n_c}{n} w_c^r, \tag{1}$$

where n_c denotes the number of samples in client c, n represents the total number of samples in all clients, and w^r and w_c^r are the global model weights and the local model weights in the c-th client at round r, respectively.

FedGNNs enable collaborative training across multiple graph data sources without compromising the privacy of individual datasets. Centralized FedGNNs rely on a central server to manage communication and coordination among clients. Depending on where the graph is stored, training can either take place on the server itself or on the individual clients. GNNs update the embedding of a given node by aggregating information from its neighboring nodes with the following function:

$$X^{l+1} = GNN(A, X^l, w^l), \tag{2}$$

where $GNN(\cdot)$ indicates the graph aggregation function, methods X^l and X^{l+1} represent the node embedding in the l-th and $(l+1)$-th layers, respectively, A

denotes the graph adjacency matrix, and w^l are the trainable model weights in the l-th layer. When the model is trained at the central server using an interclient graph, it is assumed that clients with similar models or feature embeddings are more likely to be connected. The interclient graph itself can either be predefined or dynamically generated using a self-attention mechanism during training. However, a key challenge in this framework is the simultaneous optimization of both the local models and the server-side GNN model, forming a complex bilevel learning problem [16]. In the pharmaceutical industry, FedGNNs facilitate collaborative research among organizations by allowing them to train models on distributed molecular data without sharing proprietary information [20]. This approach improves the discovery of new drugs while maintaining data confidentiality. FedGNNs are applied to disease prediction by enabling institutions to collaboratively analyze patient data represented as graphs [5].

Beyond pharmaceutical and medical research, the privacy-preserving capabilities of FedGNNs extend to cybersecurity applications, where secure, distributed learning is crucial. For instance, in intrusion detection systems (IDSes) against advanced persistent threats (APTs), the work in [19] integrates FL with GNNs, enabling distributed learning without sharing raw data. This improves adaptability and threat detection while incorporating enhanced encryption to prevent man-in-the-middle attacks and reverse engineering. In PSI research, Fed-K-PSI, a federated analytics-based PSI protocol, is introduced in [3] to allow a server to query identifiers appearing with the same value in at least K out of N parties, while ensuring privacy and anonymity. Fed-K-PSI uses a combination of symmetric-key hashing, keyed Bloom filters, and homomorphic encryption to securely compute the set intersection over k clients. Each client encodes its set using a keyed hash function and participates in a multi-round communication protocol where encrypted representations are compared to identify common elements. The protocol leverages additive homomorphism to avoid revealing the raw inputs, while optimizing the size of the exchange via probabilistic data structures. Unlike traditional PSI methods that rely on Oblivious Transfer (OT) and require complete data representation, Fed-K-PSI enables distributed privacy-preserving queries without revealing non-matching identifiers or values. Further advancements in privacy-preserving feature retrieved private set intersection (P2FRPSI) [14] allow a client to retrieve only the intersection that satisfies a hidden predicate, ensuring that neither the predicate nor additional data are exposed.

3 Threshold Multiparty Private Set Intersection via Federated Graph Neural Networks

Our proposed federated learning-enabled threshold multiparty private set intersection (FL-T-MPSI) framework is designed for m mutually untrusting parties who each hold a private dataset. The goal is to compute an intersection of elements that appear in at least t datasets while ensuring no additional information is leaked beyond what is necessary. Our approach integrates FL principles to dynamically determine the threshold t and enhance collaborative efficiency. Each

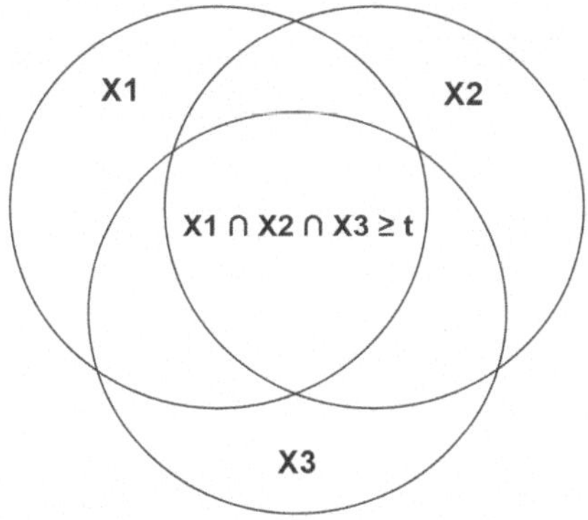

Fig. 1. Threshold Multiparty Private Set Intersection.

party P_i has a data set S_i with a maximum size of n. A predefined threshold t determines the minimum number of datasets in which an element must appear to be included in the final intersection, see Fig. 1. A semi-honest adversary model is assumed, where parties follow the protocol but may attempt to infer private information from received messages. The system is designed to tolerate collusion up to a defined number of parties.

The proposed FL-T-MPSI protocol consists of the following phases. First, in the data preprocessing and encoding phase, each participating party preprocesses its dataset to extract relevant elements and converts them into a format suitable for cryptographic operations. Specifically, our preprocessing phase converts categorical class labels into fixed-length integer encodings, followed by encryption using RSA-OAEP [12]. These integers are then serialized into strings, allowing compatibility with public-key encryption and network transmission. Before transmission, data are encrypted to prevent unauthorized access and ensure that no party learns private information outside the computed intersection. Next, the protocol proceeds to threshold-based filtering, where a dynamic thresholding mechanism determines whether an element should be included in the final intersection. The threshold is not fixed but rather adapts based on the distribution of input sets. The protocol integrates FedGNNs to aggregate model updates from participants while respecting privacy constraints, see Fig. 2. To achieve privacy preservation while enabling threshold-based set intersection, the protocol uses homomorphic encryption to allow computations on encrypted data without requiring decryption, ensuring that no party learns intermediate values during the MPSI computation.

Our architecture comprises three core modules. Data preprocessing and encoding allow clients to preprocess and encode their datasets into cryptographic tokens. A FedGNN is executed locally at each client to predict optimal threshold values, with gradient exchange secured via MPC. A central server coordinates encrypted message passing using RSA-based oblivious transfer. All PSI results are filtered through a dynamic threshold mechanism based on computed embed-

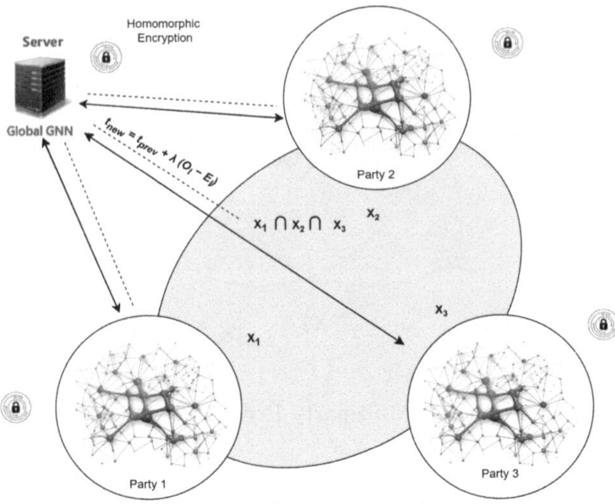

Fig. 2. FedGNN-enabled Threshold Multiparty Private Set Intersection.

dings. Communication is asynchronous and leverages a federated loop with three steps per global epoch: local update, encrypted transfer, and server-side aggregation.

A core feature of the proposed protocol is its dynamic thresholding mechanism, which eliminates the need for a pre-defined static threshold. Instead of requiring a fixed t value, the protocol dynamically adjusts t based on observed data characteristics and real-time collaboration needs. The process begins by initializing a default threshold based on prior statistical distributions of dataset overlap. Then, as the protocol progresses, a federated optimization model iteratively refines t by analyzing the proportion of elements shared across datasets. This adaptive thresholding mechanism prevents overly restrictive intersections that could exclude valuable information and ensures that collaboration occurs only when sufficient overlap exists between participants' datasets. The adjustment follows the formula:

$$t_{\text{new}} = t_{\text{prev}} + \lambda \cdot (O_l - E_l), \tag{3}$$

where λ is a learning rate parameter, O_l is the observed overlap that represents the actual number of shared elements detected across parties and E_l is the expected overlap that is estimated based on previous interactions.

In cybersecurity threat intelligence sharing, different organizations may hold datasets with overlapping threat indicators, but their relationships are not purely random-certain organizations are more likely to observe similar threats. Similarly, in biomedical research, different institutions may contribute data that is structurally related due to geographical or institutional collaborations. To capture these dependencies, we incorporate FedGNNs into the T-MPSI framework. This allows us to model interactions between datasets and their respective parties in a graph structure, leading to more informed threshold adjustments and improved learning efficiency. In our FedGNN integration, each client node represents a dataset holder, and edges represent data similarity based on partial overlaps. The node embeddings are learned from metadata features such as set cardinality, update frequency, and data modality. These embeddings are aggregated via message passing using a graph attention network to infer optimal threshold adjustment and prioritize communication paths, which directly influence the dynamic threshold t. For graph representation of multiparty interactions each party is represented as a node P_i. The edges between nodes represent data similarity, overlap in datasets. The weights of the edges capture the strength of interactions, allowing the model to infer data distribution trends across different parties. In the learning process of FedGNN, a graph $G = (V, E)$ is initialized, where V represents participating parties and E encodes relationships based on past data overlaps. Each client computes local graph embeddings based on its dataset. These embeddings are aggregated in a federated manner without sharing raw data. Node representations are updated based on message-passing from neighboring nodes, allowing each party to refine its view of dataset similarity. The final node embeddings are used to predict the optimal threshold t for intersection computation. The FedGNN model refines the dynamic threshold based on learned relationships, improving accuracy compared to simple heuristic-based adjustments.

4 Results

The implementation of FL-T-MPSI approach follows a distributed architecture, integrating cryptographic security, federated learning, and graph neural networks to optimize multiparty collaboration. On the client-side, each participating party preprocesses its dataset by extracting relevant features and converting them into an encrypted format using HE. Additionally, each client computes local graph embeddings to model data relationships within a FedGNN, enabling threshold adaptation without requiring direct data exchange. On the server-side, the protocol begins by aggregating encrypted client updates and performing set intersection computations based on the dynamic threshold mechanism.

The FL framework is implemented in Python, utilizing PyTorch Geometric for GNN computations, FedScale for FL integration, and PySEAL for cryptographic operations. Communication between clients and the server is handled through secure socket connections, ensuring encrypted data transmission. The protocol is scalable, allowing it to handle an increasing number of participants and large datasets while minimizing communication overhead. For our experiments, we utilized NVIDIA Jetson Xavier NX 16GB GPUs for training the FedGNN, enabling efficient distributed learning and threshold adaptation. Furthermore, 2 TB NVMe SSDs were employed to handle high-speed encryption and dataset preprocessing, minimizing storage bottlenecks and accelerating cryptographic operations.

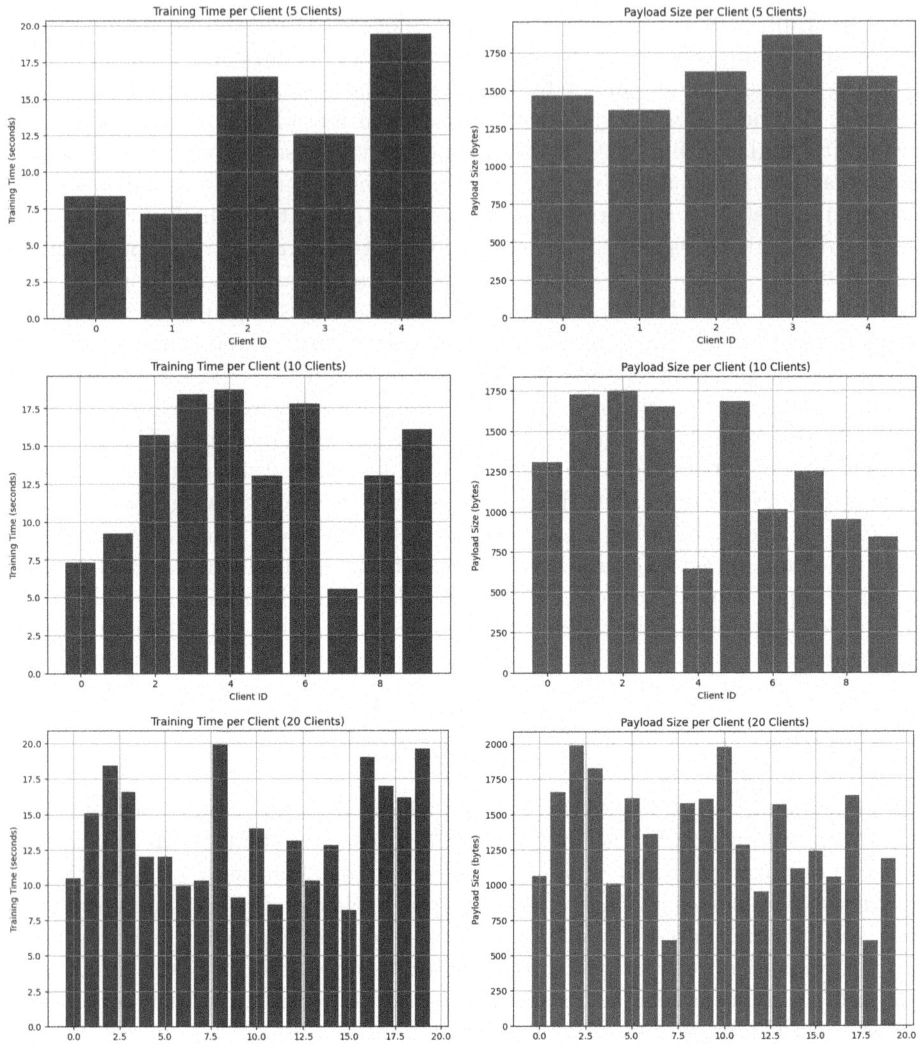

Fig. 3. Training time and payload size (N= 5, 10 and 20 clients).

For the evaluation of the applicability of our FedGNN-based protocol in real-world medical CTI settings, we simulated a multi-institutional healthcare network where hospitals and research organizations share cyber threat indicators while preserving privacy. The dataset used is the Malicious URLs dataset [26]. In this setting, multiple healthcare providers possess logs of cybersecurity incidents, including phishing attacks, malware infections, and network intrusions, but are unwilling to expose their entire datasets due to regulatory constraints (e.g., HIPAA, GDPR). The goal of this experiment was to allow institutions to identify shared threat indicators only when a predefined threshold t of hospitals

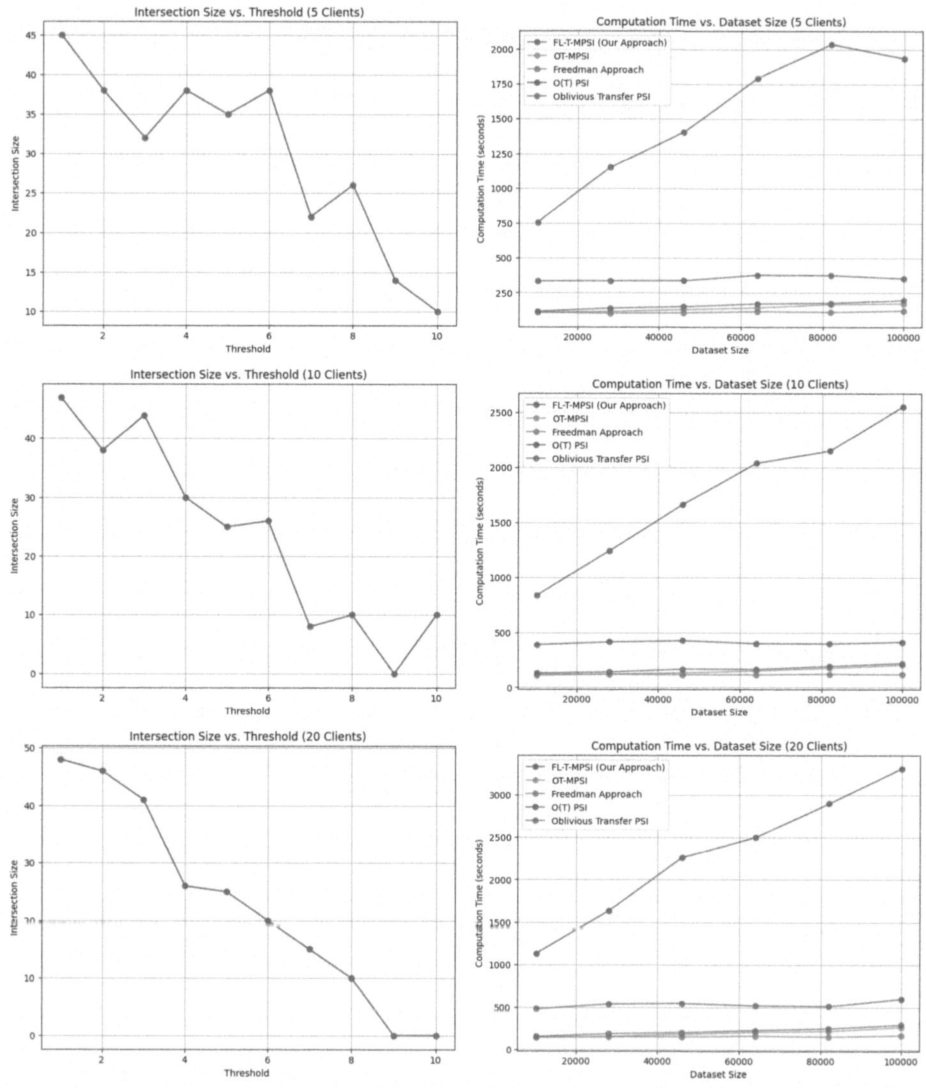

Fig. 4. Intersection size for FL-T-MPSI and its computation time against T-MPSI protocols (N= 5, 10 and 20 clients).

have observed the same indicator, ensuring that isolated or institution-specific incidents remain undisclosed.

In our experiments, we initially considered five participating parties (hospitals) that acted as independent clients, each holding a dataset of 100,000 threat indicators. These indicators included malicious URLs that are able to affect healthcare infrastructure. Our dynamic threshold-based PSI protocol was employed to compute an intersection of threat indicators appearing in at least

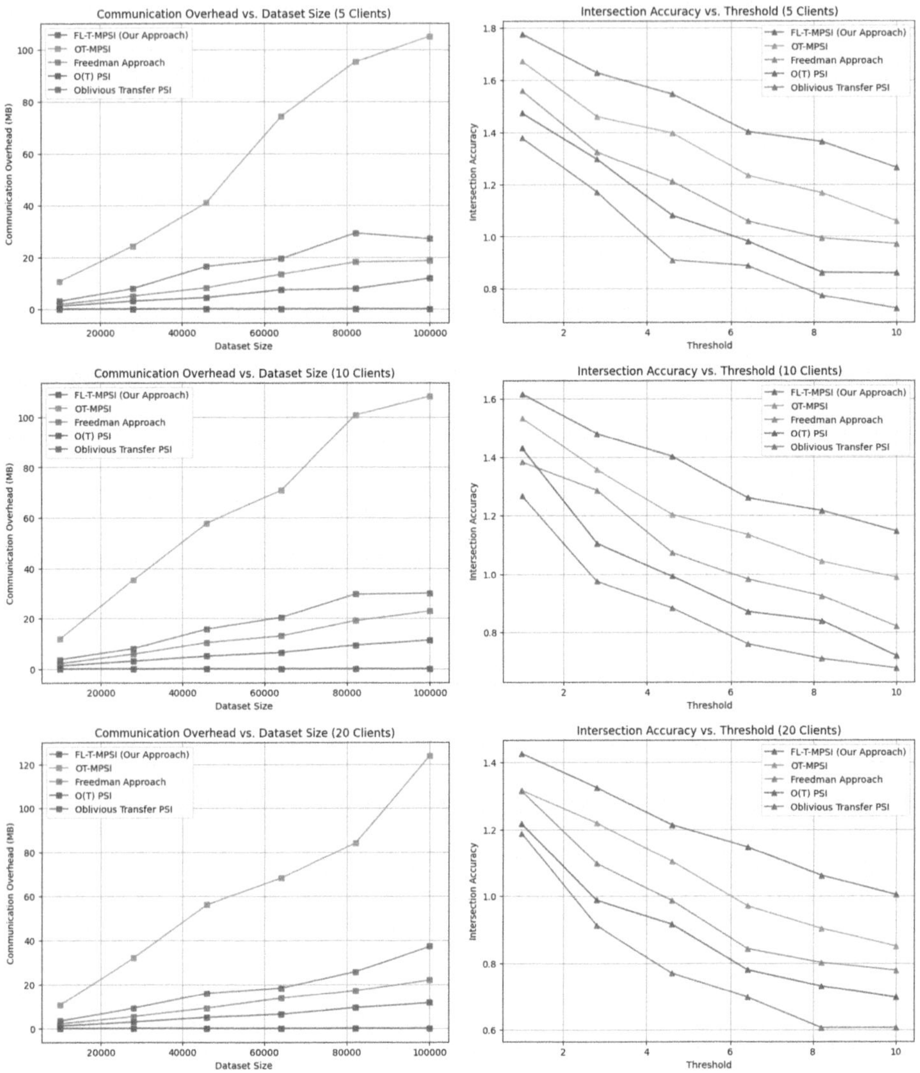

Fig. 5. Comparison of intersection accuracy and communication efficiency with T-MPSI protocols (N = 5, 10 and 20 clients).

$t = 3$ datasets while ensuring that no party learned non-intersecting elements. Then, we scaled up to 10 and 20 clients. Finally, we compared our approach against popular MPSI protocols, namely OT-MPSI [17], O(T)PSI [30], oblivious transfer PSI [3] and the Freedman et al. [4] approach. Our experiments were conducted with three different client populations N: 5 clients, 10 clients, and 20 clients, to evaluate scalability, communication efficiency, and intersection accuracy. Our results are presented in Fig. 3, 4 and 5 showcasing the training and payload time per client, the intersection size against the threshold and the corresponding communication efficiency with regards to dataset size. The comparisons considered the communication overhead and the intersection accuracy for both client population scenarios.

5 Discussion

Our results demonstrate the effectiveness of our adaptive threshold mechanism and FedGNNs in optimizing intersection computations while minimizing overhead. More specifically, they demonstrate that as the number of clients increases, computation time naturally grows across all methods highlighting improvements in processing time, scalability, and communication cost reduction as well as the protocol's robustness in real-world privacy-sensitive domains, such as cyber threat intelligence collaboration. However, our approach consistently achieves lower computation times than OT-MPSI and Oblivious Transfer PSI, both of which exhibit significantly higher computational overhead due to their complex cryptographic operation, see Fig. 4. Freedman approach and O(T) PSI perform well for smaller datasets but struggle with multiparty scalability. Our approach effectively balances efficiency and security, maintaining competitive computation times across increasing client populations.

One of the most critical factors in multiparty PSI is the communication overhead, which determines the feasibility of large-scale deployments. Our results show that our approach substantially reduces communication overhead compared to the OT-MPSI approach while the communication cost of Freedman scales proportionally but remains higher than our approach due to its lack of threshold optimization. O(T) PSI maintains a relatively low overhead, but its lack of full multiparty support limits its real-world applicability. The experiment further confirms that as the dataset size increases, T-MPSI scales efficiently, making it well-suited for federated learning and real-world deployments. The intersection accuracy results indicate that our approach dynamically adapts the threshold selection better than other methods, ensuring that meaningful intersections are retained across different client populations, see Fig. 3 and 5. As the number of clients grows, fixed-threshold approaches exhibit decreased accuracy due to rigid threshold constraints, leading to unnecessary exclusions of valid intersections. In contrast, the implementation of FedGNNs allows for threshold adjustments that maintain high accuracy while reducing unnecessary computations. This adaptability ensures that our FedGNN-enabled protocol remains robust and practical for dynamic, privacy-sensitive applications such as CTI collaboration. The experiments validate that our approach scales effectively from 5 to 20 clients while preserving efficient computation and communication costs while the introduction of a threshold-based filtering mechanism optimizes intersection size computations, leading to controlled scaling effects. Even as the size of the dataset increases, our method is compared with its counterparts in handling large-scale data intersections without excessive computational penalties. A quantitative comparison of our approach can be summarized as follows: In terms of intersection accuracy, with an average of 96.5% across all thresholds, our approach outperforms all the rest except O(T) MPSI by a minimum of ~6%. In computation overhead terms, our approach stayed below 250 s for datasets of up to 10k records, while OT-MPSI reaches 3000 s. In scalability terms, FL-T-MPSI maintained consistent performance from 5 to 20 clients with less than 20% variation in runtime. From a security standpoint, our protocol is designed to be resilient against collusion by up to $t-1$ parties, ensuring that only threshold-satisfying items are revealed. The encrypted communication overhead remains low, with payloads typically under 20 MB and transmission delays below 250 s. In terms of parameter tuning, the dynamic threshold mechanism is governed by a learning rate λ, and empirical sensitivity analysis shows that $\lambda = 0.3$ balances stability and responsiveness to client overlap [27].

While FL-T-MPSI delivers significant improvements in computation efficiency, communication overhead, and intersection accuracy, certain trade-offs must be considered. The use of FedGNNs introduces additional model training overhead, which may not be feasible in settings where computational resources are severely constrained. Additionally, while our protocol minimizes privacy leakage, extremely adversarial settings (e.g., collusion attacks from multiple clients) may still pose risks that require further mitigation techniques such as differential privacy enhancements.

6 Conclusion

This work introduces a federated learning-enabled threshold multiparty private set intersection, an advanced cryptographic protocol designed to enable privacy-preserving multiparty collaboration. Unlike traditional PSI protocols, which enforce strict exact matching, our approach leverages dynamic threshold adaptation through FedGNNs, ensuring that the intersection process remains flexible and meaningful in diverse real-world applications such as cyber threat intelligence, federated learning, and secure data analytics. By utilizing homomorphic encryption, we achieve strong privacy guarantees while minimizing communication and computational overhead.

Extensive evaluations in three different client settings demonstrate that FL-T-MPSI consistently outperforms popular TMPSI protocols in terms of computation efficiency, intersection accuracy, and scalability. The results confirm that FL-T-MPSI offers a scalable and efficient solution for privacy-preserving multiparty analytics, enabling secure data collaboration without excessive computational costs. Future work may explore further optimizations in federated threshold learning, adaptive privacy mechanisms for resisting collusion attacks, and deployments in real-world federated AI systems.

Acknowledgements. a. We would like to thank Dr. Zekeriya Erkin and Mr. Jorrit Van Assen from the Cybersecurity Technology Group, Technische Universiteit Delft for their fruitful discussions and points regarding this research study.

b. This research was supported by the European Union's Horizon Europe research and innovation programme under grant agreement No. 101094901, the SEPTON project (Security Protection Tools for Networked Medical Devices).

References

1. Alaeifar, P., Pal, S., Jadidi, Z., Hussain, M., Foo, E.: Current approaches and future directions for cyber threat intelligence sharing: a survey. J. Inf. Secur. Appl. **83**, 103786 (2024)
2. Chen, J., Yan, H., Liu, Z., Zhang, M., Xiong, H., Yu, S.: When federated learning meets privacy-preserving computation. ACM Comput. Surv. **56**(12) (2024). https://doi.org/10.1145/3679013
3. Elkordy, A.R., Ezzeldin, Y.H., Avestimehr, S.: Federated k-private set intersection. In: Proceedings of the 31st ACM International Conference on Information & Knowledge Management, CIKM 2022, dpp. 436–445. Association for Computing Machinery, New York (2022). https://doi.org/10.1145/3511808.3557321
4. Freedman, M.J., Nissim, K., Pinkas, B.: Efficient private matching and set intersection. In: International Conference on the Theory and Applications of Cryptographic Techniques, pp. 1–19. Springer (2004)
5. Fu, X., Chen, Z., Zhang, B., Chen, C., Li, J.: Federated graph learning with structure proxy alignment. In: Proceedings of the 30th ACM SIGKDD Conference on Knowle DGE Discovery and Data Mining, KDD 2024, pp. 827–838. Association for Computing Machinery, New York (2024). https://doi.org/10.1145/3637528.3671717
6. Ghosh, S., Nilges, T.: An algebraic approach to maliciously secure private set intersection. In: Annual International Conference on the Theory and Applications of Cryptographic Techniques, pp. 154–185. Springer (2019)
7. Ghosh, S., Simkin, M.: The communication complexity of threshold private set intersection. In: Annual International Cryptology Conference, pp. 3–29. Springer (2019)
8. Ghosh, S., Simkin, M.: Threshold private set intersection with better communication complexity. In: Boldyreva, A., Kolesnikov, V. (eds.) Public-Key Cryptography - PKC 2023, pp. 251–272. Springer, Cham (2023)
9. Guan, C., van Assen, J., Erkin, Z.: Collective threshold multiparty private set intersection protocols for cyber threat intelligence. In: 2024 IEEE International Workshop on Information Forensics and Security (WIFS), pp. 1–6 (2024). https://doi.org/10.1109/WIFS61860.2024.10810671
10. Hallgren, P., Orlandi, C., Sabelfeld, A.: PrivatePool: privacy-preserving ridesharing. In: 2017 IEEE 30th Computer Security Foundations Symposium (CSF), pp. 276–291. IEEE (2017)
11. Hu, J., Liu, Z., Zuo, C.: Delegated multi-party private set intersection from secret sharing. Cryptology ePrint Archive (2025)
12. Kiltz, E., O'Neill, A., Smith, A.: Instantiability of RSA-OAEP under chosen-plaintext attack. In: Rabin, T. (ed.) Advances in Cryptology – CRYPTO 2010, pp. 295–313. Springer, Heidelberg (2010)
13. Kissner, L., Song, D.: Privacy-preserving set operations. In: Annual International Cryptology Conference, pp. 241–257. Springer (2005)

14. Ling, G., Tang, F., Cai, C., Shan, J., Xue, H., Li, W., Tang, P., Huang, X., Qiu, W.: P^2frpsi: privacy-preserving feature retrieved private set intersection. IEEE Trans. Inf. Forensics Secur. **19**, 2201–2216 (2024). https://doi.org/10.1109/TIFS.2023.3343973
15. Liu, F.H., Zhang, E., Qin, L.: Efficient multiparty probabilistic threshold private set intersection. In: Proceedings of the 2023 ACM SIGSAC Conference on Computer and Communications Security, CCS 2023, pp. 2188–2201. Association for Computing Machinery, New York (2023). https://doi.org/10.1145/3576915.3623158
16. Liu, R., Xing, P., Deng, Z., Li, A., Guan, C., Yu, H.: Federated graph neural networks: overview, techniques, and challenges. IEEE Trans. Neural Netw. Learn. Syst. (2024)
17. Ma, L., et al.: Over-threshold multi-party private set operation protocols for lightweight clients. Comput. Stand. Interfaces **88**, 103781 (2024)
18. Mahmoud, N., Abdelkader, H., Aly, A.: BRIP: towards a privacy-preserving, trustworthy, and transparent ride-sharing system with semantic matching powered by blockchain and ipfs. J. Netw. Comput. Appl. **226**, 103870 (2024)
19. Mansour Bahar, A.A., Ferrahi, K.S., Messai, M.L., Seba, H., Amrouche, K.: FedHE-graph: federated learning with hybrid encryption on graph neural networks for advanced persistent threat detection. In: Proceedings of the 19th International Conference on Availability, Reliability and Security, ARES 2024, Association for Computing Machinery, New York (2024). https://doi.org/10.1145/3664476.3670466
20. Manu, D., Yao, J., Liu, W., Sun, X.: GraphGANFed: a federated generative framework for graph-structured molecules towards efficient drug discovery. IEEE/ACM Trans. Comput. Biol. Bioinform. (2024)
21. Naresh, V.S., Thamarai, M.: Privacy-preserving data mining and machine learning in healthcare: applications, challenges, and solutions. Wiley Interdisc. Rev.: Data Min. Knowl. Discov. **13**(2), e1490 (2023)
22. Pinkas, B., Schneider, T., Weinert, C., Wieder, U.: Efficient circuit-based PSI via cuckoo hashing. In: Annual International Conference on the Theory and Applications of Cryptographic Techniques, pp. 125–157. Springer (2018)
23. Qi, P., Chiaro, D., Guzzo, A., Ianni, M., Fortino, G., Piccialli, F.: Model aggregation techniques in federated learning: a comprehensive survey. Futur. Gener. Comput. Syst. **150**, 272–293 (2024). https://doi.org/10.1016/j.future.2023.09.008
24. Rahad, M., Shabab, R., Ahammad, M.S., Reza, M.M., Karmaker, A., Hossain, M.A.: Kl-FedDis: a federated learning approach with distribution information sharing using Kullback-Leibler divergence for non-iid data. Neurosci. Inform. **5**(1), 100182 (2025). https://doi.org/10.1016/j.neuri.2024.100182
25. Sheikh, Z.A., Singh, Y., Singh, P.K., Ghafoor, K.Z.: Intelligent and secure framework for critical infrastructure (CPS): current trends, challenges, and future scope. Comput. Commun. **193**, 302–331 (2022)
26. Siddhartha, M.: Malicious URLs dataset. https://www.kaggle.com/datasets/sid321axn/malicious-urls-dataset. Accessed January 2025
27. Smith, V., Chiang, C.K., Sanjabi, M., Talwalkar, A.S.: Federated multi-task learning. In: Guyon, I., et al. (eds.) Advances in Neural Information Processing Systems, vol. 30. Curran Associates, Inc. (2017). https://proceedings.neurips.cc/paper_files/paper/2017/file/6211080fa89981f66b1a0c9d55c61d0f-Paper.pdf
28. Sun, Z.: Efficient multiparty private set intersection protocol based on function secret sharing. In: Karras, D.A., Gheisari, M. (eds.) International Conference on Computer Network Security and Software Engineering (CNSSE 2024), vol. 13175,

p. 1317505. International Society for Optics and Photonics, SPIE (2024). https://doi.org/10.1117/12.3031902
29. Vos, J., Conti, M., Erkin, Z.: SoK: collusion-resistant multi-party private set intersections in the semi-honest model. In: 2024 IEEE Symposium on Security and Privacy (SP), pp. 465–483 (2024). https://doi.org/10.1109/SP54263.2024.00079
30. Wei, L., Liu, J., Zhang, L., Wang, Q., Zhang, W., Qian, X.: Efficient multi-party private set intersection protocols for large participants and small sets. Comput. Stand. Interfaces **87**, 103764 (2024). https://doi.org/10.1016/j.csi.2023.103764
31. Wen, J., Zhang, Z., Lan, Y., Cui, Z., Cai, J., Zhang, W.: A survey on federated learning: challenges and applications. Int. J. Mach. Learn. Cybern. **14**(2), 513–535 (2023)
32. Zhao, S., Zhao, C., Huang, Y., Song, X., Xu, Q.: Two-party threshold private set intersection protocols from lightweight cryptographic primitives. Cryptography **8**(4), 58 (2024)
33. Zhao, Y., Chow, S.S.: Can you find the one for me? In: Proceedings of the 2018 Workshop on Privacy in the Electronic Society, pp. 54–65 (2018)

Correction to: Analysis of Permissions and Trackers in Android Apps Targeting Children Under Thirteen

Jamie Fitzpatrick and Virginia N. L. Franqueira

Correction to: Chapter 5 in: I. Praça et al. (Eds.): *Cybersecurity*, CCIS 2500, https://doi.org/10.1007/978-3-031-94855-8_5

In the originally published version of chapter 5, Figures 1 and 2 were duplicated. The error was fixed by replacing Figure 2 with the following figure.

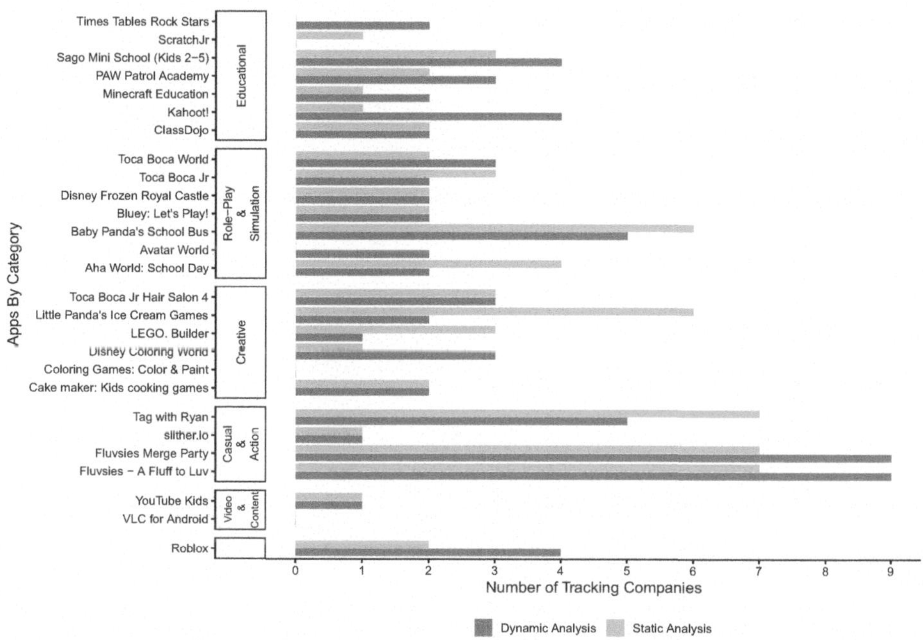

Fig. 2. Number of tracker companies from static/dynamic analysis per app by category.

The updated version of this chapter can be found at https://doi.org/10.1007/978-3-031-94855-8_5

I. Praça et al. (Eds.): EICC 2025, CCIS 2500, p. C1, 2025.
https://doi.org/10.1007/978-3-031-94855-8_24

Author Index

A

Aljohani, Mohammed 151
Androutsos, Christos 338
Apvrille, Ludovic 229
Arief, Budi 98
Arvanitis, Theodoros N. 338
Athanasiou, Lambros 338

B

Bakiris, Emmanouil 275
Bernsmed, Karin 338
Bertin, Hugo 166
Birinci, Baturay 98
Boutry, Nicolas 239
Brilingaitė, Agnė 217
Bromberg, Yérom-David 166
Bukauskas, Linas 217

C

Cabral, Bernardo 38
Carmichael, Laura 288
Carpent, Xavier 151
Castillo, Andrés 338
Çetin, Orçun 98
Chakraborty, Joheen 20
Chakraborty, Soumyo V. 20
Chimuco, Francisco T. 182
Citskovska, Elizabete 217
Criado, Regino 255

D

Dacier, Marc 166
Dias, Tiago 38
Doulamis, Nikolaos 356

E

Edwards, Matthew 3
Epiphaniou, Gregory 338
Erdogan, Gencer 288

F

Ferreira, Luís Lino 38
Fitzpatrick, Jamie 63
Fotiadis, Dimitrios I. 338
Fotos, Nikolaos 338
Franqueira, Virginia N. L. 63
Freire, Mário M. 182
Freyer, Oscar 338
Fruscione, Marco 338
Furnell, Steven 151

G

Gassmann, Luke 3
Gervassis, Nicholas 151
Gilbert, Stephen 338
Gkois, George 338
Gonçalves, José 38, 201

H

Houcke, Sebastien 52

I

Inácio, Pedro R. M. 182

J

Jaber, Majed 239
Jensen, Meiko 82

K

Kalogeras, Dimitris 356
Karas, Dimitrios S. 338
Keller, Jörg 133
Khan, Mahreen 229
Kumar, Ashwinee 324

L

Lang, Michael 116
Larsson, Emil 82
Lielbārde, Sandra 217
Litzinger, Sebastian 133

I. Praça et al. (Eds.): EICC 2025, CCIS 2500, pp. 373–374, 2025.
https://doi.org/10.1007/978-3-031-94855-8

M
Maia, Eva 38, 201
Masmoudi, Ahmed 52
Matalobos, Juan Manuel 255
Messinis, Sotirios 356
Messinis, Sotiris 338
Milojevic, Dusko 338
Mitra, Anisha 307
Moral, Santiago 255
Moukafih, Nabil 338
Mushtaq, Maria 229

N
Nasser, Yehya 52

O
Oikonomakis, George 275
Ostermann, Max 338

P
Pacalet, Renaud 229
Papadakis, Nikos 338
Pareja, Pablo 52
Parrend, Pierre 239
Pereira, Pedro 201
Pezoulas, Vasilis 338
Pirta, Rūta 217
Praça, Isabel 38, 201
Protonotarios, Nicholas E. 356

Q
Quinn, Paul 324

R
Rani, Nanda 20
Roponena, Evita 217
Roy Chowdhury, Dipanwita 307

S
Saha, Bikash 20
Senior, Samuel M. 288
Sequeiros, João B. F. 182
Severino, Ricardo 38
Shaikh, Salman 166
Shukla, Sandeep Kumar 20
Silva, Miguel 38
Singh, Divyanshu 20
Skytterholm, Andrea Neverdal 288, 338
Stevenson, James 3
Su, Ruoxin 324

T
Taylor, Steve 288, 338
Tverdal, Simeon 288

U
Uysal, Çağlar 98

V
Vitorino, João 201

Y
Yesiltuna, Oguzhan 324

The manufacturer's authorised representative in the EU is Springer Nature Customer Service Centre GmbH, Europaplatz 3, 69115 Heidelberg, Germany. If you have any concerns regarding our products, please contact ProductSafety@springernature.com

Printed and bound by CPI Group (UK) Ltd, Croydon, CR0 4YY
30/09/2026
02225560-0013